Advanced Information and Knowledge Processing

Series Editors
Professor Lakhmi Jain
Lakhmi.jain@unisa.edu.au

Professor Xindong Wu
xwu@cems.uvm.edu

For other titles published in this series, go to
www.springer.com/series/4738

Yong Shi · Yingjie Tian · Gang Kou · Yi Peng ·
Jianping Li

Optimization Based Data Mining: Theory and Applications

 Springer

Yong Shi
Research Center on Fictitious Economy and
Data Science
Chinese Academy of Sciences
Beijing 100190
China
yshi@gucas.ac.cn

and
College of Information Science &
Technology
University of Nebraska at Omaha
Omaha, NE 68182
USA
yshi@unomaka.edu

Yingjie Tian
Research Center on Fictitious Economy and
Data Science
Chinese Academy of Sciences
Beijing 100190
China
tianyingjie1213@163.com

Gang Kou
School of Management and Economics
University of Electronic Science and
Technology of China
Chengdu 610054
China
kougang@yahoo.com

Yi Peng
School of Management and Economics
University of Electronic Science and
Technology of China
Chengdu 610054
China
pengyicd@gmail.com

Jianping Li
Institute of Policy and Management
Chinese Academy of Sciences
Beijing 100190
China
ljp@casipm.ac.cn

ISSN 1610-3947
ISBN 978-1-4471-2653-9   ISBN 978-0-85729-504-0 (eBook)
DOI 10.1007/978-0-85729-504-0
Springer London Dordrecht Heidelberg New York

British Library Cataloguing in Publication Data
A catalogue record for this book is available from the British Library

Cover design: deblik

Printed on acid-free paper

Springer is part of Springer Science+Business Media (www.springer.com)

This book is dedicated to the colleagues and students who have worked with the authors

Preface

The purpose of this book is to provide up-to-date progress both in Multiple Criteria Programming (MCP) and Support Vector Machines (SVMs) that have become powerful tools in the field of data mining. Most of the content in this book are directly from the research and application activities that our research group has conducted over the last ten years.

Although the data mining community is familiar with Vapnik's SVM [206] in classification, using optimization techniques to deal with data separation and data analysis goes back more than fifty years. In the 1960s, O.L. Mangasarian formulated the principle of large margin classifiers and tackled it using linear programming. He and his colleagues have reformed his approaches in SVMs [141]. In the 1970s, A. Charnes and W.W. Cooper initiated Data Envelopment Analysis, where linear or quadratic programming is used to evaluate the efficiency of decision-making units in a given training dataset. Started from the 1980s, F. Glover proposed a number of linear programming models to solve the discriminant problem with a small-size of dataset [75]. Since 1998, the author and co-authors of this book have not only proposed and extended such a series of optimization-based classification models via Multiple Criteria Programming (MCP), but also improved a number of SVM related classification methods. These methods are different from statistics, decision tree induction, and neural networks in terms of the techniques of separating data.

When MCP is used for classification, there are two common criteria. The first one is the overlapping degree (e.g., norms of all overlapping) with respect to the separating hyperplane. The lower this degree, the better the classification. The second is the distance from a point to the separating hyperplane. The larger the sum of these distances, the better the classification. Accordingly, in linear cases, the objective of classification is either minimizing the sum of all overlapping or maximizing the sum of the distances. MCP can also be viewed as extensions of SVM. Under the framework of mathematical programming, both MCP and SVM share the same advantage of using a hyperplane for separating the data. With certain interpretation, MCP measures all possible distances from the training samples to separating hyperplane, while SVM only considers a fixed distance from the support vectors. This allows MCP approaches to become an alternative for data separation.

As we all know, optimization lies at the heart of most data mining approaches. Whenever data mining problems, such as classification and regression, are formulated by MCP or SVM, they can be reduced into different types of optimization problems, including quadratic, linear, nonlinear, fuzzy, second-order cone, semi-definite, and semi-infinite programs.

This book mainly focuses on MCP and SVM, especially their recent theoretical progress and real-life applications in various fields. Generally speaking, the book is organized into three parts, and each part contains several related chapters. Part one addresses some basic concepts and important theoretical topics on SVMs. It contains Chaps. 1, 2, 3, 4, 5, and 6. Chapter 1 reviews standard C-SVM for classification problem and extends it to problems with nominal attributes. Chapter 2 introduces LOO bounds for several algorithms of SVMs, which can speed up the process of searching for appropriate parameters in SVMs. Chapters 3 and 4 consider SVMs for multi-class, unsupervised, and semi-supervised problems by different mathematical programming models. Chapter 5 describes robust optimization models for several uncertain problems. Chapter 6 combines standard SVMs with feature selection strategies at the same time via p-norm minimization where $0 < p < 1$.

Part two mainly deals with MCP for data mining. Chapter 7 first introduces basic concepts and models of MCP, and then constructs penalized Multiple Criteria Linear Programming (MCLP) and regularized MCLP. Chapters 8, 9 and 11 describe several extensions of MCLP and Multiple Criteria Quadratic Programming (MCQP) in order to build different models under various objectives and constraints. Chapter 10 provides non-additive measured MCLP when interactions among attributes are allowed for classification.

Part three presents a variety of real-life applications of MCP and SVMs models. Chapters 12, 13, and 14 are finance applications, including firm financial analysis, personal credit management and health insurance fraud detection. Chapters 15 and 16 are about web services, including network intrusion detection and the analysis for the pattern of lost VIP email customer accounts. Chapter 17 is related to HIV-1 informatics for designing specific therapies, while Chap. 18 handles antigen and anti-body informatics. Chapter 19 concerns geochemical analyses. For the convenience of the reader, each chapter of applications is self-contained and self-explained.

Finally, Chap. 20 introduces the concept of intelligent knowledge management first time and describes in detail the theoretical framework of intelligent knowledge. The contents of this chapter go beyond the traditional domain of data mining and look for how to produce knowledge support to the end users by combing hidden patterns from data mining and human knowledge.

We are indebted to many people around the work for their encouragement and kind support of our research on MCP and SVMs. We would like to thank Prof. Nai-yang Deng (China Agricultural University), Prof. Wei-xuan Xu (Institute of Policy and Management, Chinese Academy of Sciences), Prof. Zhengxin Chen (University of Nebraska at Omaha), Prof. Ling-ling Zhang (Graduate University of Chinese Academy of Sciences), Dr. Chun-hua Zhang (RenMin University of China), Dr. Zhi-xia Yang (XinJiang University, China), and Dr. Kun Zhao (Beijing WuZi University).

In the last five years, there are a number of colleagues and graduate students at the Research Center on Fictitious Economy and Data Science, Chinese Academy of Sciences who contributed to our research projects as well as the preparation of this book. Among them, we want to thank Dr. Xiao-fei Zhou, Dr. Ling-feng Niu, Dr. Xing-sen Li, Dr. Peng Zhang, Dr. Dong-ling Zhang, Dr. Zhi-wang Zhang, Dr. Yue-jin Zhang, Zhan Zhang, Guang-li Nie, Ruo-ying Chen, Zhong-bin OuYang, Wen-jing Chen, Ying Wang, Yue-hua Zhang, Xiu-xiang Zhao, Rui Wang.

Finally, we would like acknowledge a number of funding agencies who provided their generous support to our research activities on this book. They are First Data Corporation, Omaha, USA for the research fund "Multiple Criteria Decision Making in Credit Card Portfolio Management" (1998); the National Natural Science Foundation of China for the overseas excellent youth fund "Data Mining in Bank Loan Risk Management" (#70028101, 2001–2003), the regular project "Multiple Criteria Non-linear Based Data Mining Methods and Applications" (#70472074, 2005–2007), the regular project "Convex Programming Theory and Methods in Data Mining" (#10601064, 2007–2009), the key project "Optimization and Data Mining" (#70531040, 2006–2009), the regular project "Knowledge-Driven Multi-criteria Decision Making for Data Mining: Theories and Applications" (#70901011, 2010–2012), the regular project "Towards Reliable Software: A Standardize for Software Defects Measurement & Evaluation" (#70901015, 2010–2012), the innovative group grant "Data Mining and Intelligent Knowledge Management" (#70621001, #70921061, 2007–2012); the President Fund of Graduate University of Chinese Academy of Sciences; the Global Economic Monitoring and Policy Simulation Pre-research Project, Chinese Academy of Sciences (#KACX1-YW-0906, 2009–2011); US Air Force Research Laboratory for the contract "Proactive and Predictive Information Assurance for Next Generation Systems (P2INGS)" (#F30602-03-C-0247, 2003–2005); Nebraska EPScOR, the National Science Foundation of USA for industrial partnership fund "Creating Knowledge for Business Intelligence" (2009–2010); BHP Billiton Co., Australia for the research fund "Data Mining for Petroleum Exploration" (2005–2010); Nebraska Furniture Market—a unit of Berkshire Hathaway Investment Co., Omaha, USA for the research fund "Revolving Charge Accounts Receivable Retrospective Analysis" (2008–2009); and the CAS/SAFEA International Partnership Program for Creative Research Teams "Data Science-Based Fictitious Economy and Environmental Policy Research" (2010–2012).

Chengdu, China   Yong Shi
December 31, 2010   Yingjie Tian
                                                                                        Gang Kou
                                                                                           Yi Peng
                                                                                     Jianping Li

Contents

Part I
Support Vector Machines:
Theory and Algorithms

Chapter 1
Support Vector Machines for Classification Problems

Support vector machines (SVMs), introduced by Vapnik in the early 1990's [23, 206], are powerful techniques for machine learning and data mining. Recent breakthroughs have led to advancements in the theory and applications. SVMs were developed to solve the classification problem at first, but they have been extended to the domain of regression [198], clustering problems [243, 245]. Such standard SVMs require the solution of either a quadratic or a linear programming.

The classification problem can be restricted to considering the two-class problem without loss of generality. It can be described as follows: suppose that two classes of objects are given, we are then faced a new object, and have to assign it to one of the two classes.

This problem is formulated mathematically [53]: Given a training set

$$T = \{(x_1, y_1), \ldots, (x_l, y_l)\} \in (R^n \times \{-1, 1\})^l, \tag{1.1}$$

where $x_i = ([x_i]_1, \ldots, [x_i]_n)^T$ is called an input with the attributes $[x_i]_j$, $j = 1, \ldots, n$, and $y_i = -1$ or 1 is called the corresponding output, $i = 1, \ldots, l$. The question is, for a new input $\bar{x} = ([\bar{x}_1], \ldots, [\bar{x}_n])^T$, to find its corresponding $\bar{y}$.

1.1 Method of Maximum Margin

Consider the example in Fig. 1.1. Here the problem is called linearly separable because that the set of training vectors (points) belong to two separated classes, there are many possible lines that can separate the data. Let us discuss which line is better.

Suppose that the direction of the line is given, just as the w in Fig. 1.2. We can see that line l_1 with direction w can separate the points correctly. If we put l_1 right-up and left-down until l_1 touches some points of each class, we will get two "support" lines l_2 and l_3, all the lines parallel to and between them can separate the points correctly also. Obviously the middle line l is the "best".

Y. Shi et al., *Optimization Based Data Mining: Theory and Applications*,
Advanced Information and Knowledge Processing,
DOI 10.1007/978-0-85729-504-0_1, © Springer-Verlag London Limited 2011

Fig. 1.1 Linearly separable
problem

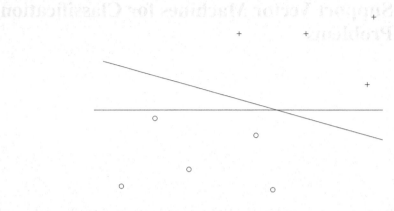

Fig. 1.2 Two support lines
with fixed direction

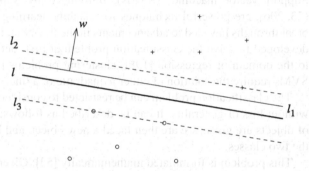

Fig. 1.3 The direction with
maximum margin

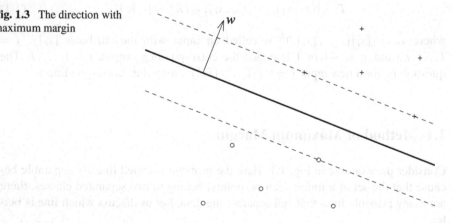

Now how to choose the direction w of the line? Just as the description above, for a
given w, we will get two support lines, the distance between them is called "margin"
corresponding to w. We can image that the direction with maximum margin should
be chosen, as in Fig. 1.3.

If the equation of the separating line is given as

$$(w \cdot x) + b = 0, \tag{1.2}$$

there is some redundancy in (1.2), and without loss of generality it is appropriate to consider a canonical hyperplane, where the parameters w, b are constrained so that the equation of line l_2 is

$$(w \cdot x) + b = 1, \tag{1.3}$$

and line l_3 is given as

$$(w \cdot x) + b = -1. \tag{1.4}$$

So the margin is given by $\frac{2}{\|w\|}$. The idea of maximizing the margin introduces the following optimization problem:

$$\min_{w,b} \frac{1}{2} \|w\|^2, \tag{1.5}$$

$$\text{s.t.} \quad y_i((w \cdot x_i) + b) \geq 1, \quad i = 1, \dots, l. \tag{1.6}$$

The above method is deduced for classification problem in 2-dimensional space, but it also works for general n dimension space, where the corresponding line becomes hyperplane.

1.2 Dual Problem

The solution to the optimization problem (1.5)–(1.6) is given by its Lagrangian dual problem,

$$\min_{\alpha} \frac{1}{2} \sum_{i=1}^{l} \sum_{j=1}^{l} y_i y_j \alpha_i \alpha_j (x_i \cdot x_j) - \sum_{j=1}^{l} \alpha_j, \tag{1.7}$$

$$\text{s.t.} \quad \sum_{i=1}^{l} y_i \alpha_i = 0, \tag{1.8}$$

$$\alpha_i \geq 0, \quad i = 1, \dots, l. \tag{1.9}$$

Theorem 1.1 *Considering the linearly separable problem. Suppose $\alpha^* = (\alpha_1^*, \dots, \alpha_l^*)^T$ is a solution of dual problem (1.7)–(1.9), so $\alpha^* \neq 0$, i.e. there is a component $\alpha_j^* > 0$, and the solution (w^*, b^*) of primal problem (1.5)–(1.6) is given by*

$$w^* = \sum_{i=1}^{l} \alpha_i^* y_i x_i, \tag{1.10}$$

$$b^* = y_j - \sum_{i=1}^{l} y_i \alpha_i^* (x_i \cdot x_j); \tag{1.11}$$

or

$$w^* = \sum_{i=1}^{l} \alpha_i^* y_i x_i, \tag{1.12}$$

$$b^* = -\left(w^* \cdot \sum_{i=1}^{l} \alpha_i^* x_i \right) \bigg/ \left(2 \sum_{y_i=1} \alpha_i^* \right). \tag{1.13}$$

After getting the solution (w^*, b^*) of primal problem, the optimal separating hyperplane is given by

$$(w^* \cdot x) + b^* = 0. \tag{1.14}$$

Definition 1.2 (Support vector)   Suppose $\alpha^* = (\alpha_1^*, \ldots, \alpha_l^*)^T$ is a solution of dual problem (1.7)–(1.9). The input x_i corresponding to $\alpha_i^* > 0$ is termed support vector (SV).

For the case of linearly separable problem, all the SVs will lie on the hyperplane $(w^* \cdot x) + b^* = 1$ or $(w^* \cdot x) + b^* = -1$, this result can be derived from the proof above, and hence the number of SV can be very small. Consequently the separating hyperplane is determined by a small subset of the training set; the other points could be removed from the training set and recalculating the hyperplane would produce the same answer.

1.3 Soft Margin

So far the discussion has been restricted to the case where the training data is linearly separable. However, in general this will not be the case if noises cause the overlap of the classes, e.g., Fig. 1.4. To accommodate this case, one introduces slack variables ξ_i for all $i = 1, \ldots, l$ in order to relax the constraints of (1.6)

$$y_i((w \cdot x_i) + b) \geq 1 - \xi_i, \quad i = 1, \ldots, l. \tag{1.15}$$

A satisfying classifier is then found by controlling both the margin term $\|w\|$ and the sum of the slacks $\sum_{i=1}^{l} \xi_i$. One possible realization of such a soft margin classifier is obtained by solving the following problem.

$$\min_{w,b,\xi} \frac{1}{2} \|w\|^2 + C \sum_{i=1}^{l} \xi_i, \tag{1.16}$$

Fig. 1.4 Linear classification
problem with overlap

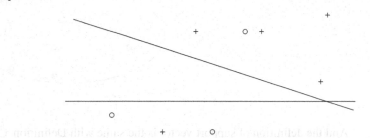

$$\text{s.t.} \quad y_i((w \cdot x_i) + b) + \xi_i \geq 1, \quad i = 1, \ldots, l, \tag{1.17}$$

$$\xi_i \geq 0, \quad i = 1, \ldots, l, \tag{1.18}$$

where the constant $C > 0$ determines the trade-off between margin maximization and training error minimization.

This again leads to the following Lagrangian dual problem

$$\min_{\alpha} \frac{1}{2} \sum_{i=1}^{l} \sum_{j=1}^{l} y_i y_j \alpha_i \alpha_j (x_i \cdot x_j) - \sum_{j=1}^{l} \alpha_j, \tag{1.19}$$

$$\text{s.t.} \quad \sum_{i=1}^{l} y_i \alpha_i = 0, \tag{1.20}$$

$$0 \leq \alpha_i \leq C, \quad i = 1, \ldots, l. \tag{1.21}$$

where the only difference from problem (1.7)–(1.9) of separable case is an upper bound C on the Lagrange multipliers α_i.

Similar with Theorem 1.1, we also get a theorem as follows:

Theorem 1.3 *Suppose* $\alpha^* = (\alpha_1^*, \ldots, \alpha_l^*)^T$ *is a solution of dual problem* (1.19)–(1.21). *If there exist* $0 < \alpha_j^* < C$, *then the solution* (w^*, b^*) *of primal problem* (1.16)–(1.18) *is given by*

$$w^* = \sum_{i=1}^{l} \alpha_i^* y_i x_i \tag{1.22}$$

and

$$b^* = y_j - \sum_{i=1}^{l} y_i \alpha_i^* (x_i \cdot x_j). \tag{1.23}$$

Fig. 1.5 Nonlinear
classification problem

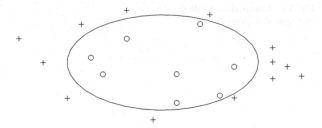

And the definition of support vector is the same with Definition 1.2.

1.4 *C*-Support Vector Classification

For the case where a linear boundary is totally inappropriate, e.g., Fig. 1.5. We can map the input x into a high dimensional feature space $x = \Phi(x)$ by introducing a mapping Φ, if an appropriate non-linear mapping is chosen a priori, an optimal separating hyperplane may be constructed in this feature space. And in this space, the primal problem and dual problem solved becomes separately

$$\min_{w,b,\xi} \frac{1}{2}\|w\|^2 + C\sum_{i=1}^{l}\xi_i, \tag{1.24}$$

$$\text{s.t.} \quad y_i((w \cdot \Phi(x_i)) + b) + \xi_i \geq 1, \quad i = 1,\ldots,l, \tag{1.25}$$

$$\xi_i \geq 0, \quad i = 1,\ldots,l. \tag{1.26}$$

$$\min_{\alpha} \frac{1}{2}\sum_{i=1}^{l}\sum_{j=1}^{l} y_i y_j \alpha_i \alpha_j (\Phi(x_i) \cdot \Phi(x_j)) - \sum_{j=1}^{l}\alpha_j, \tag{1.27}$$

$$\text{s.t.} \quad \sum_{i=1}^{l} y_i \alpha_i = 0, \tag{1.28}$$

$$0 \leq \alpha_i \leq C, \quad i = 1,\ldots,l. \tag{1.29}$$

As the mapping appears only in the dot product $(\Phi(x_i) \cdot \Phi(x_j))$, so by introducing a function $K(x,x') = (\Phi(x) \cdot \Phi(x'))$ termed kernel function, the above dual problem turns to be

$$\min_{\alpha} \frac{1}{2}\sum_{i=1}^{l}\sum_{j=1}^{l} y_i y_j \alpha_i \alpha_j K(x_i, x_j) - \sum_{j=1}^{l}\alpha_j, \tag{1.30}$$

$$\text{s.t.} \quad \sum_{i=1}^{l} y_i \alpha_i = 0, \tag{1.31}$$

$$0 \leq \alpha_i \leq C, \quad i = 1,\ldots,l. \tag{1.32}$$

Theorem 1.4 *Suppose* $\alpha^* = (\alpha_1^*, \ldots, \alpha_l^*)^{\mathrm{T}}$ *is a solution of dual problem* (1.30)–(1.32). *If there exist* $0 < \alpha_j^* < C$, *then the optimal separating hyperplane in the feature space is given by*

$$f(x) = \mathrm{sgn}((w^* \cdot x) + b^*) = \sum_{i=1}^{l} \alpha_i^* y_i K(x_i, x) + b^*, \qquad (1.33)$$

where

$$b^* = y_j - \sum_{i=1}^{l} y_i \alpha_i^* K(x_i, x_j). \qquad (1.34)$$

If $K(x, x')$ is a symmetric positive definite function, which satisfies Mercer's conditions, then the function represents an inner product in feature space and termed kernel function. The idea of the kernel function is to enable operations to be performed in the input space rather than the potentially high dimensional feature space. Hence the inner product does not need to be evaluated in the feature space. This provides a way of addressing the curse of dimensionality.

Examples of kernel function are now given [174]:

(1) linear kernels

$$K(x, x') = (x \cdot x'); \qquad (1.35)$$

(2) polynomial kernels are popular methods for non-linear modeling,

$$K(x, x') = (x \cdot x')^d, \qquad (1.36)$$

$$K(x, x') = ((x \cdot x') + 1)^d; \qquad (1.37)$$

(3) radial basis kernels have received significant attention, most commonly with a Gaussian of the form

$$K(x, x') = \exp(-\|x - x'\|^2 / \sigma^2); \qquad (1.38)$$

(4) sigmoid kernels

$$K(x, x') = \tanh(\kappa(x \cdot x') + \vartheta), \qquad (1.39)$$

where $\kappa > 0$ and $\vartheta < 0$.

Therefore, based on Theorem 1.4, the standard algorithm of Support Vector Machine for classification is given as follows:

Algorithm 1.5 (*C*-Support Vector Classification (*C*-SVC))

(1) Given a training set $T = \{(x_1, y_1), \ldots, (x_l, y_l)\} \in (R^n \times \{-1, 1\})^l$;
(2) Select a kernel $K(\cdot, \cdot)$, and a parameter $C > 0$;
(3) Solve problem (1.30)–(1.32) and get its solution $\alpha^* = (\alpha_1^*, \ldots, \alpha_l^*)^{\mathrm{T}}$;

(4) Compute the threshold b^*, and construct the decision function as

$$f(x) = \text{sgn}\left(\sum_{i=1}^{l} \alpha_i^* y_i K(x_i, x) + b^*\right).$$ (1.40)

1.5 C-Support Vector Classification with Nominal Attributes

For the classification problem, we are often given a training set like (1.1), where the attributes $[x_i]_j$ and $[\bar{x}]_j$, $j = 1, \ldots, n$, are allowed to take either continuous values or nominal values [204].

Now we consider the training set (1.1) with nominal attributes [199]. Suppose the input $x = ([x]_1, \ldots, [x]_n)^T$, where the jth nominal attribute $[x]_j$ take M_j states, $j = 1, \ldots, n$. The most popular approach in classification method is as follows: Let R^{M_j} be the M_j-dimensional space. The jth nominal attribute $[x]_j$ is represented as one of the M_j unit vectors in R^{M_j}. Thus the input space of the training set (1.1) can be embedded into a Euclidean space with the dimension $M_1 \times M_2 \times \cdots \times M_n$, and every input x is represented by n unit vectors which belong to the spaces $R^{M_1}, R^{M_2}, \ldots, R^{M_{n-1}}$ and R^{M_n} respectively.

However, the above strategy has a severe shortcoming in distance measure. The reason is that it assumes that all attribute values are of equal distance from each other. The equal distance implies that any two different attribute values have the same degree of dissimilarities. Obviously this is not always to be preferred.

1.5.1 From Fixed Points to Flexible Points

Let us improve the above most popular approach by overcoming the shortcoming pointed out in the end of the last section.

We deal with the training set (1.1) in the following way. Suppose that the jth nominal attribute $[x]_j$ takes values in M_j states

$$[x]_j \in \{v_{j1}, v_{j2}, \ldots, v_{jM_j}\}, \quad j = 1, \ldots, n.$$ (1.41)

We embed the jth nominal attribute $[x]_j$ into an $M_j - 1$ dimensional Euclidean space R^{M_j-1}: the first value v_{j1} corresponds to the point $(0, \ldots, 0)^T$, the second value v_{j2} corresponds to the point $(\sigma_1^j, 0, \ldots, 0)^T$, the third value v_{j3} correspond to the point $(\sigma_2^j, \sigma_3^j, 0, \ldots, 0)^T, \ldots$, and the last value v_{jM_j} corresponds to the point $(\sigma_{q_j+1}^j, \ldots, \sigma_{q_j+M_j-1}^j)^T$, where $q_j = \frac{(M_j-1)(M_j-2)}{2}$. Therefore for the jth nominal attribute $[x]_j$, there are p_j variables $\{\sigma_1^j, \sigma_2^j, \ldots, \sigma_{p_j}^j\}$ to be determined, where

$$p_j = \frac{M_j(M_j - 1)}{2}.$$ (1.42)

In other words, for $j = 1, \ldots, n$, the jth nominal attribute $[x]_j$ corresponds to a matrix

$$
\begin{pmatrix} v_{j1} \\ v_{j2} \\ v_{j3} \\ \vdots \\ v_{jM_j} \end{pmatrix} \longrightarrow \begin{pmatrix} 0 & 0 & 0 & \cdots & 0 \\ \sigma_1^j & 0 & 0 & \cdots & 0 \\ \sigma_2^j & \sigma_3^j & 0 & \cdots & 0 \\ \vdots & \vdots & \vdots & \vdots & \vdots \\ \sigma_{q_j+1}^j & \sigma_{q_j+2}^j & \sigma_{q_j+2}^j & \cdots & \sigma_{p_j}^j \end{pmatrix} \triangleq H_j \in R^{M_j \times (M_j - 1)}.
$$

(1.43)

Suppose an input $x = ([x]_1, \ldots, [x]_n)^{\mathrm{T}}$ taking nominal value $(v_{1k_1}, v_{2k_2}, \ldots, v_{nk_n})$, where k_j is the k_jth value in $\{v_{j1}, v_{j2}, \ldots, v_{jM_j}\}$. Then x corresponds to a vector

$$
x \to \tilde{x} = ((H_1)_{k_1}, \ldots, (H_n)_{k_n})^{\mathrm{T}},
$$

(1.44)

where $(H_j)_{k_j}$ is the k_jth row of H_j, $j = 1, \ldots, n$. Thus the training set (1.1) turns to be

$$
\tilde{T} = \{(\tilde{x}_1, y_1), \ldots, (\tilde{x}_l, y_l)\},
$$

(1.45)

where $\tilde{x}_i$ is obtained from x_i by the relationship (1.44) and (1.43).

Obviously, if we want to construct a decision function based on the training set (1.45) by *C*-SVC, the final decision function depends on the positions of the above embedded points, in other words, depends on the set

$$
\Sigma = \{\sigma_1^j, \sigma_2^j, \ldots, \sigma_{p_j}^j, \ j = 1, \ldots, n\},
$$

(1.46)

where p_j is given by (1.42).

1.5.2 *C*-SVC with Nominal Attributes

The values of $\{\sigma_1^j, \sigma_2^j, \ldots, \sigma_{p_j}^j, j = 1, \ldots, n\}$ in Σ in (1.46) can be obtained by learning. For example, it is reasonable to select them such that the LOO error for SVC is minimized, see, e.g. [207].

The definition of LOO error for Algorithm 1.5 is given as follows:

Definition 1.6 Consider Algorithm 1.5 with the training set (1.45). Let $f_{\tilde{T}|t}(x)$ be the decision function obtained by the algorithm from the training set $\tilde{T}|t = \tilde{T} \setminus \{(\tilde{x}_t, y_t)\}$, then the LOO error of the algorithm with respect to the loss function $c(x, y, f(x))$ and the training set $\tilde{T}$ is defined as

$$
R_{\mathrm{LOO}}(\tilde{T}) = \sum_{i=1}^{l} c(\tilde{x}_i, y_i, f(\tilde{x}_i)).
$$

(1.47)

In the above definition, the loss function is usually taken to be the 0–1 loss function

$$c(\tilde{x}_i, y_i, f(\tilde{x}_i)) = \begin{cases} 0, & y_i = f(\tilde{x}_i); \\ 1, & y_i \neq f(\tilde{x}_i). \end{cases} \tag{1.48}$$

Therefore, we investigate the LOO error with (1.48) below.

Obviously, the LOO error $R_{\mathrm{LOO}}(\tilde{T})$ depends on the set (1.46)

$$R_{\mathrm{LOO}}(\tilde{T}) = R_{\mathrm{LOO}}(\tilde{T}; \Sigma). \tag{1.49}$$

The basic idea of our algorithm is: First, select the values in Σ by minimizing the LOO error, i.e. by solving the optimization problem:

$$\min_{\Sigma} \, R_{\mathrm{LOO}}(\tilde{T}; \Sigma). \tag{1.50}$$

Then, using the learned values in Σ to train SVC again, and construct the final decision function. This leads to the following algorithm—C-SVC with Nominal attributes (C-SVCN):

Algorithm 1.7 (C-SVCN)

(1) Given a training set T defined in (1.1) with nominal attributes, where the jth nominal attribute $[x]_j$ takes values in M_j states (1.41));

(2) Introducing a parameter set $\Sigma = \{\sigma_1^j, \sigma_2^j, \ldots, \sigma_{p_j}^j, j = 1, \ldots, n\}$ appeared in (1.43) and turn T (1.1) to $\tilde{T}$ (1.45);

(3) Select a kernel $K(\cdot, \cdot)$ and a parameter $C > 0$;

(4) Solve problem (1.50) with replacing T by $\tilde{T}$, and get the learned values $\bar{\Sigma} = \{\bar{\sigma}_1^j, \bar{\sigma}_2^j, \ldots, \bar{\sigma}_{p_j}^j, j = 1, \ldots, n\}$;

(5) Using the parameter values in $\bar{\Sigma}$, turn T to $\bar{T} = \{(\bar{x}_1, y_1), \ldots, (\bar{x}_l, y_l)\}$ via (1.45) with replacing "the wave $\sim$" by "the bar $-$";

(6) Solve problem (1.30)–(1.32) with replacing T by $\bar{T}$ and get the solution $\alpha^* = (\alpha_1^*, \ldots, \alpha_l^*)^{\mathrm{T}}$;

(7) Compute the threshold b by KKT conditions, and construct the decision function as

$$f(\bar{x}) = \mathrm{sgn}\left(\sum_{i=1}^{l} \alpha_i^* y_i K(\bar{x}_i, \bar{x}) + b\right). \tag{1.51}$$

where $\bar{x}$ is obtained from x by the relationship (1.44) with replacing $\tilde{x}$ by $\bar{x}$.

1.5.3 Numerical Experiments

In this section, the preliminary experiments on Algorithm 1.7 are presented. As shown in Table 1.1, some standard data sets in [21] are tested by executing Algo-

Table 1.1 Data sets

Data set	♯ Nominal attributes	♯ Training patterns	♯ Test patterns
monks-1	6	300	132
monks-3	6	300	132
tic-tac-toe	9	500	458

Table 1.2 Classification errors on testing set

Data set	Popular SVCN	New SVCN
monks-1	21.2%	18.9%
monks-3	19.7%	16.7%
tic-tac-toe	22.9%	19.87%

rithm 1.7. For every set, we split it into two parts, one is for training, and the other for testing.

When we choose RBF kernel function $K(x, x') = \exp(\frac{-\|x-x'\|}{2\delta^2})$, and choose $(\delta, C) = (0.1, 10)$, $(\delta, C) = (0.7, 10)$, $(\delta, C) = (3, 0.1)$, $(\delta, C) = (3, 100)$ respectively, we compare Algorithm 1.7 with the most popular approach using unit vectors (Popular SVCN), the mean classification errors on testing sets are listed in Table 1.2. It is easy to see that Algorithm 1.7 leads to smaller classification errors.

Another simplified version of dealing with the jth nominal attribute with M_j states, $j = 1, \ldots, n$, is also another choice. Here (1.43) is replaced by

$$\begin{pmatrix} v_{j1} \\ v_{j2} \\ v_{j3} \\ \vdots \\ v_{jM_j} \end{pmatrix} \longrightarrow \begin{pmatrix} \sigma_1^j & 0 & 0 & \cdots & 0 \\ 0 & \sigma_2^j & 0 & \cdots & 0 \\ \cdots & & & & \\ 0 & 0 & 0 & \cdots & \sigma_{M_j}^j \end{pmatrix} \triangleq H_j \in R^{M_j \times M_j}. \tag{1.52}$$

Of course, extend Algorithm 1.7 to deal with the problem with both nominal and real attributes and the problem of feature selection are also interesting.

$$
\mathbf{K} \in \mathbb{R}^{M \times M}, \quad \mathbf{K} =
\begin{pmatrix}
0 & 0 & \cdots & 0 \\
0 & 0 & \cdots & 0 \\
\vdots & & \ddots & \vdots \\
0 & 0 & \cdots & 0
\end{pmatrix}
\quad (1.52)
$$

Chapter 2
LOO Bounds for Support Vector Machines

2.1 Introduction

The success of support vector machine depends on the tuning of its several parameters which affect the generalization error. For example, when given a training set, a practitioner will ask how to choose these parameters which will generalize well. An effective approach is to estimate the generalization error and then search for parameters so that this estimator is minimized. This requires that the estimators are both effective and computationally efficient. Devroye *et al.* [57] give an overview of error estimation. While some estimators (e.g., uniform convergence bounds) are powerful theoretical tools, they are of little use in practical applications, since they are too loose. Others (e.g., cross-validation, bootstrapping) give good estimates, but are computationally inefficient.

Leave-one-out (LOO) method is the extreme case of cross-validation, and in this case, a single point is excluded from the training set, and the classifier is trained using the remaining points. It is then determined whether this new classifier correctly labels the point that was excluded. The process is repeated over the entire training set, and the LOO error is computed by taking the average over these trials. LOO error provides an almost unbiased estimate of the generalization error.

However one shortcoming of the LOO method is that it is highly time consuming, thus methods are sought to speed up the process. An effective approach is to approximate the LOO error by its upper bound that is a function of the parameters. Then, we search for parameter so that this upper bound is minimized. This approach has successfully been developed for both support vector classification machine [97, 114, 119, 207] and support vector regression machine [34].

In this chapter we will introduce other LOO bounds for several algorithms of support vector machine [200, 201, 231].

Y. Shi et al., *Optimization Based Data Mining: Theory and Applications*, Advanced Information and Knowledge Processing, DOI 10.1007/978-0-85729-504-0_2, © Springer-Verlag London Limited 2011

2.2 LOO Bounds for ε-Support Vector Regression

2.2.1 Standard ε-Support Vector Regression

First, we introduce the standard ε-support vector regression (ε-SVR). Consider a regression problem with a training set

$$T = \{(x_1, y_1), \ldots, (x_l, y_l)\} \in (R^n \times \mathcal{Y})^l, \tag{2.1}$$

where $x_i \in R^n$, $y_i \in \mathcal{Y} = R$, $i = 1, \ldots, l$. Suppose that the loss function is selected to be the ε-insensitive loss function

$$c(x, y, f(x)) = |y - f(x)|_\varepsilon = \max\{0, |y - f(x)| - \varepsilon\}. \tag{2.2}$$

In support vector regression framework, the input space is first mapped to a higher dimensional space $\mathcal{H}$ by

$$x = \Phi(x), \tag{2.3}$$

and the training set T turns to be

$$\bar{T} = \{(x_1, y_1), \ldots, (x_l, y_l)\} \in (\mathcal{H} \times \mathcal{Y})^l, \tag{2.4}$$

where $x_i = \Phi(x_i) \in \mathcal{H}$, $y_i \in \mathcal{Y} = R$, $i = 1, \ldots, l$. Then in space $\mathcal{H}$, the following primal problem is constructed:

$$\min_{w,b,\xi,\xi^*} \frac{1}{2}\|w\|^2 + C \sum_{i=1}^{l}(\xi_i + \xi_i^*), \tag{2.5}$$

$$\text{s.t.} \quad (w \cdot x_i) + b - y_i \leq \varepsilon + \xi_i, \quad i = 1, \ldots, l, \tag{2.6}$$

$$y_i - (w \cdot x_i) - b \leq \varepsilon + \xi_i^*, \quad i = 1, \ldots, l, \tag{2.7}$$

$$\xi_i, \xi_i^* \geq 0, \quad i = 1, \ldots, l. \tag{2.8}$$

And ε-SVR solves this problem by introducing its dual problem

$$\max_{\alpha_T^{(*)}} J_T(\alpha^{(*)}) = -\frac{1}{2} \sum_{i,j=1}^{l} (\alpha_i^* - \alpha_i)(\alpha_j^* - \alpha_j) K(x_i, x_j)$$

$$- \varepsilon \sum_{i=1}^{l}(\alpha_i^* + \alpha_i) + \sum_{i=1}^{l} y_i(\alpha_i^* - \alpha_i), \tag{2.9}$$

$$\text{s.t.} \quad \sum_{i=1}^{l}(\alpha_i^* - \alpha_i) = 0, \tag{2.10}$$

$$0 \leq \alpha_i, \alpha_i^* \leq C, \quad i = 1, \ldots, l, \tag{2.11}$$

where $\alpha_T^{(*)} = (\alpha_1, \alpha_1^*, \ldots, \alpha_l, \alpha_l^*)^T$, and $K(x_i, x_j) = (x_i \cdot x_j) = (\Phi(x_i) \cdot \Phi(x_j))$ is the kernel function. Thus, the algorithm can be established as follows:

Algorithm 2.1 (ε-SVR)

(1) Given a training set T defined in (2.1);
(2) Select a kernel $K(\cdot, \cdot)$, and parameters $C > 0$ and $\varepsilon > 0$;
(3) Solve problem (2.9)–(2.11) and get its solution $\bar{\alpha}_T^{(*)} = (\bar{\alpha}_1, \bar{\alpha}_1^*, \ldots, \bar{\alpha}_l, \bar{\alpha}_l^*)^T$;
(4) Construct the decision function as

$$f(x) = f_T(x) = (\bar{w} \cdot x) + \bar{b} = \sum_{i=1}^{l}(\bar{\alpha}_i^* - \bar{\alpha}_i)K(x_i, x) + \bar{b}, \qquad (2.12)$$

where $\bar{b}$ is computed as follows: either choose one $\bar{\alpha}_j \in (0, C)$, then

$$\bar{b} = y_j - \sum_{i=1}^{l}(\bar{\alpha}_i^* - \bar{\alpha}_i)K(x_i, x_j) + \varepsilon; \qquad (2.13)$$

or choose one $\bar{\alpha}_k^* \in (0, C)$, then

$$\bar{b} = y_k - \sum_{i=1}^{l}(\bar{\alpha}_i^* - \bar{\alpha}_i)K(x_i, x_k) - \varepsilon. \qquad (2.14)$$

The uniqueness of the solution of primal problem (2.5)–(2.8) is shown by the following theorem [29].

Theorem 2.2 *Suppose* $\bar{\alpha}_T^{(*)} = (\bar{\alpha}_1, \bar{\alpha}_1^*, \ldots, \bar{\alpha}_l, \bar{\alpha}_l^*)^T$ *is an optimal solution of dual problem* (2.9)–(2.11), *and there exists a subscript i such that either* $0 < \bar{\alpha}_i < C$ *or* $0 < \bar{\alpha}_i^* < C$. *Then the decision function*

$$f(x) = f_T(x) = (w \cdot x) + b = \sum_{i=1}^{l}(\bar{\alpha}_i^* - \bar{\alpha}_i)K(x, x_i) + b$$

obtained by Algorithm 2.1 is unique.

2.2.2 The First LOO Bound

The kernel and the parameters in Algorithm 2.1 are reasonably selected by minimizing the LOO error or its bound. In this section, we recall the definition of this error at first, and then estimate its bound.

The definition of LOO error with respect to Algorithm 2.1 is given as follows:

Definition 2.3 For Algorithm 2.1, consider the ε-insensitive loss function (2.2) and the training set

$$T = \{(x_1, y_1), \ldots, (x_l, y_l)\} \in \{R^n \times \mathcal{Y}\}^l, \tag{2.15}$$

where $x_i \in R^n$, $y_i \in \mathcal{Y} = R$. Let $f_{T|t}(x)$ be the decision function obtained by Algorithm 2.1 from the training set $T|t = T \setminus \{(x_t, y_t)\}$, then the LOO error of Algorithm 2.1 (with respect to the loss function (2.2) and the training set T) is defined as

$$R_{\text{LOO}}(T) = \sum_{t=1}^{l} |f_{T|t}(x_t) - y_t|_\varepsilon. \tag{2.16}$$

Obviously, the computation cost of the LOO error is very expensive if l is large. In fact, for a training set including l points, the computing of the LOO error implies l times of training. So finding a more easily computed approximation of the LOO error is necessary. An interesting approach is to estimate an upper bound of the LOO error, such that this bound can be computed by training only once.

Now we derive an upper bound of the LOO error for Algorithm 2.1. Obviously, its LOO bound is related with the training set $T|t = T \setminus \{(x_t, y_t)\}$, $t = 1, \ldots, l$. The corresponding primal problem is

$$\min_{w^t, \xi^t, \xi^{t*}} \frac{1}{2} \|w^t\|^2 + C \sum_{i=1}^{l} (\xi_i^t + \xi_i^{*t}), \tag{2.17}$$

$$\text{s.t.} \quad (w^t \cdot x_i) + b^t - y_i \le \varepsilon + \xi_i^t, \quad i = 1, \ldots, t-1, t+1, \ldots, l, \tag{2.18}$$

$$y_i - (w^t \cdot x_i) - b^t \le \varepsilon + \xi_i^{*t}, \quad i = 1, \ldots, t-1, t+1, \ldots, l, \tag{2.19}$$

$$\xi_i^t, \xi_i^{*t} \ge 0, \quad i = 1, \ldots, t-1, t+1, \ldots, l. \tag{2.20}$$

Its dual problem is

$$\max_{\alpha_{T|t}^{(*)}} J_{T|t}(\alpha_{T|t}^{(*)}) = -\frac{1}{2} \sum_{i,j \ne t} (\alpha_i^* - \alpha_i)(\alpha_j^* - \alpha_j) K(x_i, x_j)$$

$$- \varepsilon \sum_{i \ne t} (\alpha_i^* + \alpha_i) + \sum_{i \ne t} y_i(\alpha_i^* - \alpha_i), \tag{2.21}$$

$$\text{s.t.} \quad \sum_{i \ne t} (\alpha_i^* - \alpha_i) = 0, \tag{2.22}$$

$$0 \le \alpha_i, \alpha_i^* \le C, \quad i = 1, \ldots, t-1, t+1, \ldots, l, \tag{2.23}$$

where $\alpha_{T|t}^{(*)} = (\alpha_1, \alpha_1^*, \ldots, \alpha_{t-1}, \alpha_{t-1}^*, \alpha_{t+1}, \alpha_{t+1}^*, \ldots, \alpha_l, \alpha_l^*)^{\text{T}}$.

Now let us introduce useful lemmas:

Lemma 2.4 *Suppose problem* (2.9)–(2.11) *has a solution* $\bar{\alpha}_T^{(*)} = (\bar{\alpha}_1, \bar{\alpha}_1^*, \ldots, \bar{\alpha}_l, \bar{\alpha}_l^*)^{\text{T}}$ *with a subscript i such that either $0 < \bar{\alpha}_i < C$ or $0 < \bar{\alpha}_i^* < C$. Suppose also that, for all any $t = 1, \ldots, l$, problem* (2.21)–(2.23) *has a solution* $\tilde{\alpha}_{T|t}^{(*)} = (\tilde{\alpha}_1, \tilde{\alpha}_1^*, \ldots, \tilde{\alpha}_{t-1}, \tilde{\alpha}_{t-1}^*, \tilde{\alpha}_{t+1}, \tilde{\alpha}_{t+1}^*, \ldots, \tilde{\alpha}_l, \tilde{\alpha}_l^*)^{\text{T}}$ *with a subscript j such that either*

$0 < \tilde{\alpha}_j < C$ or $0 < \tilde{\alpha}_j^* < C$. Let $f_T(x)$ and $f_{T|t}(x)$ be the decision functions obtained by Algorithm 2.1 respectively from the training set T and $T|t = T\backslash\{x_t, y_t\}$. Then for $t = 1, \ldots, l$, we have

  (i) If $\tilde{\alpha}_t = \tilde{\alpha}_t^* = 0$, then $|f_{T|t}(x_t) - y_t| = |f(x_t) - y_t|$;
 (ii) If $\tilde{\alpha}_t > 0$, then $f_{T|t}(x_t) \geq y_t$;
(iii) If $\tilde{\alpha}_t^* > 0$, then $f_{T|t}(x_t) \leq y_t$.

Proof Prove the case (i) first: Consider the primal problem (2.5)–(2.8) corresponding to the problem (2.9)–(2.11). Denote its solution as $(\bar{w}, \bar{b}, \bar{\xi}^{(*)})$. Note that the corresponding Lagrange multiplier vector is just the solution $\bar{\alpha}_T^{(*)} = (\bar{\alpha}_1, \bar{\alpha}_1^*, \ldots, \bar{\alpha}_l, \bar{\alpha}_l^*)^T$ of the problem (2.9)–(2.11). Therefore the KKT conditions can be represented as

$$\bar{w} = \sum_{i=1}^{l}(\bar{\alpha}_i^* - \bar{\alpha}_i)x_i, \tag{2.24}$$

$$\sum_{i=1}^{l}(\bar{\alpha}_i^* - \bar{\alpha}_i) = 0, \tag{2.25}$$

$$(\bar{w} \cdot x_i) + \bar{b} - y_i \leq \varepsilon + \bar{\xi}_i, \quad i = 1, \ldots, l, \tag{2.26}$$

$$y_i - (\bar{w} \cdot x_i) - \bar{b} \leq \varepsilon + \bar{\xi}_i^*, \quad i = 1, \ldots, l, \tag{2.27}$$

$$\bar{\xi}_i, \bar{\xi}_i^* \geq 0, \quad i = 1, \ldots, l, \tag{2.28}$$

$$((\bar{w} \cdot x_i) + \bar{b} - y_i - \varepsilon - \bar{\xi}_i)\bar{\alpha}_i = 0, \quad i = 1, \ldots, l, \tag{2.29}$$

$$((\bar{w} \cdot x_i) + \bar{b} - y_i + \varepsilon - \bar{\xi}_i^*)\bar{\alpha}_i^* = 0, \quad i = 1, \ldots, l, \tag{2.30}$$

$$(C - \bar{\alpha}_i)\bar{\xi}_i = 0, (C - \bar{\alpha}_i^*)\bar{\xi}_i^* = 0, \quad i = 1, \ldots, l, \tag{2.31}$$

$$0 \leq \bar{\alpha}_i, \bar{\alpha}_i^* \leq C, \quad i = 1, \ldots, l. \tag{2.32}$$

Define

$$\tilde{w} = \bar{w}, \qquad \tilde{b} = \bar{b}, \tag{2.33}$$

$$\tilde{\xi}^{(*)} \triangleq (\tilde{\xi}_1, \tilde{\xi}_1^*, \ldots, \tilde{\xi}_{t-1}, \tilde{\xi}_{t-1}^*, \tilde{\xi}_{t+1}, \tilde{\xi}_{t+1}^*, \ldots, \tilde{\xi}_l, \tilde{\xi}_l^*)^T$$
$$= (\bar{\xi}_1, \bar{\xi}_1^*, \ldots, \bar{\xi}_{t-1}, \bar{\xi}_{t-1}^*, \bar{\xi}_{t+1}, \bar{\xi}_{t+1}^*, \ldots, \bar{\xi}_l, \bar{\xi}_l^*)^T, \tag{2.34}$$

and

$$\tilde{\alpha}_{T|t}^{(*)} \triangleq (\tilde{\alpha}_1, \tilde{\alpha}_1^*, \ldots, \tilde{\alpha}_{t-1}, \tilde{\alpha}_{t-1}^*, \tilde{\alpha}_{t+1}, \tilde{\alpha}_{t+1}^*, \ldots, \tilde{\alpha}_l, \tilde{\alpha}_l^*)^T$$
$$= (\bar{\alpha}_1, \bar{\alpha}_1^*, \ldots, \bar{\alpha}_{t-1}, \bar{\alpha}_{t-1}^*, \bar{\alpha}_{t+1}, \bar{\alpha}_{t+1}^*, \ldots, \bar{\alpha}_l, \bar{\alpha}_l^*)^T. \tag{2.35}$$

It is not difficult to see, from (2.24)–(2.32), that $(\tilde{w}, \tilde{b}, \tilde{\xi}^{(*)})$ with the vector $\tilde{\alpha}_{T|t}^{(*)}$ satisfies the KKT conditions of problem (2.17)–(2.20). Therefore, $(\tilde{w}, \tilde{b}, \tilde{\xi}^{(*)})$ is the optimal solution of the problem (2.17)–(2.20). Noticing (2.33) and using Theorem 2.2, we claim that $f_{T|t}(x) = f(x)$, so

$$|f_{T|t}(x_t) - y_t| = |f(x_t) - y_t|. \tag{2.36}$$

Next, prove the case (ii): Consider the solution with respect to (w, b) of problem (2.5)–(2.8) and problem (2.17)–(2.20). There are two possibilities: They have respectively solution $(\bar{w}, \bar{b})$ and $(\tilde{w}, \tilde{b})$ with $(\bar{w}, \bar{b}) = (\tilde{w}, \tilde{b})$, or have no these solutions. For the former case, it is obvious, from the KKT condition (2.29), that we have

$$f_{T|t}(x_t) = (\tilde{w} \cdot x_t) + \tilde{b} = (\bar{w} \cdot x_t) + \bar{b} = y_t + \varepsilon + \bar{\xi}_t > y_t. \tag{2.37}$$

So we need only to investigate the latter case.

Let $(\bar{w}, \bar{b}, \bar{\xi}^{(*)})$ and $(\tilde{w}, \tilde{b}, \tilde{\xi}^{(*)})$ respectively be the solution of primal problem (2.5)–(2.8) and problem (2.17)–(2.20) with

$$(\bar{w}, \bar{b}) \neq (\tilde{w}, \tilde{b}). \tag{2.38}$$

We prove $f_{T|t}(x_t) \geq y_t$ by a contradiction. Suppose that $\bar{\alpha}_t > 0$, and $f_{T|t}(x_t) < y_t$. From the KKT condition (2.29), we have

$$(\bar{w} \cdot x_t) + \bar{b} = y_t + \varepsilon + \bar{\xi}_t \geq y_t + \varepsilon > y_t > f_{T|t}(x_t) = (\tilde{w} \cdot x_t) + \tilde{b}. \tag{2.39}$$

Therefore, there exists $0 < p < 1$ such that

$$(1 - p)((\bar{w} \cdot x_t) + \bar{b}) + p((\tilde{w} \cdot x_t) + \tilde{b}) = y_t. \tag{2.40}$$

Let

$$(\hat{w}, \hat{b}, \hat{\xi}^{(*)}) = (1 - p)(\bar{w}, \bar{b}, \bar{\xi}^{(*)}) + p(\tilde{w}, \tilde{b}, \check{\xi}^{(*)}), \tag{2.41}$$

where $\check{\xi}^{(*)}$ is obtained from $\tilde{\xi}^{(*)}$ by

$$\check{\xi} = (\tilde{\xi}_1, \tilde{\xi}_1^*, \ldots, \tilde{\xi}_{t-1}, \tilde{\xi}_{t-1}^*, 0, 0, \tilde{\xi}_{t+1}, \tilde{\xi}_{t+1}^*, \ldots, \tilde{\xi}_l, \tilde{\xi}_l^*)^{\mathrm{T}}. \tag{2.42}$$

Thus, $(\hat{w}, \hat{b}, \hat{\xi}^{(*)})$ with deleting the $(2t)$th and $(2t + 1)$th components of $\hat{\xi}^{(*)}$ is a feasible solution of problem (2.17)–(2.20). Therefore, noticing the convexity property, we have

$$\frac{1}{2}(\hat{w} \cdot \hat{w}) + C \sum_{i \neq t}(\hat{\xi}_i + \hat{\xi}_i^*)$$

$$\leq (1 - p)\left(\frac{1}{2}(\bar{w} \cdot \bar{w}) + C \sum_{i \neq t}(\bar{\xi}_i + \bar{\xi}_i^*)\right) + p\left(\frac{1}{2}(\tilde{w} \cdot \tilde{w}) + C \sum_{i \neq t}(\tilde{\xi}_i + \tilde{\xi}_i^*)\right)$$

$$< \frac{1}{2}(\bar{w} \cdot \bar{w}) + C \sum_{i \neq t}(\bar{\xi}_i + \bar{\xi}_i^*), \tag{2.43}$$

where the last inequality comes from the fact that $(\bar{w}, \bar{b}, \bar{\xi}^{(*)})$ with deleting the $(2t)$th and $(2t+1)$th components of $\bar{\xi}^{(*)}$ is a feasible solution of problem (2.17)–(2.20).

On the other hand, the fact $\bar{\alpha}_t > 0$, implies that $\bar{\xi}_t \geq 0$ and $\bar{\xi}_t^* = 0$. Thus, according to (2.42),

$$\hat{\xi}_t = (1-p)\bar{\xi}_t + p\check{\xi}_t = (1-p)\bar{\xi}_t \leq \bar{\xi}_t, \qquad \hat{\xi}_t^* = (1-p)\bar{\xi}_t^* + p\check{\xi}_t^* = 0. \quad (2.44)$$

From (2.40), we know that $(\hat{w}, \hat{b}, \hat{\xi}, \hat{\xi}^*)$ satisfies the constraint

$$-\varepsilon - \hat{\xi}_t^* \leq (\hat{w} \cdot x_t) + \hat{b} - y_t = 0 \leq \varepsilon + \hat{\xi}_t, \quad (2.45)$$

so it is also a feasible solution of problem (2.5)–(2.8). However from (2.43) and (2.44) we have

$$\frac{1}{2}(\hat{w} \cdot \hat{w}) + C\sum_{i=1}^{l}(\hat{\xi}_i + \hat{\xi}_i^*) = \frac{1}{2}(\hat{w} \cdot \hat{w}) + C\sum_{i \neq t}(\hat{\xi}_i + \hat{\xi}_i^*) + C(\hat{\xi}_t + \hat{\xi}_t^*)$$

$$< \frac{1}{2}(\bar{w} \cdot \bar{w}) + C\sum_{i \neq t}(\bar{\xi}_i + \bar{\xi}_i^*) + C(\bar{\xi}_t + \bar{\xi}_t^*)$$

$$= \frac{1}{2}(\bar{w} \cdot \bar{w}) + C\sum_{i=1}^{l}(\bar{\xi}_i + \bar{\xi}_i^*), \quad (2.46)$$

which is contradictive with that $(\bar{w}, \bar{\xi}, \bar{\xi}^*)$ is the solution of problem (2.5)–(2.8). Thus if $\alpha_t > 0$, there must be $f_{T|t}(x_t) \geq y_t$.

The proof of the case (iii) is similar to case (ii) and is omitted here. $\square$

Theorem 2.5 *Consider Algorithm 2.1. Suppose $\bar{\alpha}_T^{(*)} = (\bar{\alpha}_1, \bar{\alpha}_1^*, \ldots, \bar{\alpha}_l, \bar{\alpha}_l^*)^{\mathrm{T}}$ is the optimal solution of problem (2.9)–(2.11) and $f(x)$ is the corresponding decision function. Then the LOO error of this algorithm satisfies*

$$R_{\mathrm{LOO}}(T) \leq \sum_{t=1}^{l} |f(x_t) - y_t - (\bar{\alpha}_t^* - \bar{\alpha}_t)(R^2 + K(x_t, x_t))|_\varepsilon$$

$$= \sum_{t=1}^{l} \left| \sum_{i=1}^{l}(\bar{\alpha}_i^* - \bar{\alpha}_i)K(x_i, x_t) + \bar{b} - y_t - (\bar{\alpha}_t^* - \bar{\alpha}_t)(R^2 + K(x_t, x_t)) \right|_\varepsilon,$$

$$(2.47)$$

where

$$R^2 = \max\{K(x_i, x_j) \mid i, j = 1, \ldots, l\}. \quad (2.48)$$

Proof It is sufficient to prove that, for $t = 1, \ldots, l$,

$$|f(x_t) - y_t - (\bar{\alpha}_t^* - \bar{\alpha}_t)(R^2 + K(x_t, x_t))|_\varepsilon \geq |y_t - f_{T|t}(x_t)|_\varepsilon. \quad (2.49)$$

Its validity will be shown by investigating three cases separately:

(i) The case $\bar{\alpha}_t^* = \bar{\alpha}_t = 0$. In this case, by Lemma 2.4, $|f_{T|t}(x_t) - y_t| = |f(x_t) - y_t|$, it is obvious that

$$|f(x_t) - y_t - (\bar{\alpha}_t^* - \bar{\alpha}_t)(R^2 + K(x_t, x_t))| = |f_{T|t}(x_t) - y_t|, \qquad (2.50)$$

so the conclusion (2.49) is true.

(ii) The case $\bar{\alpha}_t > 0$. In this case, we have $\bar{\alpha}_t^* = 0$.

First we will construct a feasible solution of problem (2.9)–(2.11) from the solution of problem (2.21)–(2.23) to get some equality.

Suppose problem (2.21)–(2.23) has a solution $\tilde{\alpha}_{T|t}^{(*)} = (\tilde{\alpha}_1, \tilde{\alpha}_1^*, \ldots, \tilde{\alpha}_{t-1}, \tilde{\alpha}_{t-1}^*, \tilde{\alpha}_{t+1}, \tilde{\alpha}_{t+1}^*, \ldots, \tilde{\alpha}_l, \tilde{\alpha}_l^*)^T$, and there exists a subscript j such that $0 < \tilde{\alpha}_j < C$ or $0 < \tilde{\alpha}_j^* < C$. Now construct $\gamma^{(*)} = (\gamma_1, \gamma_1^*, \ldots, \gamma_l, \gamma_l^*)^T$ as follows:

$$\gamma_i^{(*)} = \begin{cases} \tilde{\alpha}_i^{(*)}, & \text{if } \alpha_i^{(*)} = 0 \text{ or } \alpha_i^{(*)} = C, \\ \tilde{\alpha}_i^{(*)} - v_i^{(*)}, & \text{if } i \in SV^t, \\ \tilde{\alpha}_i^{(*)}, & \text{if } i = t, \end{cases} \qquad (2.51)$$

where $SV^t = \{i \mid 0 < \tilde{\alpha}_i^{(*)} < C, \ i = 1, \ldots, t-1, t+1, l\}$, and $v = (v_1, v_1^*, \ldots, v_l, v_l^*)^T \geq 0$ satisfies

$$\sum_{i \in SV^t} v_i^* = \bar{\alpha}_t^* = 0, \qquad \sum_{i \in SV^t} v_i = \bar{\alpha}_t, \qquad v_i = 0 \quad \forall i \notin SV^t. \qquad (2.52)$$

It is easy to verify that

$$\sum_{i=1}^l (\gamma_i^* - \gamma_i) = 0, \quad 0 \leq \gamma_i, \gamma_i^* \leq C, \ i = 1, \ldots, l, \qquad (2.53)$$

so $\gamma^{(*)}$ is a feasible solution of problem (2.9)–(2.11),

$$J(\gamma^{(*)}) = J_{T|t}(\tilde{\alpha}_{T|t}^{(*)}) - \frac{1}{2}(\bar{\alpha}_t^* - \bar{\alpha}_t)^2 K(x_t, x_t) - \varepsilon(\bar{\alpha}_t^* + \bar{\alpha}_t) + y_t(\bar{\alpha}_t^* - \bar{\alpha}_t)$$

$$- (\bar{\alpha}_t^* - \bar{\alpha}_t) \sum_{i \neq t} (\tilde{\alpha}_i^* - \tilde{\alpha}_i) K(x_t, x_i)$$

$$- \sum_{i \in SV^t} (v_i^* - v_i)\left(y_i + \varepsilon - \sum_{j \neq t} (\tilde{\alpha}_j^* - \tilde{\alpha}_j) K(x_i, x_j) \right)$$

$$- \frac{1}{2} \sum_{i, j \in SV^t} (v_i^* - v_i)(v_j^* - v_j) K(x_i, x_j)$$

$$+ (\bar{\alpha}_t^* - \bar{\alpha}_t) \sum_{i \in SV^t} (v_i^* - v_i) K(x_i, x_t). \qquad (2.54)$$

Because there exist $0 < \tilde{\alpha}_i < C$ or $0 < \tilde{\alpha}_i^* < C$, so the solution with respect to b of problem (2.21)–(2.23) is unique, and we have

$$y_i + \varepsilon - \sum_{j \neq t} (\tilde{\alpha}_j^* - \tilde{\alpha}_j) K(x_i, x_j) = \tilde{b}; \tag{2.55}$$

furthermore, by

$$\sum_{i \in SV^T} (v_i^* - v_i) = \sum_{i \neq t} (v_i^* - v_i) = (\tilde{\alpha}_t^* - \tilde{\alpha}_t), \tag{2.56}$$

we get

$$(\tilde{\alpha}_t^* - \tilde{\alpha}_t) \left(\sum_{i \neq t} (\tilde{\alpha}_i^* - \tilde{\alpha}_i) K(x_t, x_i) + \tilde{b} \right)$$

$$= -J(\gamma^{(*)}) + J_{T|t}(\tilde{\alpha}_{T|t}^{(*)}) - \frac{1}{2}(\tilde{\alpha}_t^* - \tilde{\alpha}_t)^2 K(x_t, x_t) - \varepsilon(\tilde{\alpha}_t^* + \tilde{\alpha}_t) + y_t(\tilde{\alpha}_t^* - \tilde{\alpha}_t)$$

$$- \frac{1}{2} \sum_{i,j \in SV^t} (v_i^* - v_i)(v_j^* - v_j) K(x_i, x_j)$$

$$+ (\tilde{\alpha}_t^* - \tilde{\alpha}_t) \sum_{i \in SV^t} (v_i^* - v_i) K(x_i, x_t). \tag{2.57}$$

Next, we will construct a feasible solution of problem (2.21)–(2.23) from the solution of problem (2.9)–(2.11) to get another equality.

Similarly, we construct $\beta_{T|t}^{(*)} = (\beta_1, \beta_1^*, \ldots, \beta_{t-1}, \beta_{t-1}^*, \beta_{t+1}, \beta_{t+1}^*, \ldots, \beta_l, \beta_l^*)^\mathrm{T}$ from the solution $\bar{\alpha}^{(*)}$ of problem (2.9)–(2.11) as follows:

$$\beta_i^{(*)} = \begin{cases} \bar{\alpha}_i^{(*)}, & \text{if } \bar{\alpha}_i^{(*)} = 0 \text{ or } \bar{\alpha}_i^{(*)} = C, \\ \bar{\alpha}_i^{(*)} + \eta_i^{(*)}, & \text{if } i \in SV \setminus \{t\}, \end{cases} \tag{2.58}$$

where $SV = \{i \mid 0 < \bar{\alpha}_i^{(*)} < C, i = 1, \ldots, l\}$, and $\eta_{T|t}^{(*)} = (\eta_1, \eta_1^*, \ldots, \eta_{t-1}, \eta_{t-1}^*, \eta_{t+1}, \eta_{t+1}^*, \ldots, \eta_l, \eta_l^*)^\mathrm{T} \geq 0$ satisfies

$$\sum_{i \in SV \setminus \{t\}} \eta_i^* = \bar{\alpha}_t^* = 0, \qquad \sum_{i \in SV \setminus \{t\}} \eta_i = \bar{\alpha}_t. \tag{2.59}$$

It is easy to verify that

$$\sum_{i \neq t} (\beta_i^* - \beta_i) = 0, \quad 0 \leq \beta_i, \beta_i^* \leq C, \ i = 1, \ldots, t-1, t+1, \ldots, l. \tag{2.60}$$

So $\beta_{T|t}^{(*)}$ is a feasible solution of problem (2.21)–(2.23), and we have

$$J_{T|t}(\beta_{T|t}^{(*)}) = J(\bar{\alpha}^{(*)}) + \frac{1}{2}(\bar{\alpha}_t^* - \bar{\alpha}_t)^2 K(x_t, x_t) + \varepsilon(\bar{\alpha}_t^* + \bar{\alpha}_t) - y_t(\bar{\alpha}_t^* - \bar{\alpha}_t)$$

$$+ (\bar{\alpha}_t^* - \bar{\alpha}_t) \sum_{i \neq t} (\bar{\alpha}_i^* - \bar{\alpha}_i) K(x_t, x_i)$$

$$+ \sum_{i \in SV \setminus \{t\}} (\eta_i^* - \eta_i) \left(y_i + \varepsilon - \sum_{j \neq t} (\bar{\alpha}_j^* - \bar{\alpha}_j) K(x_i, x_j) \right)$$

$$- \frac{1}{2} \sum_{i,j \in SV \setminus \{t\}} (\eta_i^* - \eta_i)(\eta_j^* - \eta_j) K(x_i, x_j)$$

$$+ (\bar{\alpha}_t^* - \bar{\alpha}_t) \sum_{i \in SV \setminus \{t\}} (\eta_i^* - \eta_i) K(x_i, x_t). \qquad (2.61)$$

Because there exist $0 < \bar{\alpha}_i < C$ or $0 < \bar{\alpha}_i^* < C$, so the solution with respect to b of problem (2.9)–(2.11) is unique, and we have

$$y_i + \varepsilon - \sum_{j \neq t} (\bar{\alpha}_j^* - \bar{\alpha}_j) K(x_i, x_j) = \bar{b}; \qquad (2.62)$$

furthermore,

$$-J(\bar{\alpha}^{(*)}) = -J_{T|t}(\beta_{T|t}^{(*)}) + \frac{1}{2}(\bar{\alpha}_t^* - \bar{\alpha}_t)^2 K(x_t, x_t) + \varepsilon(\bar{\alpha}_t^* + \bar{\alpha}_t) - y_t(\bar{\alpha}_t^* - \bar{\alpha}_t)$$

$$+ (\bar{\alpha}_t^* - \bar{\alpha}_t) \left(\sum_{i \neq t} (\bar{\alpha}_i^* - \bar{\alpha}_i) K(x_t, x_i) + \bar{b} \right)$$

$$- \frac{1}{2} \sum_{i,j \in SV \setminus \{t\}} (\eta_i^* - \eta_i)(\eta_j^* - \eta_j) K(x_i, x_j)$$

$$+ (\bar{\alpha}_t^* - \bar{\alpha}_t) \sum_{i \in SV \setminus \{t\}} (\eta_i^* - \eta_i) K(x_i, x_t). \qquad (2.63)$$

Now, by the above two equalities (2.57) and (2.63), and the obvious facts $J(\gamma^{(*)}) \leq J(\bar{\alpha}^*)$ and $J_{T|t}(\beta_{T|t}^{(*)}) \leq J_{T|t}(\alpha_{T|t}^{(*)})$, we get

$$(\bar{\alpha}_t^* - \bar{\alpha}_t) \left(\sum_{i \neq t} (\bar{\alpha}_i^* - \bar{\alpha}_i) K(x_t, x_i) + \bar{b} \right)$$

$$\geq (\bar{\alpha}_t^* - \bar{\alpha}_t) \left(\sum_{i \neq t} (\bar{\alpha}_i^* - \bar{\alpha}_i) K(x_t, x_i) + \bar{b} \right)$$

$$- \frac{1}{2} \sum_{i,j \in SV^t} (v_i^* - v_i)(v_j^* - v_j) K(x_i, x_j)$$

$$-\frac{1}{2}\sum_{i,j\in SV\setminus\{t\}}(\eta_i^* - \eta_i)(\eta_j^* - \eta_j)K(x_i, x_j)$$

$$+(\bar{\alpha}_t^* - \bar{\alpha}_t)\sum_{i\in SV^t}(v_i^* - v_i)K(x_i, x_t)$$

$$+(\bar{\alpha}_t^* - \bar{\alpha}_t)\sum_{i\in SV\setminus\{t\}}(\eta_i^* - \eta_i)K(x_i, x_t). \tag{2.64}$$

By (2.52) and (2.59),

$$(\bar{\alpha}_t^* - \bar{\alpha}_t)\sum_{i\in SV^t}(v_i^* - v_i)K(x_i, x_t) \geq 0, \tag{2.65}$$

$$(\bar{\alpha}_t^* - \bar{\alpha}_t)\sum_{i\in SV\setminus\{t\}}(\eta_i^* - \eta_i)K(x_i, x_t) \geq 0. \tag{2.66}$$

Reminding the definition (2.48), we have

$$\frac{1}{2}\sum_{i,j\in SV^t}(v_i^* - v_i)(v_j^* - v_j)K(x_i, x_j) \leq \frac{1}{2}(\bar{\alpha}_t^* - \bar{\alpha}_t)^2 R^2, \tag{2.67}$$

$$\frac{1}{2}\sum_{i,j\in SV\setminus\{t\}}(\eta_i^* - \eta_i)(\eta_j^* - \eta_j)K(x_i, x_j) \leq \frac{1}{2}(\bar{\alpha}_t^* - \bar{\alpha}_t)^2 R^2, \tag{2.68}$$

therefore

$$\sum_{i\neq t}(\tilde{\alpha}_i^* - \tilde{\alpha}_i)K(x_t, x_i) + \tilde{b}$$

$$\leq \left(\sum_{i\neq t}(\bar{\alpha}_i^* - \bar{\alpha}_i)K(x_t, x_i) + \bar{b}\right) - (\bar{\alpha}_t^* - \bar{\alpha}_t)R^2. \tag{2.69}$$

By Lemma 2.4, the fact $\bar{\alpha}_t > 0$ implies that $f_{T|t}(x_t) \geq y_t$. Therefore

$$0 \leq \left(\sum_{i\neq t}(\tilde{\alpha}_i^* - \tilde{\alpha}_i)K(x_t, x_i) + \tilde{b}\right) - y_t$$

$$\leq \left(\sum_{i\neq t}(\bar{\alpha}_i^* - \bar{\alpha}_i)K(x_t, x_i) + \bar{b}\right) - (\bar{\alpha}_t^* - \bar{\alpha}_t)R^2 - y_t$$

$$= f(x_t) - y_t - (\bar{\alpha}_t^* - \bar{\alpha}_t)(R^2 + K(x_t, x_t)), \tag{2.70}$$

that is,

$$\left|\left(\sum_{i \neq t}(\tilde{\alpha}_i^* - \tilde{\alpha}_i)K(x_t, x_i) + \tilde{b}\right) - y_t\right|$$

$$\leq |f(x_t) - y_t - (\bar{\alpha}_t^* - \bar{\alpha}_t)(R^2 + K(x_t, x_t))| \tag{2.71}$$

and because the function $|\cdot|_\varepsilon$ is monotonically increasing, so the conclusion (2.49) is true.

(iii) The case $\bar{\alpha}_t^* > 0$. Similar with the discussion of case (ii), the conclusion (2.49) is true.   □

2.2.3 A Variation of ε-Support Vector Regression

If we consider the decision function with the formulation $f(x) = (w \cdot x)$ in ε-SVR, we will get the primal problem

$$\min_{w,\xi,\xi^*} \frac{1}{2}\|w\|^2 + C\sum_{i=1}^{l}(\xi_i + \xi_i^*), \tag{2.72}$$

$$\text{s.t.} \quad (w \cdot x_i) - y_i \leq \varepsilon + \xi_i, \quad i = 1,\ldots,l, \tag{2.73}$$

$$y_i - (w \cdot x_i) \leq \varepsilon + \xi_i^*, \quad i = 1,\ldots,l, \tag{2.74}$$

$$\xi_i, \xi_i^* \geq 0, \quad i = 1,\ldots,l. \tag{2.75}$$

The corresponding dual problem is

$$\max_{\alpha_T^{(*)}} J_T(\alpha_T^{(*)}) = -\frac{1}{2}\sum_{i,j=1}^{l}(\alpha_i^* - \alpha_i)(\alpha_j^* - \alpha_j)K(x_i, x_j)$$

$$- \varepsilon\sum_{i=1}^{l}(\alpha_i^* + \alpha_i) + \sum_{i=1}^{l}y_i(\alpha_i^* - \alpha_i), \tag{2.76}$$

$$\text{s.t.} \quad 0 \leq \alpha_i, \alpha_i^* \leq C, \quad i = 1,\ldots,l, \tag{2.77}$$

where $\alpha_T^{(*)} = (\alpha_1, \alpha_1^*, \ldots, \alpha_l, \alpha_l^*)^T$, and $K(x_i, x_j) = (x_i \cdot x_j) = (\Phi(x_i) \cdot \Phi(x_j))$ is the kernel function.

Thus the algorithm can be established as follows:

Algorithm 2.6 (A variation of the standard ε-SVR)

(1) Given a training set T defined in (2.1);
(2) Select a kernel $K(\cdot, \cdot)$, and parameters C and ε;
(3) Solve problem (2.76)–(2.77) and get its solution $\bar{\alpha}_T^{(*)} = (\bar{\alpha}_1, \bar{\alpha}_1^*, \ldots, \bar{\alpha}_l, \bar{\alpha}_l^*)^T$;

(4) Construct the decision function

$$f(x) = f_T(x) = (\mathbf{w} \cdot \mathbf{x}) = \sum_{i=1}^{l} (\bar{\alpha}_i^* - \bar{\alpha}_i) K(x, x_i). \tag{2.78}$$

Because the objective function of problem (2.72) is strictly convex with respect to w, so the solution of problem (2.72)–(2.75) with respect to w is unique. Therefore we have the following theorem.

Theorem 2.7 *The decision function*

$$f(x) = f_T(x) = (\mathbf{w} \cdot \mathbf{x}) = \sum_{i=1}^{l} (\bar{\alpha}_i^* - \bar{\alpha}_i) K(x, x_i)$$

obtained by Algorithm 2.6 is unique.

2.2.4 The Second LOO Bound

Now we derive an upper bound of the LOO error for Algorithm 2.6. Obviously, its LOO bound is related with the training set $T|t = T \setminus \{(x_t, y_t)\}$, $t = 1, \ldots, l$. The corresponding primal problem should be

$$\min_{\mathbf{w}^t, \xi^t, \xi^{t*}} \frac{1}{2} \|\mathbf{w}^t\|^2 + C \sum_{i=1}^{l} (\xi_i^t + \xi_i^{*t}), \tag{2.79}$$

$$\text{s.t.} \quad (\mathbf{w}^t \cdot x_i) - y_i \leq \varepsilon + \xi_i^t, \quad i = 1, \ldots, t-1, t+1, \ldots, l, \tag{2.80}$$

$$y_i - (\mathbf{w}^t \cdot x_i) \leq \varepsilon + \xi_i^{*t}, \quad i = 1, \ldots, t-1, t+1, \ldots, l, \tag{2.81}$$

$$\xi_i^t, \xi_i^{*t} \geq 0, \quad i = 1, \ldots, t-1, t+1, \ldots, l. \tag{2.82}$$

Its dual problem is

$$\max_{\alpha_{T|t}^{(*)}} J_{T|t}(\alpha_{T|t}^{(*)}) \triangleq -\frac{1}{2} \sum_{i,j \neq t} (\alpha_i^* - \alpha_i)(\alpha_j^* - \alpha_j) K(x_i, x_j)$$

$$- \varepsilon \sum_{i \neq t} (\alpha_i^* + \alpha_i) + \sum_{i \neq t} y_i (\alpha_i^* - \alpha_i), \tag{2.83}$$

$$\text{s.t.} \quad 0 \leq \alpha_i, \alpha_i^* \leq C, \quad i = 1, \ldots, t-1, t+1, \ldots, l, \tag{2.84}$$

where $\alpha_{T|t}^{(*)} = (\alpha_1, \alpha_1^*, \ldots, \alpha_{t-1}, \alpha_{t-1}^*, \alpha_{t+1}, \alpha_{t+1}*, \ldots, \alpha_l, \alpha_l^*)^{\mathrm{T}}$.

Now let us introduce two useful lemmas:

Lemma 2.8 *Suppose problem* (2.76)–(2.77) *has a solution* $\bar{\alpha}_T^{(*)} = (\bar{\alpha}_1, \bar{\alpha}_1^*, \ldots, \bar{\alpha}_l, \bar{\alpha}_l^*)^{\mathrm{T}}$. *Let* $f_T(x)$ *and* $f_{T|t}(x)$ *be the decision functions obtained by Algorithm* 2.6 *respectively from the training set* T *and* $T|t = T \setminus \{x_t, y_t\}$. *Then for* $t = 1, \ldots, l$, *we have*

(i) *If* $\bar{\alpha}_t = \bar{\alpha}_t^* = 0$, *then* $|f_{T|t}(x_t) - y_t| = |f(x_t) - y_t|$;
(ii) *If* $\bar{\alpha}_t > 0$, *then* $f_{T|t}(x_t) \geq y_t$;
(iii) *If* $\bar{\alpha}_t^* > 0$, *then* $f_{T|t}(x_t) \leq y_t$.

Proof It's proof is similar with Lemma 2.4, and is omitted here.   □

Lemma 2.9 *Suppose problem* (2.76)–(2.77) *has a solution* $\bar{\alpha}_T^{(*)} = (\bar{\alpha}_1, \bar{\alpha}_1^*, \ldots, \bar{\alpha}_l, \bar{\alpha}_l^*)^{\mathrm{T}}$. *Let* $f_{T|t}(x)$ *be the decision function obtained by Algorithm* 2.6 *from the training set* $T|t = T - \{(x_t, y_t)\}$. *Then for* $t = 1, \ldots, l$, *we have*

$$-(\bar{\alpha}_t^* - \bar{\alpha}_t) \sum_{i \neq t} (\bar{\alpha}_i^* - \bar{\alpha}_i) K(x_i, x_t) \geq -(\bar{\alpha}_t^* - \bar{\alpha}_t) f_{T|t}(x_t). \qquad (2.85)$$

Proof Obviously problem (2.76)–(2.77) can be expressed as

$$\max_{\alpha_T^{(*)}} J_T(\alpha_T^{(*)}) = J_{T|t}(\alpha_{T|t}^{(*)}) - (\alpha_t^* - \alpha_t) \sum_{i \neq t} (\alpha_i^* - \alpha_i) K(x_i, x_t)$$

$$- \frac{1}{2}(\alpha_t^* - \alpha_t)^2 K(x_t, x_t) - \varepsilon(\alpha_t^* + \alpha_t) + y_t(\alpha_t^* - \alpha_t), \quad (2.86)$$

$$\text{s.t.} \quad 0 \leq \alpha_i, \alpha_i^* \leq C, \quad i = 1, \ldots, l, \qquad (2.87)$$

where $J_{T|t}(\alpha_{T|t}^{(*)})$ is given by (2.83). Because $\bar{\alpha}_T^{(*)} = (\bar{\alpha}_1, \bar{\alpha}_1^*, \ldots, \bar{\alpha}_l, \bar{\alpha}_l^*)^{\mathrm{T}}$ is the solution of problem (2.76)–(2.77) or (2.86)–(2.87), then problem (2.86)–(2.87) can be rewritten as

$$\max_{\alpha_T^{(*)}} J_{T|t}(\alpha_{T|t}^{(*)}) = (\alpha_t^* - \alpha_t) \sum_{i \neq t} (\alpha_i^* - \alpha_i) K(x_i, x_t)$$

$$- \frac{1}{2}(\alpha_t^* - \alpha_t)^2 K(x_t, x_t) - \varepsilon(\alpha_t^* + \alpha_t) + y_t(\alpha_t^* - \alpha_t), \quad (2.88)$$

$$\text{s.t.} \quad 0 \leq \alpha_i, \alpha_i^* \leq C, \quad i = 1, \ldots, t-1, t+1, \ldots, l, \qquad (2.89)$$

$$\alpha_t^{(*)} = \bar{\alpha}_t^{(*)}. \qquad (2.90)$$

Substitute the equality (2.90) into the objective function directly, the problem (2.88)–(2.90) turns to be

$$\max_{\alpha_{T|t}^{(*)}} \hat{J}_T(\alpha_{T|t}^{(*)}) \triangleq J_{T|t}(\alpha_{T|t}^{(*)}) - (\bar{\alpha}_t^* - \bar{\alpha}_t) \sum_{i \neq t} (\alpha_i^* - \alpha_i) K(x_i, x_t), \quad (2.91)$$

$$\text{s.t.} \quad 0 \leq \alpha_i, \alpha_i^* \leq C, \quad i = 1, \ldots, t-1, t+1, \ldots, l. \qquad (2.92)$$

It is easy to see that

$$\hat{\alpha}_{T|t}^{(*)} \triangleq (\hat{\alpha}_1, \hat{\alpha}_1^*, \ldots, \hat{\alpha}_{t-1}, \hat{\alpha}_{t-1}^*, \hat{\alpha}_{t+1}, \hat{\alpha}_{t+1}^*, \ldots, \hat{\alpha}_l, \hat{\alpha}_l^*)^{\mathrm{T}}$$

$$= (\bar{\alpha}_1, \bar{\alpha}_1^*, \ldots, \bar{\alpha}_{t-1}, \bar{\alpha}_{t-1}^*, \bar{\alpha}_{t+1}, \bar{\alpha}_{t+1}^*, \ldots, \bar{\alpha}_l, \bar{\alpha}_l^*)^{\mathrm{T}} \qquad (2.93)$$

is an optimal solution of problem (2.91)–(2.92). Because $\tilde{\alpha}_{T|t}^{(*)}$ is the optimal solution of problem (2.83)–(2.84), and is also a feasible solution of problem (2.91)–(2.92), we have

$$\widehat{J}_T(\hat{\alpha}_{T|t}^{(*)}) \geq \widehat{J}_T(\tilde{\alpha}_{T|t}^{(*)}). \qquad (2.94)$$

Therefore by (2.91) and (2.93),

$$J_{T|t}(\hat{\alpha}_{T|t}^{(*)}) - (\bar{\alpha}_t^* - \bar{\alpha}_t) \sum_{i \neq t} (\hat{\alpha}_i^* - \hat{\alpha}_i) K(x_i, x_t)$$

$$\geq J_{T|t}(\tilde{\alpha}_{T|t}^{(*)}) - (\bar{\alpha}_t^* - \bar{\alpha}_t) \sum_{i \neq t} (\tilde{\alpha}_i^* - \tilde{\alpha}_i) K(x_i, x_t), \qquad (2.95)$$

that is

$$-(\bar{\alpha}_t^* - \bar{\alpha}_t) \sum_{i \neq t} (\tilde{\alpha}_i^* - \tilde{\alpha}_i) K(x_i, x_t)$$

$$\geq J_{T|t}(\tilde{\alpha}_{T|t}^{(*)}) - J_{T|t}(\hat{\alpha}_{T|t}^{(*)}) - (\bar{\alpha}_t^* - \bar{\alpha}_t) \sum_{i \neq t} (\hat{\alpha}_i^* - \hat{\alpha}_i) K(x_i, x_t). \qquad (2.96)$$

Because $\tilde{\alpha}_{T|t}^{(*)}$ maximizes the objective function $J_{T|t}(\alpha_{T|t}^{(*)})$, we have

$$J_{T|t}(\tilde{\alpha}_{T|t}^{(*)}) - J_{T|t}(\hat{\alpha}_{T|t}^{(*)}) \geq 0. \qquad (2.97)$$

So the conclusion comes from (2.96) and (2.97). □

Now we are in a position to show our main conclusion.

Theorem 2.10 *Consider Algorithm* 2.6. *Suppose* $\bar{\alpha}_T^{(*)} = (\bar{\alpha}_1, \bar{\alpha}_1^*, \ldots, \bar{\alpha}_l, \bar{\alpha}_l^*)^{\mathrm{T}}$ *is the optimal solution of problem* (2.76)–(2.77) *and* $f(x)$ *is the decision function. Then the LOO error of this algorithm satisfies*

$$R_{\mathrm{LOO}}(T) \leq \frac{1}{l} \sum_{t=1}^{l} |y_t - f(x_t) + (\bar{\alpha}_t^* - \bar{\alpha}_t) K(x_t, x_t)|_\varepsilon. \qquad (2.98)$$

Proof It is sufficient to prove that, for $t = 1, \ldots, l$,

$$|y_t - f(x_t) + (\bar{\alpha}_t^* - \bar{\alpha}_t) K(x_t, x_t)|_\varepsilon \geq |y_t - f_{T|t}(x_t)|_\varepsilon. \qquad (2.99)$$

We complete the proof by investigating three cases separately:

(i) The case $\bar{\alpha}_t^* > 0$. In this case, we have $\bar{\alpha}_t = 0$, so $(\bar{\alpha}_t^* - \bar{\alpha}_t) > 0$. Thus by Lemma 2.9,

$$-\sum_{i \neq t}(\bar{\alpha}_i^* - \bar{\alpha}_i)K(x_i, x_t) \geq -f_{T|t}(x_t). \tag{2.100}$$

Furthermore, by Lemma 2.8, the fact $\bar{\alpha}_t^* > 0$ implies that $f_{T|t}(x_t) \leq y_t$. Therefore, inequality (2.100) leads to

$$y_t - \sum_{i \neq t}(\bar{\alpha}_i^* - \bar{\alpha}_i)K(x_i, x_t) \geq y_t - f_{T|t}(x_t) \geq 0, \tag{2.101}$$

and because the function $| \cdot |_\varepsilon$ is monotonically increasing, so the conclusion (2.99) is true.

(ii) The case $\bar{\alpha}_t > 0$. In this case, we have $\bar{\alpha}_t^* = 0$, so $-(\bar{\alpha}_t^* - \bar{\alpha}_t) > 0$. Thus, by Lemma 2.9,

$$\sum_{i \neq t}(\bar{\alpha}_i^* - \bar{\alpha}_i)K(x_i, x_t) \geq f_{T|t}(x_t). \tag{2.102}$$

Furthermore, by Lemma 2.8, the fact $\bar{\alpha}_t > 0$ implies that $f_{T|t}(x_t) \geq y_t$. Therefore, inequality (2.102) leads to

$$\sum_{i \neq t}(\bar{\alpha}_i^* - \bar{\alpha}_i)K(x_i, x_t) - y_t \geq f_{T|t}(x_t) - y_t \geq 0, \tag{2.103}$$

and because the function $| \cdot |_\varepsilon$ is monotonically increasing, so the conclusion (2.99) is true.

(iii) The validity of the conclusion (2.99) is obvious for the case $\bar{\alpha}_t^* = \bar{\alpha}_t = 0$ by Lemma 2.9. □

2.2.5 Numerical Experiments

In this section, we will compare the proposed first LOO bound and second LOO bound with the true corresponding LOO errors. Consider the real dataset—"Boston Housing Data", which is a standard regression testing problem. This dataset includes 506 instances, each instance has 13 attributes and a real-valued output.

Here we randomly choose 50 instances for training, and the Radial Basis Kernel

$$K(x, x') = \exp\left(\frac{-\|x - x'\|^2}{\sigma^2}\right) \tag{2.104}$$

is applied, where σ is the kernel parameter. So the parameters to be chosen in Algorithm 2.1 and Algorithm 2.6 include C, ε, σ, and in our experiments, we choose these three parameters from the following sets:

$$C \in S_C = \{0.01, 0.1, 0.2, 0.5, 1, 2, 5, 10, 20, 50, 100, 200, 500,$$

$$1000, 10000\}, \tag{2.105}$$

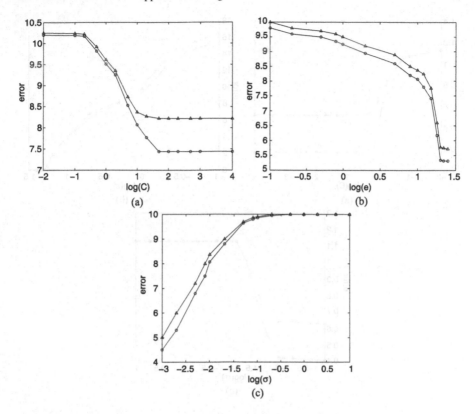

Fig. 2.1 LOO errors and LOO bounds of Algorithm 2.1

$$\varepsilon \in S_\varepsilon = \{0.1, 0.2, 0.5, 0.8, 1, 2, 5, 8, 10, 12, 15, 18, 20, 22, 25\}, \qquad (2.106)$$

$$\sigma \in S_\sigma = \{0.001, 0.002, 0.005, 0.008, 0.01, 0.02, 0.05, 0.08, 0.1, 0.2, 0.5, 1, 2,$$
$$5, 10\}. \qquad (2.107)$$

However, we do not consider their all combinations. We will only perform three experiments for each algorithm. In the first experiment, we fix $\varepsilon = 10$, $\sigma = 0.01$, and choose C from S_C. Applying these parameters in Algorithm 2.1 and Algorithm 2.6, and using Definition 2.3, the two LOO errors are computed. On the other hand, according to Theorem 2.5 and Theorem 2.10, the two corresponding LOO error bounds are obtained. Both the LOO errors and the LOO bounds are showed in Fig. 2.1(a) and Fig. 2.2(a), where "o" denotes LOO error and "$\triangle$" denotes the corresponding LOO bound.

Similarly, in the second experiment, let $C = 10$, $\sigma = 0.01$, and choose ε from S_ε, the compared result is showed in Fig. 2.1(b) and Fig. 2.2(b). At last, in the third experiment let $C = 10$, $\varepsilon = 10$, and choose σ from S_σ, the compared result is showed in Fig. 2.1(c) and Fig. 2.2(c). Note that in order to be visible clearly, the

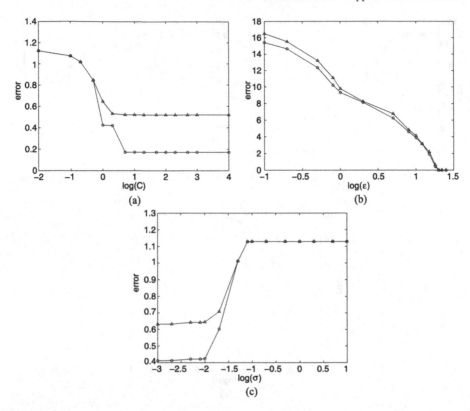

Fig. 2.2 LOO errors and LOO bounds of Algorithm 2.6

values of LOO errors and LOO bounds in the figures are all be divided by 10, and the values of $[C, \varepsilon, \sigma]$ are all changed to $[\log_{10}(C), \log_{10}(\varepsilon), \log_{10}(\sigma)]$.

From Figs. 2.1 and 2.2, we see that our two LOO bounds are really upper bounds of the corresponding true LOO errors, and more important, they almost have the same trend with the corresponding true LOO errors when the parameters are changing. So in order to choose the optimal parameters in Algorithm 2.1 and Algorithm 2.6, we only need to minimize the proposed LOO bound instead of LOO error itself. Obviously it must cost much less time.

2.3 LOO Bounds for Support Vector Ordinal Regression Machine

This section will focus on LOO bounds for support vector ordinal regression machine (SVORM) proposed in [177] which solves ordinal regression problem. Problem of ordinal regression arises in many fields, e.g., in information retrieval [109], in econometric models [194], and in classical statistics [7]. It is complementary to classification problem and metric regression problem due to its discrete and ordered outcome space. Several methods corresponding with SVM have been proposed to solve this problem, such as in [110] which is based on a mapping from objects to scalar utility values and enforces large margin rank boundaries. SVORM was constructed by applying the large margin principle used in SVM to the ordinal regression problem, and outperforms existing ordinal regression algorithms [177].

Selecting appropriate parameters in SVORM is also an important problem, techniques such as cross-validation and LOO error can also be applied except for their inefficient computation. Therefore, we will present two LOO error bounds for SVORM. The first one corresponds to an upper bound for the C-SVC in [207] by Vapnik and Chapelle, while the second one to an upper bound in [119] by Joachims. Obviously, the derivation of our two bounds are more complicated because multiclass classification, instead of 2-class classification, is solved by SVORM.

2.3.1 Support Vector Ordinal Regression Machine

Ordinal regression problem can be described as follows: Suppose a training set is given by

$$T = \{(x_i^j, y_i^j)\}_{i=1,\dots,l^j}^{j=1,\dots,k} \in (R^n \times \mathcal{Y})^l, \qquad (2.108)$$

where $x_i^j \in R^n$ is the input, $y_i^j = j \in \mathcal{Y} = \{1, \dots, k\}$ is the output or the class label, $i = 1, \dots, l^j$ is the index with each class and $l = \sum_{j=1}^k l^j$ is the number of sample points. We need to find $k - 1$ parallel hyperplanes represented by vector w and an orderly real sequence $b_1 \le \dots \le b_{k-1}$ defining the hyperplanes $(w, b_1), \dots, (w, b_{k-1})$ such that the data are separated by dividing the space into equally ranked regions by the decision function

$$f(x) = \min_{r \in \{1,\dots,k\}} \{r : (w \cdot x) - b_r < 0\}, \qquad (2.109)$$

where $b_k = +\infty$. In other words, all input points x satisfying $b_{r-1} < (w \cdot x) < b_r$ are assigned the rank r, where $b_0 = -\infty$.

Now we briefly introduce the fixed margin version of SVORM as a direct generalization of C-SVM [206]. Figure 2.3 gives out the geometric interpretation of this strategy.

For the training set (2.108), the input is mapped into a Hilbert space by a function $x = \Phi(x) : x \in R^n \to x \in \mathcal{H}$, where $\mathcal{H}$ is the Hilbert space. Then the primal problem of SVORM is the following optimization problem:

$$\min_{w,b,\xi^{(*)}} \frac{1}{2} \|w\|^2 + C \sum_{j=1}^k \sum_{i=1}^{l^j} (\xi_i^j + \xi_i^{*j}), \qquad (2.110)$$

$$\text{s.t.} \quad (w \cdot \Phi(x_i^j)) - b_j \le -1 + \xi_i^j, \quad j = 1, \dots, k, \ i = 1, \dots, l^j, \qquad (2.111)$$

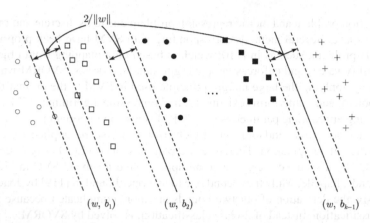

Fig. 2.3 Fixed-margin policy for ordinal problem: the margin to be maximized is the one defined by the closest (neighboring) pair of classes. Formally, let w, b_j be the hyperplane separating the two pairs of classes which are the closest among all the neighboring pairs of classes. Let w, b_j be scaled such the distance of the boundary points from the hyperplane is 1, i.e., the margin between the classes $j, j + 1$ is $1/\|w\|$. Thus, the fixed margin policy for ranking learning is to find the direction w and the scalars $b_1, \ldots, b_{k-1}$ such that $\|w\|$ is minimized (i.e., the margin between classes $j, j + 1$ is maximized) subject to the separability constraints (modulo margin errors in the non-separable case)

$$(w \cdot \Phi(x_i^j)) - b_{j-1} \geq 1 - \xi_i^{*j}, \quad j = 1, \ldots, k, \ i = 1, \ldots, l^j, \quad (2.112)$$

$$\xi_i^j \geq 0, \quad \xi_i^{*j} \geq 0, \quad j = 1, \ldots, k, \ i = 1, \ldots, l^j, \quad (2.113)$$

where $w \in \mathcal{H}$, $b = (b_1, \ldots, b_{k-1})^T$, $b_0 = -\infty$, $b_k = +\infty$, $\xi^{(*)} = (\xi_1^1, \ldots, \xi_{l^1}^1, \ldots, \xi_1^k, \ldots, \xi_{l^k}^k, \xi_1^{*1}, \ldots, \xi_{l^1}^{*1}, \ldots, \xi_1^{*k}, \ldots, \xi_{l^k}^{*k})$ and the penalty parameter $C > 0$. The dual of the problem (2.110)–(2.113) can be expressed as [42]:

$$\min_{\alpha^{(*)}} \frac{1}{2} \sum_{j,i} \sum_{j',i'} (\alpha_i^{*j} - \alpha_i^j)(\alpha_{i'}^{*j'} - \alpha_{i'}^{j'}) K(x_i^j, x_{i'}^{j'}) - \sum_{j,i} (\alpha_i^j + \alpha_i^{*j}), \quad (2.114)$$

$$\text{s.t.} \quad \sum_{i=1}^{l^j} \alpha_i^j = \sum_{i=1}^{l^{j+1}} \alpha_i^{*j+1}, \quad j = 1, \ldots, k - 1, \quad (2.115)$$

$$0 \leq \alpha_i^j, \alpha_i^{*j} \leq C, \quad j = 1, \ldots, k, \ i = 1, \ldots, l^j, \quad (2.116)$$

where $\alpha^{(*)} = (\alpha_1^1, \ldots, \alpha_{l^1}^1, \ldots, \alpha_1^k, \ldots, \alpha_{l^k}^k, \alpha_1^{*1}, \ldots, \alpha_{l^1}^{*1}, \ldots, \alpha_1^{*k}, \ldots, \alpha_{l^k}^{*k})^T, \alpha_i^{*1} = 0, i = 1, \ldots, l^1, \alpha_i^k = 0, i = 1, \ldots, l^k$.

For optimal solutions w and $\alpha^{(*)}$, the primal–dual relationship shows

$$w = \sum_{j=1}^{k} \sum_{i=1}^{l^j} (\alpha_i^{*j} - \alpha_i^j) \Phi(x_i^j). \quad (2.117)$$

So in the decision function (2.109) the real value function $g(x)$ is given by

$$g(x) = (\mathbf{w} \cdot \mathbf{x}) = \sum_{j=1}^{k} \sum_{i=1}^{l^j} (\alpha_i^{*j} - \alpha_i^j) K(x_i^j, x), \qquad (2.118)$$

where $K(x_i^j, x) = (\Phi(x_i^j) \cdot \Phi(x))$ is the kernel function. The scalars $b_1, \ldots, b_{k-1}$ can be obtained from the KKT conditions of primal problem (2.110)–(2.113).

This leads to the following algorithm:

Algorithm 2.11 (SVORM)

(1) Given a training set (2.108);
(2) Select a scalar $C > 0$ and a kernel function $K(x, x')$. Solve the dual problem (2.114)–(2.116), and get its optimal solution $\alpha^{(*)} = (\alpha_1^1, \ldots, \alpha_{l^1}^1, \ldots, \alpha_1^k, \ldots, \alpha_{l^k}^k, \alpha_1^{*1}, \ldots, \alpha_{l^1}^{*1}, \ldots, \alpha_1^{*k}, \ldots, \alpha_{l^k}^{*k})^{\mathrm{T}}$;
(3) Compute

$$g(x) = (\mathbf{w} \cdot \mathbf{x}) = \sum_{j=1}^{k} \sum_{i=1}^{l^j} (\alpha_i^{*j} - \alpha_i^j) K(x_i^j, x); \qquad (2.119)$$

(4) For $j = 1, \ldots, k - 1$, execute the following steps:

    (4.1) Choose a component $\alpha_i^j \in (0, C)$ in $\alpha^{(*)}$. If we get such subscript i, set

$$b_j = 1 + \sum_{j'=1}^{k} \sum_{i'=1}^{l^{j'}} (\alpha_{i'}^{*j'} - \alpha_{i'}^{j'}) K(x_{i'}^{j'}, x_i^j);$$

    otherwise go to step (4.2);

    (4.2) Choose a component $\alpha_i^{*j+1} \in (0, C)$ in $\alpha^{(*)}$. If we get such subscript i, set

$$b_j = \sum_{j'=1}^{k} \sum_{i'=1}^{l^{j'}} (\bar{\alpha}_{i'}^{*j'} - \bar{\alpha}_{i'}^{j'}) K(x_{i'}^{j'}, x_i^{j+1}) - 1;$$

    otherwise go to step (4.3);

    (4.3) Set

$$b_j = \frac{1}{2}(b_j^{\mathrm{dn}} + b_j^{\mathrm{up}}),$$

    where

$$b_j^{\mathrm{dn}} = \max\left\{ \max_{i \in I_1^j}(g(x_i^j) + 1), \max_{i \in I_4^j}(g(x_i^{j+1}) - 1) \right\},$$

$$b_j^{\mathrm{up}} = \min\left\{ \min_{i \in I_3^j}(g(x_i^j) + 1), \min_{i \in I_2^j}(g(x_i^{j+1}) - 1) \right\},$$

and

$$I_1^j = \{i \in \{1, \ldots, l^j\} \mid \alpha_i^j = 0\}, \qquad I_2^j = \{i \in \{1, \ldots, l^{j+1}\} \mid \alpha_i^{*j+1} = 0\},$$

$$I_3^j = \{i \in \{1, \ldots, l^j\} \mid \alpha_i^j = C\},$$

$$I_4^j = \{i \in \{1, \ldots, l^{j+1}\} \mid \alpha_i^{*j+1} = C\};$$

(5) If there exists $j \in \{1, \ldots, k\}$ such that $b_j \leq b_{j-1}$, stop or go to step (2);
(6) Define $b_k = +\infty$, construct the decision function

$$f(x) = \min_{r \in \{1, \ldots, k\}} \{r : g(x) - b_r < 0\}. \qquad (2.120)$$

In addition, in order to derive the LOO error bounds for SVORM, we firstly give the definitions of its support vector and its LOO error in the LOO procedure.

Definition 2.12 (Support vector) Suppose that $\alpha^{(*)}$ is the optimal solution of the dual problem (2.114)–(2.116) for the training set T (2.108). Then

(i) The input x_i^j is called *non-margin support vector about* $\alpha = (\alpha_1^1, \ldots, \alpha_{l^1}^1, \ldots,$ $\alpha_1^k, \ldots, \alpha_{l^k}^k)^T$, if the corresponding component α_i^j of $\alpha^{(*)}$ is equal to C. For $j = 1, \ldots, k$ define the index set

$$N(\alpha, j) = \{(j, i) \mid i = 1, \ldots, l^j, \ \alpha_i^j = C\}. \qquad (2.121)$$

The input x_i^j is called *non-margin support vector about* $\alpha^* = (\alpha_1^{*1}, \ldots, \alpha_{l^1}^{*1},$ $\ldots, \alpha_1^{*k}, \ldots, \alpha_{l^k}^{*k})^T$, if the corresponding component α_i^{*j} of $\alpha^{(*)}$ is equal to C. For $j = 1, \ldots, k$ define the index set

$$N(\alpha^*, j) = \{(j, i) \mid i = 1, \ldots, l^j, \ \alpha_i^{*j} = C\}. \qquad (2.122)$$

The input x_i^j is called *non-margin support vector about* $\alpha^{(*)}$, if x_i^j is either non-margin support vector about α or non-margin support vector about α^*. For $j = 1, \ldots, k$ define the index set

$$N(\alpha^{(*)}, j) = N(\alpha, j) \cup N(\alpha^*, j). \qquad (2.123)$$

(ii) The input x_i^j is called *margin support vector about* $\alpha = (\alpha_1^1, \ldots, \alpha_{l^1}^1, \ldots,$ $\alpha_1^k, \ldots, \alpha_{l^k}^k)^T$, if the corresponding component α_i^j of $\alpha^{(*)}$ is in the interval $(0, C)$ and the component α_i^{*j} of $\alpha^{(*)}$ is not equal to C. For $j = 1, \ldots, l$ define the index set

$$M(\alpha, j) = \{(j, i) \mid i = 1, \ldots, l^j, \ \alpha_i^j \in (0, C)\} \setminus N(\alpha^*, j). \qquad (2.124)$$

The input x_i^j is called *margin support vector about* $\alpha^* = (\alpha_1^{*1}, \ldots, \alpha_{l^1}^{*1}, \ldots,$ $\alpha_1^{*k}, \ldots, \alpha_{l^k}^{*k})^T$, if the corresponding component α_i^{*j} of $\alpha^{(*)}$ is in the interval

$(0, C)$ and the component α_i^j of $\alpha^{(*)}$ is not equal to C. For $j = 1, \ldots, l$ define the index set

$$M(\alpha^*, j) = \{(j, i) \mid i = 1, \ldots, l^j, \ \alpha_i^{*j} \in (0, C)\} \setminus N(\alpha, j). \quad (2.125)$$

The input x_i^j is called *margin support vector about* $\alpha^{(*)}$, if x_i^j is either margin support vector about α or margin support vector about α^*. For $j = 1, \ldots, k$ define the index set

$$M(\alpha^{(*)}, j) = M(\alpha, j) \cup M(\alpha^*, j). \quad (2.126)$$

(iii) The input x_i^j is called *support vector about* $\alpha = (\alpha_1^1, \ldots, \alpha_{l^1}^1, \ldots, \alpha_1^k, \ldots, \alpha_{lk}^k)^T$, if x_i^j is either non-margin support vector about α or margin support vector about α. For $j = 1, \ldots, k$ define the index set

$$V(\alpha, j) = M(\alpha, j) \cup N(\alpha, j). \quad (2.127)$$

The input x_i^j is called *support vector about* $\alpha^* = (\alpha_1^{*1}, \ldots, \alpha_{l^1}^{*1}, \ldots, \alpha_1^{*k}, \ldots, \alpha_{lk}^{*k})^T$, if x_i^j is either non-margin support vector about α^* or margin support vector about α^*. For $j = 1, \ldots, k$ define the index set

$$V(\alpha^*, j) = M(\alpha^*, j) \cup N(\alpha^*, j). \quad (2.128)$$

The input x_i^j is called *support vector about* $\alpha^{(*)}$, if x_i^j is either non-margin support vector about $\alpha^{(*)}$ or margin support vector about $\alpha^{(*)}$. For $j = 1, \ldots, k$ define the index set

$$V(\alpha^{(*)}, j) = V(\alpha, j) \cup V(\alpha^*, j). \quad (2.129)$$

Definition 2.13 (LOO error) Consider the training set $T_p^q = T \setminus \{(x_p^q, y_p^q)\}$, $q = 1, \ldots, k$, $p = 1, \ldots, l^q$, where T is given by (2.108). Suppose that $f_{T_p^q}(x)$ is the decision function obtained by executing Algorithm 2.11 for T_p^q, then the leave-one-out error, or LOO error for short, is defined as

$$R_{\text{LOO}}(T) = \frac{1}{l} \sum_{q=1}^{k} \sum_{p=1}^{l^q} c(x_p^q, y_p^q, f_{T_p^q}(x_p^q)), \quad (2.130)$$

where c is the 0–1 loss function

$$c(x, y, f(x)) = \hat{c}(y - f(x)),$$

with

$$\hat{c}(\xi) = \begin{cases} 0, & \text{if } \xi = 0, \\ 1, & \text{otherwise.} \end{cases}$$

From Definition 2.13 we can see that the computation of LOO error for SVORM is time-consuming and inefficient. So researching for LOO error bounds for SVORM will be necessary.

2.3.2 The First LOO Bound

In this section, we study the derivation of our first LOO bound for Algorithm 2.11 by the concept of a span.

Definition and Existence of Span

We now define an S-span of a margin support vector about α and α^* respectively.

Definition 2.14 (S-span about α) For any margin support vector x_p^q about α, define its S-span by

$$S^2(q, p) := \min\{\|x_p^q - \tilde{x}_p^q\|^2 | \tilde{x}_p^q \in \Lambda_p^q\}, \qquad (2.131)$$

where Λ_p^q is

$$\Lambda_p^q := \left\{ \sum_{i \in M_p^q(\alpha,q)} \lambda_i^q x_i^q + \sum_{i \in M_p^q(\alpha^*,q+1)} \lambda_i^{q+1} x_i^{q+1} \right\}, \qquad (2.132)$$

with the following conditions:

$$0 \leq \alpha_i^q + \lambda_i^q \alpha_p^q \leq C, \qquad 0 \leq \alpha_i^{*q} + \lambda_i^q \alpha_p^{*q} \leq C, \qquad (2.133)$$

$$0 \leq \alpha_i^{q+1} - \lambda_i^{q+1} \alpha_p^{*q} \leq C, \qquad 0 \leq \alpha_i^{*q+1} - \lambda_i^{q+1} \alpha_p^q \leq C, \qquad (2.134)$$

$$\sum_{i \in M_p^q(\alpha,q)} \lambda_i^q + \sum_{i \in M_p^q(\alpha^*,q+1)} \lambda_i^{q+1} = 1, \qquad \lambda_p^q = -1, \qquad (2.135)$$

and

$$M_p^q(\alpha, j) = M(\alpha, j)\backslash\{(q, p)\}, \qquad M_p^q(\alpha^*, j) = M(\alpha^*, j)\backslash\{(q, p)\}. \qquad (2.136)$$

Definition 2.15 (S-span about α^*) For any margin support vector x_p^q about α^*, define its S-span by

$$S^{*2}(q, p) := \min\{\|x_p^q - \hat{x}_p^q\|^2 | \hat{x}_p^q \in \Lambda_p^{*q}\}, \qquad (2.137)$$

where Λ_p^{*q} is

$$\Lambda_p^{*q} := \left\{ \sum_{i \in M_p^q(\alpha,q-1)} \lambda_i^{q-1} x_i^{q-1} + \sum_{i \in M_p^q(\alpha^*,q)} \lambda_i^q x_i^q \right\}, \qquad (2.138)$$

with the following conditions:

$$0 \le \alpha_i^{q-1} - \lambda_i^{q-1}\alpha_p^{*q} \le C, \qquad 0 \le \alpha_i^{*q-1} - \lambda_i^{q-1}\alpha_p^{q} \le C, \qquad (2.139)$$

$$0 \le \alpha_i^{q} + \lambda_i^{q}\alpha_p^{q} \le C, \qquad 0 \le \alpha_i^{*q} + \lambda_i^{q}\alpha_p^{*q} \le C, \qquad (2.140)$$

$$\sum_{i \in M_p^q(\alpha, q-1)} \lambda_i^{q-1} + \sum_{i \in M_p^q(\alpha^*, q)} \lambda_i^{q} = 1, \qquad \lambda_p^{q} = -1, \qquad (2.141)$$

and

$$M_p^q(\alpha, j) = M(\alpha, j)\backslash\{(q, p)\}, \qquad M_p^q(\alpha^*, j) = M(\alpha^*, j)\backslash\{(q, p)\}. \qquad (2.142)$$

For the S-span $S^2(q, p)$ and $S^{*2}(q, p)$ defined above, it is necessary to show that the set Λ_p^q and Λ_p^{*q} are non-empty. To this end, we make use of the following lemma.

Lemma 2.16 *The both sets Λ_p^q and Λ_p^{*q} defined by (2.132) and (2.138) are non-empty.*

The proof is omitted here, which can be referred to [230] and [231].

According to the above Definition 2.14 and Definition 2.15, we have the following two lemmas.

Lemma 2.17 *Suppose that $\alpha^{(*)}$ is an optimal solution of the dual problem (2.114)–(2.116) for the training set T (2.108) and x_p^q is a margin support vector about α. Then we can construct a feasible solution $\tilde{\alpha}^{(*)}$ of the dual problem (2.114)–(2.116) for the training set $T_p^q = T \backslash \{(x_p^q, y_p^q)\}$ by*

$$\tilde{\alpha}_i^{q} = \alpha_i^{q} + \lambda_i^{q}\alpha_p^{q}, \quad \tilde{\alpha}_i^{*q} = \alpha_i^{*q} + \lambda_i^{q}\alpha_p^{*q}, \quad i \in M_p^q(\alpha, q), \qquad (2.143)$$

$$\tilde{\alpha}_i^{q+1} = \alpha_i^{q+1} - \lambda_i^{q+1}\alpha_p^{*q}, \quad \tilde{\alpha}_i^{*q+1} = \alpha_i^{*q+1} - \lambda_i^{q+1}\alpha_p^{q},$$

$$i \in M_p^q(\alpha^*, q+1), \qquad (2.144)$$

$$\tilde{\alpha}_i^{q} = \alpha_i^{q}, \tilde{\alpha}_i^{*q} = \alpha_i^{*q}, \quad i \notin M_p^q(\alpha, q), \qquad (2.145)$$

$$\tilde{\alpha}_i^{q+1} = \alpha_i^{q+1}, \quad \tilde{\alpha}_i^{*q+1} = \alpha_i^{*q+1}, \quad i \notin M_p^q(\alpha^*, q+1), \qquad (2.146)$$

$$\tilde{\alpha}_i^{j} = \alpha_i^{j}, \quad \tilde{\alpha}_i^{*j} = \alpha_i^{*j}, \quad j = 1, \dots, q-1, q+2, \dots, k, \; i = 1, \dots, l^j, \qquad (2.147)$$

and

$$\sum_{i \in M_p^q(\alpha, q)} \lambda_i^{q}x_i^{q} + \sum_{i \in M_p^q(\alpha^*, q+1)} \lambda_i^{q+1}x_i^{q+1} \in \Lambda_p^q.$$

The proof is omitted here, which can be referred to [230] and [231].

Lemma 2.18 *Suppose that $\alpha^{(*)}$ is an optimal solution of the dual problem* (2.114)–
(2.116) *for the training set T* (2.108) *and x_p^q is a margin support vector about α^*.*
Then we can construct a feasible solution $\tilde{\alpha}^{(*)}$ of the dual problem (2.114)–(2.116)
for the training set $T_p^q = T \setminus \{(x_p^q, y_p^q)\}$ by

$$\hat{\alpha}_i^{q-1} = \alpha_i^{q-1} - \lambda_i^{q-1}\alpha_p^{*q}, \quad \hat{\alpha}_i^{*q-1} = \alpha_i^{*q-1} - \lambda_i^{q-1}\alpha_p^q,$$

$$i \in M_p^q(\alpha, q-1), \tag{2.148}$$

$$\hat{\alpha}_i^q = \alpha_i^q + \lambda_i^q\alpha_p^q, \quad \hat{\alpha}_i^{*q} = \alpha_i^{*q} + \lambda_i^q\alpha_p^{*q}, \quad i \in M_p^q(\alpha^*, q), \tag{2.149}$$

$$\hat{\alpha}_i^{q-1} = \alpha_i^{q-1}, \quad \hat{\alpha}_i^{*q-1} = \alpha_i^{*q-1}, \quad i \notin M_p^q(\alpha, q-1), \tag{2.150}$$

$$\hat{\alpha}_i^q = \alpha_i^q, \quad \hat{\alpha}_i^{*q} = \alpha_i^{*q}, \quad i \notin M_p^q(\alpha^*, q), \tag{2.151}$$

$$\hat{\alpha}_i^j = \alpha_i^j, \quad \hat{\alpha}_i^{*j} = \alpha_i^{*j}, \quad j = 1, \ldots, q-2, q+1, \ldots, k, \; i = 1, \ldots, l^j, \tag{2.152}$$

and

$$\sum_{i \in M_p^q(\alpha, q-1)} \lambda_i^{q-1}x_i^{q-1} + \sum_{i \in M_p^q(\alpha^*, q)} \lambda_i^q x_i^q \in \Lambda_p^{*q}.$$

The proof is omitted here, which can be referred to [230] and [231].

The Bound

Now we are in a position to introduce our first LOO error bound:

Lemma 2.19 *Suppose that $\alpha^{(*)}$ is the optimal solution the dual problem* (2.114)–
(2.116) *for the training set T* (2.108) *and $f_{T_p^q}$ is the decision function obtained by*
Algorithm 2.11 for the training set $T_p^q = T \setminus \{(x_p^q, y_p^q)\}$. For a margin support vector
x_p^q about $\alpha^{(*)}$, we have

(1) *If x_p^q is a margin support vector about α and is recognized incorrectly by the*
decision function f_p^q, then the following inequality holds

$$(\alpha_p^{*q} - \alpha_p^q)^2 S^2(p, q) \geq \min\left(C, \frac{1}{D_{q,q+1}^2}\right), \tag{2.153}$$

where $D_{q,q+1}$ is diameter of the smallest sphere containing the qth class train-
ing points and the $(q+1)$th class training points in the training set T (2.108),
and we have the following expression

$$D_{q,q+1} = \min_{D_{q,q+1},c}\left\{D_{q,q+1} \| x_i^j - c \|^2 \leq \frac{D_{q,q+1}^2}{4}, \; j = q, q+1, \; i = 1, \ldots, l^j\right\}; \tag{2.154}$$

(2) *If x_p^q is a margin support vector about α^* and is recognized incorrectly by the decision function f_p^q, then the following inequality holds*

$$(\alpha_p^{*q} - \alpha_p^q)^2 S^{*2}(p, q) \geq \min\left(C, \frac{1}{D_{q-1,q}^2}\right),$$ (2.155)

where $D_{q-1,q}$ is diameter of the smallest sphere containing the $(q-1)$th class training points and the qth class training points in the training set T (2.108), and we have the following expression

$$D_{q-1,q} = \min_{D_{q-1,q},c}\left\{D_{q-1,q}\|x_i^j - c\|^2 \leq \frac{D_{q-1,q}^2}{4},\right.$$

$$\left. j = q-1, q, \ i = 1, \ldots, l^j\right\}.$$ (2.156)

The proof is omitted here, which can be referred to [230] and [231].
The above lemma leads to the following theorem:

Theorem 2.20 *For Algorithm 2.11, the bound of LOO error is estimated by*

$$R_{\text{LOO}}(T) \leq \frac{1}{l}\sum_{q=1}^{k}\sum_{p=1}^{l^q}\left[\left|\left\{(q, p) : (\alpha_p^{*q} - \alpha_p^q)^2 S^2(q, p) \geq \min\left(C, \frac{1}{D_{q,q+1}^2}\right) \text{ or }\right.\right.\right.$$

$$\left.\left.\left.(\alpha_p^{*q} - \alpha_p^q)^2 S^{*2}(q, p) \geq \min\left(C, \frac{1}{D_{q-1,q}^2}\right)\right\}\right| + |N(\alpha^{(*)}, q)|\right],$$ (2.157)

where $D_{q,q+1}$ and $D_{q-1,q}$ are given by (2.154) and (2.156) respectively, $N(\alpha^{(*)}, q)$ is defined by (2.123) and $|\cdot|$ is the number of elements in the set.

Proof Considering the Definition 2.13 of LOO error. Denote the number of error made by the LOO procedure as $\mathcal{L}(T)$

$$\mathcal{L}(T) = \sum_{q=1}^{k}\sum_{p=1}^{l^q} c(x_p^q, y_p^q, f_{T_p^q}(x_p^q)) = \sum_{q=1}^{k}\sum_{p=1}^{l^q}\mathbb{I}_{(w_p^q \cdot x_p^q) - b_q > 0 \text{ or } (w_p^q \cdot x_p^q) - b_{q-1} < 0},$$ (2.158)

where

$$\mathbb{I}_P = \begin{cases} 1, & P \text{ is true}; \\ 0, & P \text{ is false}. \end{cases}$$

In order to estimate $\mathcal{L}(T)$, define two index sets

$$I_{\text{LOO}}^q \triangleq \{(q, p) : (w_p^q \cdot x_p^q) - b_q > 0 \text{ or } (w_p^q \cdot x_p^q) - b_{q-1} < 0\} \cup N(\alpha^{(*)}, q),$$

$$\tilde{I}_{\text{LOO}}^q \triangleq \{(q, p) : (w_p^q \cdot x_p^q) - b_q > -1 \text{ or } (w_p^q \cdot x_p^q) - b_{q-1} < 1\} \cup N(\alpha^{(*)}, q).$$

It is easy to see that

$$\mathcal{L}(T) = |I_{\text{LOO}}^q| \le |\hat{I}_{\text{LOO}}^q|. \tag{2.159}$$

By the Lemma 2.19, we have

$$|\hat{I}_{\text{LOO}}^q| = \left| \left\{ (q, p) : (\alpha_p^{*q} - \alpha_p^q)^2 S^2(q, p) \ge \min\left(C, \frac{1}{D_{q,q+1}^2}\right) \right. \right.$$

$$\left. \left. \text{or } (\alpha_p^{*q} - \alpha_p^q)^2 S^2(q, p) \ge \min\left(C, \frac{1}{D_{q-1,q}^2}\right) \right\} \right| + |N(\alpha^{(*)}, q)|. \tag{2.160}$$

So the LOO error bound (2.157) is obtained from (2.159) and (2.160). □

2.3.3 The Second LOO Bound

In this section, we study the derivation our second LOO error bound.

Remind that the dual problem for the training set T (2.108) is presented in (2.114)–(2.116). For the training set $T_p^q = T/\{(x_p^q, y_p^q)\}$, the dual problem is

$$\max_{\alpha_p^{(*)q}} W_p^q(\alpha^{(*)})$$

$$= \sum_{(j,i)\in I\backslash\{(q,p)\}} (\alpha_i^j + \alpha_i^{*j})$$

$$- \frac{1}{2} \sum_{(j,i)\in I\backslash\{(q,p)\}} \sum_{(j',i')\in I\backslash\{(q,p)\}} (\alpha_i^{*j} - \alpha_i^j)(\alpha_{i'}^{*j'} - \alpha_{i'}^{j'}) K(x_i^j, x_{i'}^{j'})$$

$$\text{s.t.} \quad \sum_{i=1}^{l^j} \alpha_i^j = \sum_{i=1}^{l^{j+1}} \alpha_i^{*j+1}, \quad j = 1, \dots, k-1, \ (j,i) \ne (q, p), \tag{2.161}$$

$$0 \le \alpha_i^j, \alpha_i^{*j} \le C, \quad (j,i) \in I \backslash \{(q, p)\}, \tag{2.162}$$

where $\alpha_p^{(*)q} = (\alpha_p^{qT}, \alpha_p^{*qT})^T$, $\alpha_p^q = (\alpha_1^1, \dots, \alpha_{l^1}^1, \dots, \alpha_1^q, \dots, \alpha_{p-1}^q, \alpha_{p+1}^q, \dots, \alpha_{l^q}^q, \dots, \alpha_1^k, \dots, \alpha_{l^k}^k)^T$, $\alpha_p^{*q} = (\alpha_1^{*1}, \dots, \alpha_{l^1}^{*1}, \dots, \alpha_1^q, \dots, \alpha_{p-1}^q, \alpha_{p+1}^q, \dots, \alpha_{l^q}^q, \dots, \alpha_1^{*k}, \dots, \alpha_{l^k}^{*k})^T$; $\alpha_i^{*1} = 0, \ i = 1, 2, \dots, l^1, \ \alpha_i^k = 0, \ i = 1, 2, \dots, l^k, \ I = \{(j,i) \mid j = 1, \dots, k, \ i = 1, \dots, l^j\}$.

In order to derive the second LOO error bound, we give the following lemma firstly.

Lemma 2.21 *Suppose that $\alpha^{(*)}$ is the optimal solution of the dual problem (2.114)–(2.116) for the training set T (2.108), and $f_{T_p^q}$ is the decision function obtained by*

Algorithm 2.11 for the training set $T_p^q = T \setminus \{(x_p^q, y_p^q)\}$. For the components of optimal solution $\alpha^{(*)}$: $\alpha_i^q, \alpha_i^{*q}, i = 1, \ldots, l^q$,

(1) *If in $\{\alpha_i^q \mid i = 1, \ldots, l^q\}$, there exists some $\alpha_i^q \in (0, C)$ and x_p^q is recognized incorrectly by the decision function f_p^q, then the following inequality holds*

$$\left[\sum_{j,i}(\alpha_i^{*j} - \alpha_i^j)K(x_p^q, x_i^j) - b_q\right] - (\alpha_p^{*q} - \alpha_p^q)(K(x_p^q, x_p^q) + R^2) \geq 0, \quad (2.163)$$

where $R^2 = \max\{K(x_i^j, x_i^j) \mid j = 1, \ldots, k, i = 1, \ldots, l^j\}$.

(2) *If in $\{\alpha_i^{*q} \mid i = 1, \ldots, l^q\}$, there exists $\alpha_i^{*q} \in (0, C)$ and x_p^q is recognized incorrectly by the decision function f_p^q, then the following inequality holds*

$$-\left[\sum_{j,i}(\alpha_i^{*j} - \alpha_i^j)K(x_p^q, x_i^j) - b_{q-1}\right] + (\alpha_p^{*q} - \alpha_p^q)(K(x_p^q, x_p^q) + R^2) \geq 0,$$

$$(2.164)$$

where $R^2 = \max\{K(x_i^j, x_i^j) \mid j = 1, \ldots, k, i = 1, \ldots, l^j\}$.

The proof is omitted here, which can be referred to [230] and [231].
The above lemma leads to the following theorem for Algorithm 2.11:

Theorem 2.22 *For Algorithm 2.11 the bound of LOO error is estimated by*

$$R_{LOO}(T) \leq \frac{1}{l}\left\{\sum_{q \in I_1}\sum_{p=1}^{l^q}\left|\left[\sum_{j,i}(\alpha_i^{*j} - \alpha_i^j)K(x_p^q, x_i^j) - b_q\right]\right.\right.$$

$$- (\alpha_p^{*q} - \alpha_p^q)(K(x_p^q, x_p^q) + R^2) \geq 0$$

$$or -\left[\sum_{j,i}(\alpha_i^{*j} - \alpha_i^j)K(x_p^q, x_i^j) - b_{q-1}\right]$$

$$\left.\left.+ (\alpha_p^{*q} - \alpha_p^q)(K(x_p^q, x_p^q) + R^2) \geq 0\right| + \sum_{q \in I_2}\sum_{p=1}^{l^q}|N(\alpha^{(*)}, q)|\right\},$$

$$(2.165)$$

where

$$I_1 = \{q \mid in\ (\alpha_1^q, \ldots, \alpha_{l^q}^q),\ (\alpha_1^{*q+1}, \ldots, \alpha_{l^{q+1}}^{*q+1})\ there\ exists\ \alpha_i^q \in (0, C)$$

$$or\ \alpha_i^{*q+1} \in (0, C)\},$$

$$I_2 = \{q \mid in\ (\alpha_1^q, \ldots, \alpha_{l^q}^q),\ (\alpha_1^{*q+1}, \ldots, \alpha_{l^{q+1}}^{*q+1})\ there\ exist\ not\ \alpha_i^q \in (0, C)$$

$$and\ \alpha_i^{*q+1} \in (0, C)\},$$

$$R^2 = \max\{K(x_i^j, x_i^j) \mid j = 1, \ldots, k, i = 1, \ldots, l^j\},$$

$N(\alpha^{(*)}, q)$ is defined by (2.123) and $|\cdot|$ is the number of elements in the set.

Proof Assume that the point x_p^q belongs to the class $q \in I_1 = \{q \mid \text{in } (\alpha_1^q, \ldots, \alpha_{lq}^q),$
$(\alpha_1^{*q+1}, \ldots, \alpha_{lq+1}^{*q+1})$ there exists $\alpha_i^q \in (0, C)$ or $\alpha_i^{*q+1} \in (0, C)\}$. Then according to
Lemma 2.21, when the LOO error procedure commits an error at the point x_p^q, the
following one of two inequalities holds

$$\left[\sum_{j,i} (\alpha_i^{*j} - \alpha_i^j) K(x_p^q, x_i^j) - b_q \right] - (\alpha_p^{*q} - \alpha_p^q)(K(x_p^q, x_p^q) + R^2) \geq 0, \quad (2.166)$$

$$-\left[\sum_{j,i} (\alpha_i^{*j} - \alpha_i^j) K(x_p^q, x_i^j) - b_{q-1} \right] + (\alpha_p^{*q} - \alpha_p^q)(K(x_p^q, x_p^q) + R^2) \geq 0,$$
$$(2.167)$$

where $R^2 = \max\{K(x_i^j, x_i^j) \mid j = 1, \ldots, k, i = 1, \ldots, l^j\}$.

If being left out the point x_p^q belongs to the class $q \in I_2 = \{q \mid \text{in } (\alpha_1^q, \ldots, \alpha_{lq}^q),$
$(\alpha_1^{*q+1}, \ldots, \alpha_{lq+1}^{*q+1})$ there does not exist $\alpha_i^q \in (0, C)$ and $\alpha_i^{*q+1} \in (0, C)\}$, then the
number of error made by the LOO error procedure is $|N(\alpha^{(*)}, q)|$, where $N(\alpha^{(*)}, q)$
is defined by (2.123) and $|\cdot|$ is the number of elements in the set.

So we get the LOO error bound (2.165) for Algorithm 2.11.   □

2.3.4 Numerical Experiments

In this section, we describe the performance of the two LOO error bounds with four
ordinal regression datasets [10]. The datasets are (1) "Employee Rejection\Accept-
ance" (ERA), (2) "Employee Selection" (ESL), (3) "Lecturers Evaluation" (LEV),
(4) "Social Workers Decisions" (SWD). A summary of the characteristics of these
datasets is presented in Table 2.1.

In our experiment, because the cost of computing LOO error is very high, we
select randomly only 60 training points from each dataset and merge these 4 multi-
class problems into 3-class problems. For each problem, we choose randomly 20
points from each class and get training set expressed as

$$T = \{(x_1^1, y_1^1), \ldots, (x_{20}^1, y_{20}^1), (x_1^2, y_1^2), \ldots, (x_{20}^2, y_{20}^2), (x_1^3, y_1^3), \ldots, (x_{20}^3, y_{20}^3)\},$$
$$(2.168)$$

where x_i^j is the input, $y_i^j = j$ is the output. In this way, corresponding to ERA, ESL,
LEV and SWD, we obtain the following training sets

$$T_{\text{ERA}}, \quad T_{\text{ESL}}, \quad T_{\text{LEV}} \quad \text{and} \quad T_{\text{SWD}}, \quad (2.169)$$

which are tested in our experiments.

Gaussian kernel function

$$K(x, x') = \exp\left(\frac{-\|x - x'\|^2}{\sigma^2} \right) \quad (2.170)$$

Table 2.1 Characteristics of the selected datasets from the ordinal datasets

Dataset	Features	Classes	Patterns
ERA	4	9	1000
ESL	4	9	488
LEV	4	5	1000
SWD	10	4	1000

is selected in our experiment, while the parameters C and σ are selected respectively from the following two sequences

$$C = \text{logspace}(-2, 4, 12), \qquad (2.171)$$

and

$$\sigma = \text{logspace}(-4, 4, 10), \qquad (2.172)$$

where logspace is a logarithmically spaced vector in MATLAB. More precisely, firstly, we find the optimal parameters C^*, σ^{*2} in (2.171) and (2.172) to our four training sets T_{ERA}, T_{ESL}, T_{LEV} and T_{SWD} by minimizing the LOO error respectively. Secondly, either C or σ is fixed to be its optimal value obtained, while the other one takes the values in (2.171) or (2.172). Figure 2.4 shows the performance of two LOO error bounds and LOO error itself. For example, the top-left figure corresponds to the training set T_{ERA} with $\sigma = \sigma^* = 10^{-4}$, and C take the value in (2.171). "LOO error" stands for the actual LOO error, "First LOO error bound" is the bound given by Theorem 2.20 and "Second LOO error bound" by Theorem 2.22.

By and large, it can be observed from Fig. 2.4 that changing trend of both LOO error bounds is almost consistent with that of LOO error itself. Concretely, when the penalty parameter C is fixed and the kernel parameter σ^2 is changed, our proposed both LOO error bounds are good performance. In other words, the lowest points of both LOO error bounds are close to those of LOO error, some difference of only one step length. So it is reasonable that the optimal parameters can be selected by minimizing these LOO error bounds instead of LOO error itself. Obviously, this strategy is highly efficient.

In this section, we derive two LOO error bounds for SVORM. The second LOO error bound is more effective than the first LOO error bound, because if we will compute the first LOO error bound, we must solve some quadratic programming problems, whereas the second LOO error bound doesn't need. Experiments demonstrate that these bounds are valid and it is hopeful to get the optimal parameter by minimizing the proposed bounds instead of the LOO error itself. In the further, we improve our proposed both LOO error bounds by smart way handling non-margin support vectors, due to the assumption that all non-margin support vectors are leave-one-out errors. In addition, an interesting study is to apply the proposed bounds on feature selection.

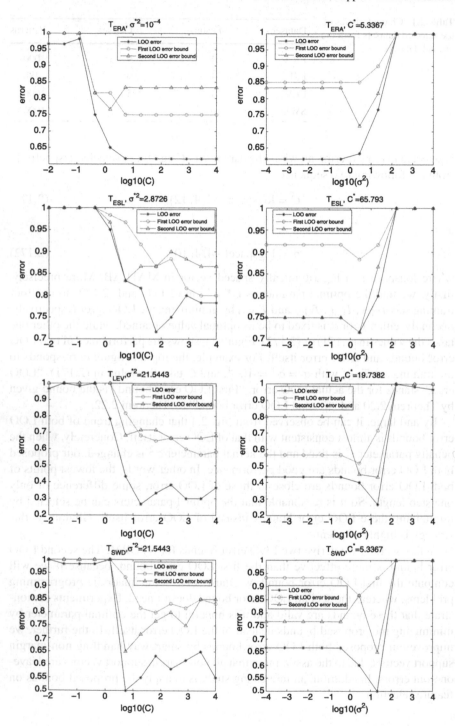

Fig. 2.4 Results of two LOO error bounds and LOO error

Chapter 3
Support Vector Machines for Multi-class Classification Problems

Multi-class classification refers to constructing a decision function $f(x)$ defined from an input space R^n onto an unordered set of classes $\mathcal{Y} = \{\Theta_1, \Theta_2, \ldots, \Theta_K\}$ based on independently and identically distributed (i.i.d) training set

$$T = \{(x_1, y_1), \ldots, (x_l, y_l)\} \in (R^n \times \mathcal{Y})^l. \tag{3.1}$$

Currently, there are roughly two types of SVC algorithms to solve the multi-class classification problems. One is the "decomposition-reconstruction" architecture approach which mainly makes direct use of binary SVC to tackle the tasks of multi-class classification, such as the "one versus the rest" method [24], the "one versus one" method [106, 130], the "error-correcting output code" method [3, 61], and the "one versus one versus rest" method [8, 249]. The other is the "all together" approach [49, 95, 136, 215], in other words, it solves the multi-class classification through only one optimization formulation. In this chapter, we first introduce two algorithms which follow the idea of [8, 249], and then construct another multi-class classification algorithm which is based on support vector ordinal regression [168, 202, 232].

3.1 K-Class Linear Programming Support Vector Classification Regression Machine (K-LPSVCR)

First we briefly review SVC and SVR in linear programming formulation.

Consider a binary classification problem with a training set

$$T = \{(x_1, y_1), \ldots, (x_l, y_l)\} \in (R^n \times \mathcal{Y})^l, \tag{3.2}$$

where $x_i \in R^n$, $y_i \in \mathcal{Y} = \{-1, 1\}$, $i = 1, \ldots, l$. The linear programming formulation of SVC is given as follows:

Y. Shi et al., *Optimization Based Data Mining: Theory and Applications*,
Advanced Information and Knowledge Processing,
DOI 10.1007/978-0-85729-504-0_3, © Springer-Verlag London Limited 2011

$$\min_{\alpha,\alpha^*,\xi,b} \sum_{i=1}^{l}(\alpha_i + \alpha_i^*) + C\sum_{i=1}^{l}\xi_i, \tag{3.3}$$

$$\text{s.t.} \quad y_i\left(\sum_{j=1}^{l}(\alpha_j - \alpha_j^*)K(x_j,x_i)+b\right) \geq 1 - \xi_i, \quad i=1,\ldots,l, \tag{3.4}$$

$$\alpha_i,\alpha_i^*,\xi_i \geq 0, \quad i=1,\ldots,l, \tag{3.5}$$

where the constant $C > 0$ and the kernel function $K(x,x')$ are chosen prior. Suppose $(\bar{\alpha}, \bar{\alpha}^*, \bar{\xi}, \bar{b})$ is the optimal solution of problem (3.3)–(3.5), then the decision function is constructed as

$$f(x) = \text{sgn}\left(\sum_{i=1}^{l}(\bar{\alpha}_i - \bar{\alpha}_i^*)K(x_i,x)+\bar{b}\right). \tag{3.6}$$

Consider a regression problem with a training set

$$T = \{(x_1,y_1),\ldots,(x_l,y_l)\} \in (R^n \times \mathcal{Y})^l, \tag{3.7}$$

where $x_i \in R^n$, $y_i \in \mathcal{Y} = R$, $i=1,\ldots,l$. By using the ε-insensitive loss function

$$|f(x) - y|_\varepsilon = \max\{0, |y - f(x)| - \varepsilon\}, \tag{3.8}$$

the linear programming formulation of SVR is given as follows:

$$\min_{\beta,\beta^*,\eta,\eta^*,b} \sum_{i=1}^{l}(\beta_i + \beta_i^*) + D\sum_{i=1}^{l}(\eta_i + \eta_i^*), \tag{3.9}$$

$$\text{s.t.} \quad \sum_{j=1}^{l}(\beta_j - \beta_j^*)K(x_j,x_i)+b - y_i \leq \varepsilon + \eta_i, \quad i=1,\ldots,l, \tag{3.10}$$

$$y_i - \sum_{j=1}^{l}(\beta_j - \beta_j^*)K(x_j,x_i) - b \leq \varepsilon + \eta_i^*, \quad i=1,\ldots,l, \tag{3.11}$$

$$\beta_i,\beta_i^*,\eta_i,\eta_i^* \geq 0, \quad i=1,\ldots,l, \tag{3.12}$$

where $D > 0$, $\varepsilon > 0$ and the kernel function $K(x,x')$ are chosen prior. Suppose $(\bar{\beta}, \bar{\beta}^*, \bar{\eta}, \bar{\eta}^*, \bar{b})$ is the optimal solution of problem (3.9)–(3.12), then the decision function is constructed as

$$f(x) = \sum_{i=1}^{l}(\bar{\beta}_i - \bar{\beta}_i^*)K(x_i,x)+\bar{b}. \tag{3.13}$$

3.1.1 K-LPSVCR

Given the training set T defined by (3.1). For an arbitrary pair $(\Theta_j, \Theta_k) \in (\mathcal{Y} \times \mathcal{Y})$ of classes with $j < k$, we will construct a decision function $f_{\Theta_{jk}}(x)$ which divides the inputs into three classes. In other words, it separates the two classes Θ_j and Θ_k as well as the remaining classes. The corresponding training set is denoted as

$$\tilde{T} = \{(\tilde{x}_1, \tilde{y}_1), \ldots, (\tilde{x}_{l_1}, \tilde{y}_{l_1}), (\tilde{x}_{l_1+1}, \tilde{y}_{l_1+1}), \ldots,$$
$$(\tilde{x}_{l_1+l_2}, \tilde{y}_{l_1+l_2}), (\tilde{x}_{l_1+l_2+1}, \tilde{y}_{l_1+l_2+1}), \ldots, (\tilde{x}_l, \tilde{y}_l)\}, \quad (3.14)$$

which is obtained from T in (3.1) by the following way:

$$\{\tilde{x}_1, \ldots, \tilde{x}_{l_1}\} = \{x_i \mid y_i = \Theta_j\},$$
$$\{\tilde{x}_{l_1+1}, \ldots, \tilde{x}_{l_1+l_2}\} = \{x_i \mid y_i = \Theta_k\} \quad (3.15)$$

and

$$\tilde{y}_i = \begin{cases} +1, & i = 1, \ldots, l_1 \\ -1, & i = l_1+1, \ldots, l_1+l_2, \\ 0, & i = l_1+l_2+1, \ldots, l. \end{cases} \quad (3.16)$$

For $C > 0$, $D > 0$, $\varepsilon > 0$ and $K(\cdot, \cdot)$ chosen prior, combining SVC and SVR in linear programming formulations, we thus obtain a linear programming problem

$$\min_{\alpha, \alpha^*, \xi, \eta, \eta^*, b} \sum_{i=1}^{l} (\alpha_i + \alpha_i^*) + C \sum_{i=1}^{l_1+l_2} \xi_i + D \sum_{i=l_1+l_2+1}^{l} (\eta_i + \eta_i^*), \quad (3.17)$$

$$\text{s.t.} \quad \tilde{y}_i \left(\sum_{j=1}^{l} (\alpha_j - \alpha_j^*) K(\tilde{x}_j, \tilde{x}_i) + b \right) \geq 1 - \xi_i,$$

$$i = 1, \ldots, l_1 + l_2, \quad (3.18)$$

$$\sum_{j=1}^{l} (\alpha_j - \alpha_j^*) K(\tilde{x}_j, \tilde{x}_i) + b \leq \varepsilon + \eta_i,$$

$$i = l_1 + l_2 + 1, \ldots, l, \quad (3.19)$$

$$\sum_{j=1}^{l} (\alpha_j - \alpha_j^*) K(\tilde{x}_j, \tilde{x}_i) + b \geq -\varepsilon - \eta_i^*,$$

$$i = l_1 + l_2 + 1, \ldots, l, \quad (3.20)$$

$$\alpha, \alpha^*, \xi, \eta, \eta^* \geq 0. \quad (3.21)$$

Suppose $(\bar{\alpha}, \bar{\alpha}^*, \bar{\xi}, \bar{\eta}, \bar{\eta}^*, \bar{b})$ is an optimal solution of problem (3.17)–(3.21), the decision function can be expressed as

$$
f_{\Theta_{j,k}}(x) = \begin{cases} +1, & \text{if } \sum_{i=1}^{l}(\bar{\alpha}_i - \bar{\alpha}_i^*)K(\tilde{x}_i, x) + \bar{b} \geq \varepsilon; \\ -1, & \text{if } \sum_{i=1}^{l}(\bar{\alpha}_i - \bar{\alpha}_i^*)K(\tilde{x}_i, x) + \bar{b} \leq -\varepsilon; \\ 0, & \text{otherwise.} \end{cases} \tag{3.22}
$$

Thus, in a K-class problem, for each pair (Θ_j, Θ_k), we have a classifier (3.22) to separate them as well as the remaining classes. So we have $K(K - 1)/2$ classifiers in total. Hence, for a new input x, $K(K - 1)/2$ outputs are obtained. Now we give the algorithm as follows

Algorithm 3.1 (K-Linear Programming Support Vector Classification Regression (K-LPSVCR))

(1) Given the training set T (3.1), construct the corresponding training set $\tilde{T}$ as (3.14);
(2) Select appropriate kernel $K(x, x')$ and parameters $C > 0$, $D > 0$, $\varepsilon > 0$;
(3) Solve problem (3.17)–(3.21), get the solution $(\bar{\alpha}, \bar{\alpha}^*, \bar{\xi}, \bar{\eta}, \bar{\eta}^*, \bar{b})$ and construct decision function as (3.22);
(4) For a new input x, translate its $K(K - 1)/2$ outputs as follows:
    (4.1) When $f_{\Theta_{j,k}}(x) = +1$, a positive vote is added to Θ_j, and no votes are added on the other classes;
    (4.2) When $f_{\Theta_{j,k}}(x) = -1$, a positive vote is added to Θ_k, and no votes are added on the other classes;
    (4.3) When $f_{\Theta_{j,k}}(x) = 0$ a negative vote is added to both Θ_j and Θ_k, and no votes are added to the other classes.
(5) After translating all of the $K(K - 1)/2$ outputs, we get the total votes of each class by adding the positive and negative votes on this class. Finally, x will be assigned to the class that gets the most votes.

The K-LPSVCR can be considered to include the SVC with $y_i = \pm 1$ (3.18) and the SVR for regression with 0 being the target value (3.19)–(3.20). It makes the fusion of the standard structures, "one versus one" and "one versus the rest", employed in the decomposition scheme of a multi-class classification procedure. On one hand, each (Θ_j, Θ_k)-K-LPSVCR classifier is trained to focus on the separation between two classes, as "one versus one" does. On the other hand, the same time, it also gives useful information about the other classes that are labeled 0, as "one versus the rest" does.

3.1.2 Numerical Experiments

In this section, experiments are performed to compare K-LPSVCR with the counterpart in [8, 249] on the same artificial and benchmark data sets. Experiments are

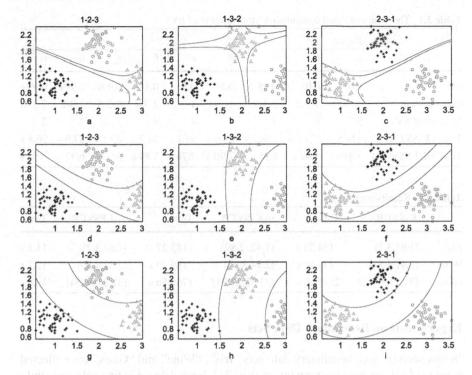

Fig. 3.1 The effect of parameter ε

carried out using Matlab v7.0 on Intel Pentium III 800 MHz PC with 256 MHz of RAM.

Experiments on Artificial Data Sets

The training set T is generated from a Gaussian distribution on R^2, it contains 150 examples in $K = 3$ classes, each of which has 50 examples [8, 249]. In this experiment, parameters $C = 30$, $D = 400$, and polynomial kernel $K(x, x') = \exp(\frac{-\|x-x'\|^2}{2\sigma^2})$ with $\sigma = 3$ is employed. The numerical results are summarized in Fig. 3.1 and Table 3.1.

In Fig. 3.1, $\frac{K(K-1)}{2} = 3$ 'one-one-rest' decision functions with $\varepsilon = 0.05, 0.5,$ 0.999 respectively are displayed.

From Fig 3.1, it can be observed that the behavior of algorithm K-LPSVCR is similar to that in [8]. For example, the parameter ε has a high influence on the optimal hyperplane determination. In addition, this similarity is also represented in the numbers of the support vectors shown in Table 3.1. However, Table 3.1 shows a great difference between K-LPSVCR and that in [8]: the time consumed here is much less than in [8]. This point will be strengthened in the following experiments.

Table 3.1 The values of ε and computation time, number of SVs

	Figure								
	a	b	c	d	e	f	g	h	i
ε	0.05	0.05	0.05	0.5	0.5	0.5	0.999	0.999	0.999
One-one-rest	1-2-3	1-3-2	2-3-1	1-2-3	1-3-2	2-3-1	1-2-3	1-3-2	2-3-1
Number of SVs	5	6	4	4	4	4	3	2	2
Time (K-SVCR)	114.8 s	121.0 s	138.2 s	86.0 s	98.1 s	96.2 s	89.1 s	87.9 s	96.8 s
Time (K-LPSVCR)	8.41 s	8.44 s	8.35 s	7.90 s	7.78 s	7.89 s	6.97 s	6.97 s	6.96 s

Table 3.2 Results comparison

	K-SVCR	Time	v-K-SVCR	Time	K-LPSVCR	Time
Iris	[1.97,3.2]	154.23 s	[1.42, 2.89]	145.27 s	[1.64,3.2]	11.1 s
Wine	[2.41,4.54]	178.11 s	[2.74,3.98]	189.29 s	[2.32,4.50]	13.2 s
Glass	[31.23,37.42]	2577.20 s	[33.48,37.21]	1765.56 s	[31.23,36.34]	18.8 s

Experiments on Benchmark Data Sets

In this section three benchmark data sets "Iris", "Wine" and "Glass" are collected from the UCI machine learning repository [21]. Each data set is first split randomly into ten subsets. Then one of these subsets is reserved as a test set and the others are summarized as the training set; this process is repeated ten times.

For data sets "Iris", "Wine", the polynomial kernels $K(x, x') = (x \cdot x')^d$ with degree $d = 4$ and $d = 3$ are employed respectively. For data set "Glass" the Gaussian kernel $K(x, x') = \exp(\frac{-\|x-x'\|^2}{2\sigma^2})$ with $\sigma = 0.2236$ is employed. Compared results between K-LPSVCR, K-SVCR in [8] and v-K-SVCR in [249] are listed in Table 3.2.

In Table 3.2, [·, ·] refers to two kinds of error percentage. The first number is the percentage of error when examples are finally assigned to the wrong classes, the second number is the percentage of error when examples are assigned to the wrong classes by any decision function $f_{\Theta_{jk}}(x)$, $j, k = 1, \ldots, l$.

More important, Table 3.2 shows that the time consumed in K-LPSVCR is much less than that both in K-SVCR and v-K-SVCR. Generally speaking, algorithm K-LPSVCR is faster than both of them over ten times. Furthermore, because K-LPSVCR has the same structure with the algorithm K-SVCR, it can also be proved to have good robustness.

3.1.3 v-K-LPSVCR

Follow the idea in Sect. 3.1.2, we can generalize algorithm K-LPSVCR to a new multi-class algorithm: v-K-class Linear Programming Support Vector Classification–Regression Machine (ν-K-LPSVCR), which can be realized by introducing ν-SVC and ν-SVR in linear programming formulations described in [95] and [187] respectively.

For fixed $C > 0$, $D > 0$, $\nu_1, \nu_2 \in (0, 1]$ and $K(\cdot, \cdot)$ chosen prior, a linear programming for multi-class problem is given as

$$\min_{\alpha, \alpha^*, \xi, \eta, \eta^*, b, \rho, \varepsilon} \frac{1}{l} \sum_{i=1}^{l} (\alpha_i + \alpha_i^*) + C\left(\frac{1}{l_1 + l_2} \sum_{i=1}^{l_1 + l_2} \xi_i - \nu_1 \rho\right)$$

$$+ D\left(\frac{1}{l - l_1 - l_2} \sum_{i=l_1 + l_2 + 1}^{l} (\eta_i + \eta_i^*) + \nu_2 \varepsilon\right) \tag{3.23}$$

$$\text{s.t.} \quad \frac{1}{l} \sum_{j=1}^{l} (\alpha_j + \alpha_j^*) = 1, \tag{3.24}$$

$$\tilde{y}_i \left(\sum_{j=1}^{l} (\alpha_j - \alpha_j^*) K(\tilde{x}_j, \tilde{x}_i) + b\right) \geq \rho - \xi_i, \quad i = 1, \ldots, l_1 + l_2, \tag{3.25}$$

$$\sum_{j=1}^{l} (\alpha_j - \alpha_j^*) K(\tilde{x}_j, \tilde{x}_i) + b \leq \varepsilon + \eta_i, \quad i = l_1 + l_2 + 1, \ldots, l, \tag{3.26}$$

$$\sum_{j=1}^{l} (\alpha_j - \alpha_j^*) K(\tilde{x}_j, \tilde{x}_i) + b \geq -\varepsilon - \eta_i^*, \quad i = l_1 + l_2 + 1, \ldots, l, \tag{3.27}$$

$$\alpha, \alpha^*, \xi, \eta, \eta^* \geq 0, \qquad \rho \geq \varepsilon \geq 0, \tag{3.28}$$

where α, α^*, ξ, η and η^* are the vectors respectively with the components α_i, α_i^*, ξ_i, η_i, η_i^*. Suppose $(\bar{\alpha}, \bar{\alpha}^*, \bar{\xi}, \bar{\eta}, \bar{\eta}^*, \bar{b}, \bar{\rho}, \bar{\varepsilon})$ is an optimal solution of problem (3.23)–(3.28), the decision function can be expressed as

$$f_{\Theta_{j,k}}(x) = \begin{cases} +1, & \text{if } g(x) \geq \varepsilon; \\ -1, & \text{if } g(x) \leq -\varepsilon; \\ 0, & \text{otherwise,} \end{cases} \tag{3.29}$$

where

$$g(x) = \sum_{i=1}^{l} (\bar{\alpha}_i - \bar{\alpha}_i^*) K(\tilde{x}_i, x) + \bar{b}. \tag{3.30}$$

According to (3.29) and (3.30), it is natural to define an input x_i in the training set to be a support vector (SV) if $(\bar{\alpha}_i - \bar{\alpha}_i^*) \neq 0$.

Table 3.3 Results comparison

	K-SVCR		v-K-SVCR		v-K-LPSVCR	
	Error	Time	Error	Time	Error	Time
Iris	[1.97,3.2]	154.23 s	[1.42, 2.89]	145.27 s	[1.62,3.2]	10.3 s
Wine	[2.41,4.54]	178.11 s	[2.74,3.98]	189.29 s	[2.40,4.50]	14.6 s
Glass	[31.23,37.42]	2577.20 s	[33.48,37.21]	1765.56 s	[31.34,36.51]	17.8 s

It should be pointed out that the parameter v in the (Θ_i, Θ_k) classifier has a similar meaning to that in v-SVM [249].

For a K-class classification problem, follow the same voting strategy as in K-LPSVCR, algorithm v-K-LPSVCR is proposed as follows: for each pair (Θ_j, Θ_k) with $j < k$, we construct a (Θ_j, Θ_k) classifier (3.29). So we have $K(K-1)/2$ classifiers in total. Hence, for a new input x, $K(K-1)/2$ outputs are obtained. We translate these outputs as follows: When $f_{\Theta_{j,k}}(x) = +1$, a positive vote is added on Θ_j, and no votes are added on the other classes; when $f_{\Theta_{j,k}}(x) = -1$, a positive vote is added on Θ_k, and no votes are added on the other classes; when $f_{\Theta_{j,k}}(x) = 0$ a negative vote is added on both Θ_j and Θ_k, and no votes are added on the other classes. After translating all of the $K(K-1)/2$ outputs, we get the total votes of each class by adding its positive and negative votes. Finally, x will be assigned to the class that gets the most votes.

Experiments are carried out to compare algorithm v-K-LPSVCR, K-SVCR in [8] and algorithm v-K-SVCR in [249] on benchmark data sets: "Iris", "Wine" and "Glass", detailed description of experiments can be found in [168], the main results is given in Table 3.3. Remarkably, Table 3.3 shows that the time consumed by algorithm v-K-LPSVCR is much less than the others while their errors are in the same level. Generally speaking, algorithm v-K-LPSVCR is faster than both of them over ten times, therefore it is suitable for solving large-scale data sets.

3.2 Support Vector Ordinal Regression Machine for Multi-class Problems

3.2.1 Kernel Ordinal Regression for 3-Class Problems

We have introduced Support vector Ordinal Regression in Sect. 2.3.1. Now, we first consider a 3-class classification problem with a training set

$$T = \{(x_1, y_1), \ldots, (x_l, y_l)\}$$
$$= \{(x_1, 1), \ldots, (x_{l_1}, 1), (x_{l_1+1}, 2), \ldots, (x_{l_1+l_2}, 2),$$
$$(x_{l_1+l_2+1}, 3), \ldots, (x_{l_1+l_2+l_3}, 3)\} \in (R^n \times \mathcal{Y})^l, \tag{3.31}$$

where the inputs $x_i \in R^n$, the outputs $y_i \in \mathcal{Y} = \{1, 2, 3\}$, $i = 1, \ldots, l$.

Let us turn to the ordinal regression described in Sect. 2.3.1. Its basic idea is: For linearly separable case (see Fig. 2.3), the margin to be maximized is associated with the two closest neighboring classes. As in conventional SVM, the margin is equal to $2/\|w\|$, thus maximizing the margin is achieved by minimizing $\|w\|$. This leads to the following algorithm through a rather tedious procedure.

Algorithm 3.2 (Kernel Ordinal Regression for 3-class problem)

(1) Given a training set (3.31);
(2) Select $C > 0$, and a kernel function $K(x, x')$;
(3) Define a symmetry matrix

$$H = \begin{pmatrix} H_1 & H_2 \\ H_3 & H_4 \end{pmatrix}, \tag{3.32}$$

where

$$H_1 = \begin{pmatrix} (K(x_i, x_j))_{l_1 \times l_1} & (K(x_i, x_j))_{l_1 \times l_2} \\ (K(x_i, x_j))_{l_2 \times l_1} & (K(x_i, x_j))_{l_2 \times l_2} \end{pmatrix},$$

$$H_2 = H_3^T = \begin{pmatrix} -(K(x_i, x_j))_{l_1 \times l_2} & -(K(x_i, x_j))_{l_1 \times l_3} \\ -(K(x_i, x_j))_{l_2 \times l_2} & -(K(x_i, x_j))_{l_2 \times l_3} \end{pmatrix},$$

$$H_4 = \begin{pmatrix} (K(x_i, x_j))_{l_2 \times l_2} & (K(x_i, x_j))_{l_2 \times l_3} \\ (K(x_i, x_j))_{l_3 \times l_2} & (K(x_i, x_j))_{l_3 \times l_3} \end{pmatrix},$$

and $(K(x_i, x_j))_{l_n \times l_m}$ denotes an $l_n \times l_m$ matrix with x_i belonging to the nth class and x_j belonging to the mth class ($n, m = 1, 2, 3$).
Solve the following optimization problem:

$$\min_{\mu} \frac{1}{2} \mu^T H \mu - \sum_{i=1}^{N} \mu_i, \tag{3.33}$$

$$\text{s.t.} \quad 0 \le \mu_i \le C, \quad i = 1, \ldots, N, \tag{3.34}$$

$$\sum_{i=1}^{l_1} \mu_i = \sum_{i=l_1+l_2+1}^{l_1+2l_2} \mu_i, \tag{3.35}$$

$$\sum_{i=l_1+1}^{l_1+l_2} \mu_i = \sum_{i=l_1+2l_2+1}^{N} \mu_i, \tag{3.36}$$

where $N = 2l - l_1 - l_3$, and get its solution $\mu = (\mu_1, \ldots, \mu_N)^T$;
(4) Construct the function

$$g(x) = -\sum_{i=1}^{l_1+l_2} \mu_i K(x_i, x) + \sum_{i=l_1+l_2+1}^{N} \mu_i K(x_i, x); \tag{3.37}$$

(5) Solve the following linear program problem:

$$\min_{b_1,b_2,\xi,\xi^*} \sum_{i=1}^{l_1+l_2} \xi_i + \sum_{i=1}^{l_2+l_3} \xi_i^*, \tag{3.38}$$

$$\text{s.t.} \quad g(x_i) - b_1 \le -1 + \xi_i, \quad i = 1, \ldots, l_1, \tag{3.39}$$

$$g(x_i) - b_2 \le -1 + \xi_i, \quad i = l_1 + 1, \ldots, l_1 + l_2, \tag{3.40}$$

$$g(x_i) - b_1 \ge 1 - \xi_{i-l_1}^*, \quad i = l_1 + 1, \ldots, l_1 + l_2, \tag{3.41}$$

$$g(x_i) - b_2 \ge 1 - \xi_{i-l_1}^*, \quad i = l_1 + l_2 + 1, \ldots, l, \tag{3.42}$$

$$\xi_i \ge 0, \quad i = 1, \ldots, l_1 + l_2, \tag{3.43}$$

$$\xi_i^* \ge 0, \quad i = 1, \ldots, l_2 + l_3, \tag{3.44}$$

where $\xi = (\xi_1, \ldots, \xi_{l_1+l_2})^{\mathrm{T}}$, $\xi^* = (\xi_1^*, \ldots, \xi_{l_2+l_3}^*)^{\mathrm{T}}$, and get its solution $(\bar{b}_1, \bar{b}_2, \bar{\xi}, \bar{\xi}^*)^{\mathrm{T}}$;

(6) Construct the decision function

$$f(x) = \begin{cases} 1, & \text{if } g(x) \le \bar{b}_1; \\ 2, & \text{if } \bar{b}_1 < g(x) \le \bar{b}_2; \\ 3, & \text{if } g(x) > \bar{b}_2, \end{cases} \tag{3.45}$$

where $g(x)$ is given by (3.37).

3.2.2 Multi-class Classification Algorithm

In this section, we propose the multi-class classification algorithm based on ordinal regression described in above section. The question is, how to use 3-class classifiers to solve a multi-class classification problem. There exist many reconstruction schemes. Remind that, when 2-class classifiers are used, there are "one versus one", "one versus rest" and other architectures. When 3-class classifiers are used, it is obvious that more architectures are possible. In order to compare this new algorithm with the ones in [8] and [249], here we also apply "one versus one versus rest" architecture.

Let the training set T be given by (3.1). For an arbitrary pair $(\Theta_j, \Theta_k) \in (\mathcal{Y} \times \mathcal{Y})$ of classes with $j < k$, Algorithm 3.2 will be applied to separate the two classes Θ_j and Θ_k as well as the remaining classes. To do so, we respectively consider the classes Θ_j, Θ_k and $\mathcal{Y} \setminus \{\Theta_j, \Theta_k\}$ as the classes 1, 3 and 2 in ordinal regression, i.e., the class Θ_j corresponds to the inputs x_i $(i = 1, \ldots, l_1)$, the class Θ_k corresponds to the inputs x_i $(i = l_1 + l_2 + 1, \ldots, l)$ and the remaining classes correspond to the inputs x_i $(i = l_1 + 1, \ldots, l_1 + l_2)$.

Thus, in the K-class problem, for each pair (Θ_j, Θ_k), we get by Algorithm 3.2 a classifier $f_{\Theta_{j,k}}(x)$ to separate them as well as the remaining classes. So we get

$K(K-1)/2$ classifiers in total. Hence, for a new input $\bar{x}$, $K(K-1)/2$ outputs are obtained. We adopt voting schemes, i.e., it is assumed that all the classifiers outputs are in the same rank, and the 'winner-takes-all' scheme is applied. In other words, $\bar{x}$ will be assigned to the class that gets the most votes.

Now we give the detail of the multi-class classification algorithm as follows:

Algorithm 3.3 (Multi-class classification algorithm based on ordinal regression)

(1) Given the training set (3.1), construct a set P with $K(K-1)/2$ elements:

$$P = \{(\Theta_j, \Theta_k) \mid (\Theta_j, \Theta_k) \in (\mathcal{Y} \times \mathcal{Y}), \ j < k\}, \qquad (3.46)$$

where $\mathcal{Y} = \{\Theta_1, \ldots, \Theta_K\}$. Set $m = 1$;

(2) If $m = 1$, take the first pair in the set P; otherwise take its next pair. Denote the pair just taken as (Θ_j, Θ_k);

(3) Transform the training set (3.1) into the following form:

$$\tilde{T} = \{(\tilde{x}_1, 1), \ldots, (\tilde{x}_{l_1}, 1), (\tilde{x}_{l_1+1}, 2), \ldots,$$
$$(\tilde{x}_{l_1+l_2}, 2), (\tilde{x}_{l_1+l_2+1}, 3), \ldots, (\tilde{x}_l, 3)\}, \qquad (3.47)$$

where

$$\{\tilde{x}_1, \ldots, \tilde{x}_{l_1}\} = \{x_i \mid y_i = \Theta_j\},$$
$$\{\tilde{x}_{l_1+1}, \ldots, \tilde{x}_{l_1+l_2}\} = \{x_i \mid y_i \in Y \setminus \{\Theta_j, \Theta_k\}\},$$
$$\{\tilde{x}_{l_1+l_2+1}, \ldots, \tilde{x}_l\} = \{x_i \mid y_i = \Theta_k\};$$

(4) Algorithm 3.2 is applied to the training set (3.47). Denote the decision function obtained as $f_{\Theta_{j,k}}(x)$;

(5) Vote: For an input $\bar{x}$, when $f_{\Theta_{j,k}}(\bar{x}) = 1$, a positive vote is added on Θ_j, and no votes are added on the other classes; when $f_{\Theta_{j,k}}(\bar{x}) = 3$, a positive vote is added on Θ_k, and no votes are added on the other classes; when $f_{\Theta_{j,k}}(\bar{x}) = 2$, a negative vote is added on both Θ_j and Θ_k, and no votes are added on the other classes;

(6) If $m = K(K-1)/2$, go to the next step; otherwise set $m = m + 1$, return to the step (3);

(7) Calculate the total votes of each class by adding the positive and negative votes on this class. The input $\bar{x}$ will be assigned to the class that gets the most votes.

3.2.3 Numerical Experiments

In this section, we present two types of experiments to evaluate Algorithm 3.3. All test problems are taken from [8] and [249].

Fig. 3.2 One-dimensional case. Class 1: 'o'; Class 2: '△'; Class 3: '◇'

Example in the Plane

According to Algorithm 3.3, for a K-class classification problem, Algorithm 3.2 is used $K(K-1)/2$ times and $K(K-1)/2$ classifiers are constructed. In order to examine the validity of these classifiers, we study an artificial example in plane which can be visualized. Note that this validity is not clear at a glance. For example, consider 3-class classification problem. The "one versus one versus rest" architecture implies that there are 3 training sets:

$$T_{1-2-3} = \{\text{Class 1 (one), Class 3 (rest), Class 2 (one)}\}, \qquad (3.48)$$

$$T_{1-3-2} = \{\text{Class 1 (one), Class 2 (rest), Class 3 (one)}\}, \qquad (3.49)$$

and

$$T_{2-3-1} = \{\text{Class 2 (one), Class 1 (rest), Class 3 (one)}\}, \qquad (3.50)$$

where Classes 3, 2, and 1 are respectively in the "middle" for ordinal regression. Each of them is expected in our mind to be separated by 2 parallel hyperplanes in feature space nicely. However, observe the simplest case when the input space is one-dimensional, see Fig. 3.2.

It is easy to see that it is impossible to separate all of the 3 training sets properly if the kernel in Algorithm 3.2 is linear. It can be imagine that similar difficulty will also happen to the low dimensional case. But as pointed in [8], these separations turn to be practical when nonlinear kernel is introduced. They construct the parallel separation hyperplanes in feature space by a combination of SVC and SVR with a parameter δ. Its success is shown by their numerical experiments when their parameter δ is chosen carefully. In the following, we shall show that this is also implemented successfully by Algorithm 3.2. Note that there is no any parameter corresponding to δ in our algorithm.

In our experiment, the training set T is generated following a Gaussian distribution on R^2 with 150 training data equally distributed in three classes, marking three classes by the following signs: Class 1 with '+', Class 2 with '△' and Class 3 with '◇'. The parameter C and the kernel $K(\cdot, \cdot)$ in Algorithm 3.2 are selected as follows: $C = 1000$, and polynomial kernel $K(x, x') = ((x \cdot x') + 1)^d$ with degree $d = 2, 3, 4$. For each group parameters we have three decision functions corresponding to the training sets: T_{1-2-3}, T_{1-3-2} and T_{2-3-1} in (3.48)–(3.50), and the training sets are marked as 1–2–3, 1–3–2 and 2–3–1. The numerical results are shown in Fig. 3.3.

From Fig. 3.3, it can be observed that the behavior of kernel ordinal regression algorithm is reasonable. The training data are separated explicitly into arbitrary order of classes. We get validity classifiers choosing different parameters of kernel. And it is also similar to that in [8] and [249], when their parameters δ and ε are chosen properly. Note that we need not make this choice.

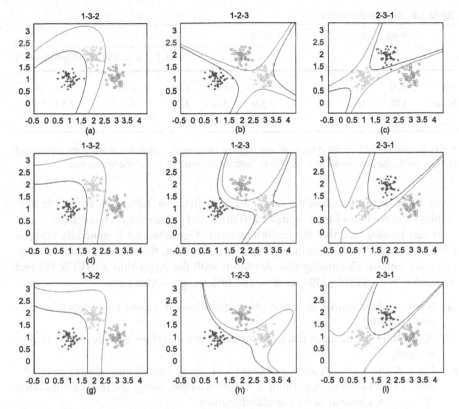

Fig. 3.3 Different kernel parameter d on artificial data set with $C = 1000$. (**a**) 1–3–2, $d = 2$; (**b**) 1–2–3, $d = 2$; (**c**) 2–3–1, $d = 2$; (**d**) 1–3–2, $d = 3$; (**e**) 1–2–3, $d = 3$; (**f**) 2–3–1, $d = 3$; (**g**) 1–3–2, $d = 4$; (**h**) 1–2–3, $d = 4$; (**i**) 2–3–1, $d = 4$

Experiments on Benchmark Data Sets

In this subsection, we test our multi-class classification algorithm on a collection of three benchmark problems same with Sect. 3.1.2, 'Iris', 'Wine' and 'Glass', from the UCI machine learning repository [21]. Each data set is first split randomly into ten subsets. Then one of these subsets is reserved as a test set and the others are summarized as the training set; this process is repeated ten times. Finally we compute the average error of the ten times, where the error of each time is described by the percentage of examples finally assigned to the wrong classes.

For data sets 'Iris' and 'Wine', the parameter $C = 10$ and the polynomial kernels $K(x, x') = ((x \cdot x') + 1)^d$ with degree $d = 2$ and $d = 3$ are employed respectively in our algorithm. For data set 'Glass', the parameter $C = 1000$ and the Gaussian kernel $K(x, x') = \exp(-\frac{\|x - x'\|^2}{2\sigma^2})$ with $\sigma = 2$ is employed. The averages errors are listed in Table 3.4. In order to compare our algorithm with Algorithm K-SVCR in [8], Algorithm ν-K-SVCR in [249], Algorithm '1-v-r' in [24] and Algorithm '1-v-1' in

Table 3.4 Experiment results

	# pts	# atr	# class	err%				
				1-v-r	1-v-1	K-SVCR	v-K-SVCR	Algorithm 3.3
Iris	150	4	3	1.33	1.33	1.93	1.33	2
Wine	178	13	3	5.6	5.6	2.29	3.3	2.81
Glass	214	9	6	35.2	36.4	30.47	32.47	26.17

pts: the number of training data, # atr: the number of example attributes, # class: the number of classes, err%: the percentage of error, 1-v-r: "one versus rest" , 1-v-1: "one versus one"

[106, 130], their corresponding errors are also listed in Table 3.4. This table also displays the number of training data, attributes and classes for each database.

It can be observed that the performance of Algorithm 3.3 is generally comparable to the other ones. Specifically, for the 'Glass' set, this algorithm outperforms other algorithms. Comparing this algorithm with the Algorithm K-SVCR [8] and Algorithm v-K-SVCR [249], we have the following conclusions:

(a) Noticing their similar structure, all of them have the same nice properties, e.g. good robustness;
(b) Algorithm 3.3 includes fewer parameters and therefore is easier to be implemented;
(c) For a K-class classification problem (3.1), many variant algorithms can be obtained by extending Algorithm 3.3, for example, using p-class classification ($2 \leq p \leq K$) instead of 3-class classification.

Chapter 4
Unsupervised and Semi-supervised Support Vector Machines

As an important branch in unsupervised learning, clustering analysis aims at partitioning a collection of objects into groups or clusters so that members within each cluster are more closely related to one another than objects assigned to different clusters [105]. Clustering algorithms provide automated tools to help identify a structure from an unlabeled set, in a variety of areas including bio-informatics, computer vision, information retrieval and data mining. There is a rich resource of prior works on this subject.

As we all know, efficient convex optimization techniques have had a profound impact on the field of machine learning, such as quadratic programming and linear programming techniques to Support Vector Machine and other kernel machine training [174]. Furthermore, Semi-definite Programming (SDP) extends the toolbox of optimization methods used in machine learning, beyond the current unconstrained, linear and quadratic programming techniques, which has provided effective algorithms to cope with the difficult computational problem in optimization and obtain high approximate solutions.

Semi-definite Programming has showed its utility in machine learning. Lanckreit *et al.* show how the kernel matrix can be learned from data via semi-definite programming techniques [134]. De Bie and Cristanini develop a new method for two-class transduction problem based on semi-definite relaxation technique [20]. Xu *et al.* based on [20, 134] develop methods to two-class unsupervised and semi-supervised classification problems in virtue of relaxation to Semi-definite Programming [227].

Given $C \in \mathcal{M}^n$, $A_i \in \mathcal{M}^n$ and $b \in \mathcal{R}^m$, where $\mathcal{M}^n$ is the set of symmetric matrix. The standard Semi-definite Programming problem is to find a matrix $X \in \mathcal{M}^n$ for the optimization problem [25]

$$\min \ C \bullet X \tag{4.1}$$

$$\text{(SDP)} \quad \text{s.t.} \quad A_i \bullet X = b_i, \quad i = 1, 2, \ldots, m, \tag{4.2}$$

$$X \succeq 0, \tag{4.3}$$

Y. Shi et al., *Optimization Based Data Mining: Theory and Applications*,
Advanced Information and Knowledge Processing,
DOI 10.1007/978-0-85729-504-0_4, © Springer-Verlag London Limited 2011

where the operation '•' is the matrix inner product $A \bullet B = \mathrm{tr}(A^T B)$, the notation $X \succeq 0$ means that X is a positive semi-definite matrix. The dual problem to SDP can be written as:

$$\max \; b^T \lambda \tag{4.4}$$

$$(SDD) \quad \text{s.t.} \quad C - \sum_{i=1}^{m} \lambda_i A_i \succeq 0. \tag{4.5}$$

Here $\lambda \in \mathcal{R}^m$. Interior point method has good effect for Semi-definite Programming, moreover there exists several softwares such as SeDuMi [191] and SDP3.

Now, we will propose several algorithms for unsupervised and semi-supervised problems based on SDP.

4.1 Unsupervised and Semi-supervised v-Support Vector Machine

In this section we construct unsupervised and semi-supervised classification algorithms which are based on Bounded v-Support Vector Machine (Bv-SVM) [78], which have a univocal parameter.

4.1.1 Bounded v-Support Vector Machine

Considering the supervised two-class classification problem with the training set

$$T = \{(x_1, y_1), \ldots, (x_l, y_l)\}, \tag{4.6}$$

where $x_i \in R^n$, $y \in \{-1, 1\}$, Bv-SVM is to find the linear discriminant $f(x) = (\mathrm{w} \cdot \phi(x)) + b$ by constructing the primal problem in some Hilbert space $\mathcal{H}$

$$\min_{\mathrm{w},b,\xi,\rho} \; \frac{1}{2}\|\mathrm{w}\|^2 + \frac{1}{2}b^2 - v\rho + \frac{1}{l}\sum_{i=1}^{l}\xi_i, \tag{4.7}$$

$$\text{s.t.} \quad y_i((\mathrm{w} \cdot \phi(x_i)) + b) \geq \rho - \xi_i, \tag{4.8}$$

$$\xi_i \geq 0, \quad i = 1, \ldots, l, \tag{4.9}$$

$$\rho \geq 0. \tag{4.10}$$

Its dual problem is

$$\max_{\alpha} \; -\frac{1}{2}\sum_{i=1}^{l}\sum_{j=1}^{l} y_i y_j \alpha_i \alpha_j K(x_i \cdot x_j) - \frac{1}{2}\sum_{i=1}^{l}\sum_{j=1}^{l} y_i y_j \alpha_i \alpha_j, \tag{4.11}$$

$$\text{s.t.} \quad 0 \leq \alpha_i \leq \frac{1}{l}, \quad i = 1, \ldots, l, \tag{4.12}$$

$$\sum_{i=1}^{l} \alpha_i \geq v, \tag{4.13}$$

where $K(x, x') = (\phi(x) \cdot \phi(x'))$ is the kernel function. Algorithm Bv-SVM can be described as follows:

Algorithm 4.1 (Bv-SVM)

(1) Given a training set $T = \{(x_1, y_1), \ldots, (x_l, y_l)\} \in (R^n \times \mathcal{Y})^l$, $x_i \in R^n$, $y_i \in \mathcal{Y} = \{-1, 1\}$, $i = 1, \ldots, l$;
(2) Select a kernel $K(x, x')$ and a parameter $C > 0$;
(3) Solve problem (4.11)–(4.13) and get its solution $\alpha^* = (\alpha_1^*, \ldots, \alpha_l^*)^T$;
(4) Construct the decision function $f(x) = \text{sgn}(\sum_{i=1}^{l} \alpha_i^* y_i (K(x, x_i) + 1))$.

4.1.2 v-SDP for Unsupervised Classification Problems

Now we will construct an algorithm for unsupervised classification problems based on Algorithm Bv-SVM. For unsupervised two-class classification problem, the training set is usually given as

$$T = \{x_1, \ldots, x_l\}, \tag{4.14}$$

where $x_i \in R^n$, the task is to assign the inputs x_i with labels -1 or 1.

Xu $et\ al.$ based on C-SVM with $b = 0$ get the optimization problem [227] that can solve unsupervised classification problem. In fact the Support Vector Machine they used is Bounded C-Support Vector Machine (BC-SVM) [141]. However, in BC-SVM, although the parameter C has explicit meaning in qualitative analysis, it has no specific in quantification. Therefore we consider the qualified SVM, which is Bv-SVM with the meaningful parameter v [78].

The parameter v varied between 0 and 1 places a lower bound on the sum of the α_i^*, which causes the linear term to be dropped from the objective function. It can be shown that the proportion of the training set that are margin errors is upper bounded by v, while v provides a lower bound on the total number of support vectors. Therefore v gives a more transparent parameter of the problem which does not depend on the scaling of the feature space, but only on the noise level in the data [51].

Now, we use the same method in [20, 134] to get the optimization problem based on Bv-SVM

$$\min_{y_i \in \{-1,+1\}^l} \min_{w,b,\xi,\rho} \frac{1}{2}\|w\|^2 + \frac{1}{2}b^2 - v\rho + \frac{1}{l}\sum_{i=1}^{l} \xi_i, \tag{4.15}$$

$$\text{s.t.} \quad y_i((w \cdot \phi(x_i)) + b) \geq \rho - \xi_i, \quad i = 1, \ldots, l, \tag{4.16}$$

$$-\varepsilon \leq \sum_{i=1}^{l} y_i \leq \varepsilon, \tag{4.17}$$

$$\xi_i \geq 0, \quad i = 1, \ldots, l, \tag{4.18}$$

$$\rho \geq 0. \tag{4.19}$$

It is difficult to solve problem (4.15)–(4.19), so we will consider to get its approximate solutions. Since Semi-definite Programming can provide effective algorithms to cope with the difficult computational problems and obtain high approximate solutions, it seems better to relax problem (4.15)–(4.19) to Semi-definite Programming. In order to relax it to SDP, we will change the form of primal Bν-SVM in use of duality, that means finding its dual problem's dual. Set $y = (y_1, \ldots, y_l)^{\mathrm{T}}$ and $M = yy^{\mathrm{T}}$, moreover $A \circ B$ denotes component-wise matrix multiplication. Relax $M = yy^{\mathrm{T}}$ to $M \succeq 0$ and $\mathrm{diag}(M) = e$, then the dual of Bν-SVM is

$$\max_{\alpha} \quad -\frac{1}{2}\alpha^{\mathrm{T}}((K + ee^{\mathrm{T}}) \circ M)\alpha, \tag{4.20}$$

$$\text{s.t.} \quad 0 \leq \alpha_i \leq \frac{1}{l}, \quad i = 1, \ldots, l, \tag{4.21}$$

$$\sum_{i=1}^{l} \alpha_i \geq \nu. \tag{4.22}$$

Now we will find the dual of the problem above for relaxation to SDP, problem (4.20)–(4.22)'s Lagrange function is

$$L(\alpha, u, r, \beta) = -\frac{1}{2}\alpha^{\mathrm{T}}((K + ee^{\mathrm{T}}) \circ M)\alpha$$
$$+ u(e^{\mathrm{T}}\alpha - \nu) + r^{\mathrm{T}}\alpha + \beta^{\mathrm{T}}\left(\frac{1}{l} - \alpha\right)$$
$$= -\frac{1}{2}\alpha^{\mathrm{T}}((K + ee^{\mathrm{T}}) \circ M)\alpha$$
$$+ (ue + r - \beta)^{\mathrm{T}}\alpha - u\nu + \frac{1}{l}\beta^{\mathrm{T}}e. \tag{4.23}$$

Suppose $K + ee^{\mathrm{T}} \succ 0$,

$$\nabla L_{\alpha} = -((K + ee^{\mathrm{T}}) \circ M)\alpha + ue + r - \beta, \tag{4.24}$$

and at the optimum, we have

$$\alpha = ((K + ee^{\mathrm{T}}) \circ M)^{-1}(ue + r - \beta). \tag{4.25}$$

Therefore the dual problem of problem (4.20)–(4.22) is

$$\min_{u,r,\beta} \quad \frac{1}{2}(ue + r - \beta)^{\mathrm{T}}((K + ee^{\mathrm{T}}) \circ M)^{-1}(ue + r - \beta) - u\nu + \frac{1}{l}\beta^{\mathrm{T}}e, \tag{4.26}$$

$$\text{s.t.} \quad u \geq 0, \quad r \geq 0, \quad \beta \geq 0, \tag{4.27}$$

Using the same method in [134], letting

$$\frac{1}{2}(ue + r - \beta)^{\mathrm{T}}((K + ee^{\mathrm{T}}) \circ M)^{-1}(ue + r - \beta) - uv + \frac{1}{l}\beta^{\mathrm{T}}e \leq \delta, \qquad (4.28)$$

then we will get linear matrix inequality by Schur Complement lemma,

$$\begin{pmatrix} (K + ee^{\mathrm{T}}) \circ M & ue + r - \beta \\ (ue + r - \beta)^{\mathrm{T}} & 2(\delta + uv - \frac{1}{l}\beta^{\mathrm{T}}e) \end{pmatrix} \succeq 0, \qquad (4.29)$$

finally we obtain the optimization problem below as SDP

$$\min_{M,\delta,u,r,\beta} \quad \delta, \qquad (4.30)$$

$$\text{s.t.} \quad \begin{pmatrix} (K + ee^{\mathrm{T}}) \circ M & ue + r - \beta \\ (ue + r - \beta)^{\mathrm{T}} & 2(\delta + uv - \frac{1}{l}\beta^{\mathrm{T}}e) \end{pmatrix} \succeq 0, \qquad (4.31)$$

$$-\varepsilon e \leq Me \leq \varepsilon e, \qquad (4.32)$$

$$M \succeq 0, \quad \mathrm{diag}(M) = e, \qquad (4.33)$$

$$r \geq 0, \quad u \geq 0, \quad \beta \geq 0. \qquad (4.34)$$

When we get the solution M^*, set $y^* = \mathrm{sgn}(t_1)$, where t_1 is eigenvector corresponding to the maximal eigenvalue of M^*, therefore we classify the inputs of training set T into two classes -1 and 1. So we construct the following algorithm:

Algorithm 4.2 (ν-SDP for Unsupervised Classification problems)

(1) Given the training set $T = \{x_1, \ldots, x_l\}$, where $x_i \in R^n$, $i = 1, \ldots, l$;
(2) Select an appropriate kernel $K(x, x')$, and parameters ν, ε, construct and solve the problem (4.30)–(4.34) and get the solution M^*, δ^*, u^*, r^* and β^*;
(3) Construct $y^* = \mathrm{sgn}(t_1)$, where t_1 is eigenvector corresponding to the maximal eigenvalue of M^*.

4.1.3 ν-SDP for Semi-supervised Classification Problems

It is easy to extend Algorithm 4.2 to semi-supervised classification algorithm. For semi-supervised two-class classification problem, the training set is usually given as

$$T = \{(x_1, y_1), \ldots, (x_l, y_l)\} \cup \{x_{l+1}, \ldots, x_{l+N}\}, \qquad (4.35)$$

where $x_i \in R^n$, $i = 1, \ldots, l + N$, $y_i \in \{-1, 1\}$, $i = 1, \ldots, l$, the task is to assign the inputs x_i, $i = l+1, \ldots, l+N$ with labels -1 or 1, and also to assign an unknown input x with label -1 or 1. We only need to add a constraint $M_{ij} = y_i y_j$, $i, j = 1, \ldots, l$

to the problem (4.30)–(4.34), and construct the programming for semi-supervised classification problem.

$$\min_{M,\delta,u,r,\beta} \quad \delta, \tag{4.36}$$

$$\text{s.t.} \quad \begin{pmatrix} (K + ee^{\mathrm{T}} \circ M) & ue + r - \beta \\ (ue + r - \beta)^{\mathrm{T}} & 2(\delta + uv - \frac{1}{l+N}\beta^{\mathrm{T}}e) \end{pmatrix} \succeq 0, \tag{4.37}$$

$$-\varepsilon e \le Me \le \varepsilon e, \tag{4.38}$$

$$M \succeq 0, \qquad \mathrm{diag}(M) = e, \tag{4.39}$$

$$M_{ij} = y_i y_j, \quad i, j = 1, \dots, l, \tag{4.40}$$

$$r \ge 0, \qquad u \ge 0, \qquad \beta \ge 0. \tag{4.41}$$

Algorithm 4.3 (v-SDP for Semi-supervised Classification problems)

(1) Given the training set (4.35);
(2) Select an appropriate kernel $K(x, x')$, and parameters v, ε, construct and solve the problem (4.36)–(4.41) and get the solution M^*, δ^*, u^*, r^* and β^*;
(3) Construct $y^* = \mathrm{sgn}(t_1)$, where t_1 is eigenvector corresponding to the maximal eigenvalue of M^*, assign the inputs $x_{l+1}, \dots, x_{l+N}$ with labels $y_{l+1}^*, \dots, y_{l+N}^*$;
(4) Construct decision function

$$f(x) = \mathrm{sgn}\left(\sum_{i=1}^{l} \alpha_i^* y_i (K(x, x_i) + 1) + \sum_{i=l+1}^{l+N} \alpha_i^* y_i^* (K(x, x_i) + 1) \right), \tag{4.42}$$

where $\alpha^* = ((K + ee^{\mathrm{T}}) \circ M^*)^{-1}(u^*e + r^* - \beta^*)$.

4.2 Numerical Experiments

4.2.1 Numerical Experiments of Algorithm 4.2

We will test Algorithm 4.2 on various data sets using SeDuMi library. In order to evaluate the influence of the meaningful parameter v, we will set value of v from 0.1 to 1 with increment 0.1 on four synthetic data sets including data set 'AI', 'Gaussian', 'circles' and 'joined-circles' [225], which every data set has sixty points. $\varepsilon = 2$ and Gaussian kernel with appropriate parameter $\sigma = 1$ are selected. Results are showed in Table 4.1, in which the number is the misclassification percent.

From Table 4.1 we can find that the result is better when v is in $[0.3, 0.8]$ than in other intervals. So the parameter v in v-SDP has the same meaning as in v-SVM. The value of v is too large or too small also results badly.

Furthermore, in order to evaluate the performance of Algorithm 4.2, we will compared it with Algorithm C-SDP [225], straightforward k-means algorithm and

Table 4.1 Algorithm 4.2 with changing ν

ν	AI	Gaussian	Circles	Joined-circles
0.1	8.2	3.3	13.3	18.03
0.2	8.2	5	1.67	14.75
0.3	9.83	1.67	10	9.83
0.4	11.48	1.67	1.67	11.48
0.5	3.28	1.67	1.67	6.67
0.6	9.83	1.67	1.67	8.1
0.7	22.95	1.67	1.67	8.1
0.8	11.48	1.67	1.67	11.48
0.9	3, 27	1.67	3.3	9.83
1.0	9.83	1.67	3.3	11.48

Table 4.2 Compared results

Algorithm	AI	Gaussian	Circles	Joined-circles
ν-SDP	11.48	1.67	1.67	11.48
C-SDP	16.7	1.67	11.67	28.33
K-means	8.20	0	48.33	48.33
DBSCAN	21.3	23.3	0	9.8

DBSCAN [69]. Both ν-SDP and C-SDP use the same parameters, and $C = 100$ in C-SDP which is same to [225]. The parameter k (number of objects in a neighborhood of an object) in DBSCAN is 3. Results are listed in Table 4.2, in which the number is the misclassification percent. And Fig. 4.1 illustrates some result of ν-SDP.

Table 4.2 shows that even the worst result of ν-SDP is not worse than that of C-SDP, then the best result is better than that of C-SDP, K-means and DBSCAN.

We also conduct ν-SDP on the real data sets including Face and Digits data sets [225]. Table 4.3 listed the results of changing ν, and Table 4.4 listed the compared results with other algorithms, in which the number is the misclassification percent. Figures 4.2 and 4.3 illustrate some results derived by ν-SDP.

4.2.2 Numerical Experiments of Algorithm 4.3

Here we test Algorithm 4.3 only on the real data sets including Face and Digits data sets. At first, we separate the data into two parts: labeled and unlabeled, then run Algorithm 4.2 to reclassify the inputs in the unlabeled part, eventually measured the misclassification error on the original dataset. The results are showed in Table 4.5 and Table 4.6, in which the number is the misclassification percent.

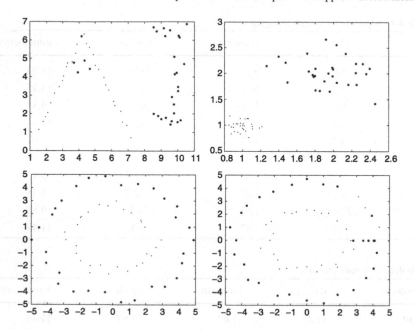

Fig. 4.1 Results of ν-SDP on data set 'AI', 'Gaussian', 'circles' and 'joined-circles'

Table 4.3 Results of changing ν

ν	Face12	Face34	Face56	Digit09	Digit17	Digit23
0.1	20	10	10	10	20	10
0.2	10	20	10	10	20	10
0.3	10	15	10	10	10	10
0.4	10	10	10	15	10	10
0.5	10	15	10	10	15	10
0.6	10	10	10	10	10	10
0.7	10	10	10	10	10	15
0.8	10	10	10	10	10	10
0.9	10	20	10	10	20	10
1.0	20	10	10	20	10	10

Table 4.4 Compared results

Algorithm	Face12	Face34	Face56	Digit09	Digit17	Digit23
ν-SDP	10	15	10	10	15	10
C-SDP	10	20	10	10	10	20
K-means	10	30	15	20	25	25
DBSCAN	50	50	50	50	50	50

Fig. 4.2 Sampling of the handwritten digits ('2' and '3'). Every *row* shows a random sampling of images from a class by ν-SDP

Fig. 4.3 Sampling of face. Every *row* shows a random sampling of images from a class by ν-SDP

Table 4.5 Results of changing ν

ν	Face12	Face34	Face56	Digit09	Digit17	Digit23
0.1	20	10	10	10	20	10
0.2	10	10	10	20	20	10
0.3	10	10	20	10	10	10
0.4	10	10	10	10	10	15
0.5	15	10	10	10	10	10
0.6	10	10	10	10	10	10
0.7	10	15	10	10	15	10
0.8	10	10	15	15	10	10
0.9	20	10	10	10	20	10
1.0	20	10	30	20	10	20

Table 4.6 Compared results

Algorithm	Face12	Face34	Face56	Digit09	Digit17	Digit23
semi-ν-SDP	10	15	10	10	15	10
semi-C-SDP	25	10	45	25	25	25

4.3 Unsupervised and Semi-supervised Lagrange Support Vector Machine

In this section we construct a new unsupervised and semi-supervised classification algorithms which is based on Lagrangian Support Vector Machine (LSVMs) [142].

Considering the supervised two-class classification problem with training set (4.6), Lagrangian Support Vector Machine construct the primal problem as

$$\min_{w,b,\xi} \frac{1}{2}(\|w\|^2 + b^2) + \frac{C}{2}\sum_{i=1}^{n}\xi_i^2, \qquad (4.43)$$

$$\text{s.t.} \quad y_i((w \cdot \phi(x_i)) - b) + \xi_i \geq 1, \quad i = 1,\ldots,l. \qquad (4.44)$$

Its dual problem is

$$\max_{\alpha} \ -\frac{1}{2}\sum_{i=1}^{l}\sum_{j=1}^{n}y_iy_j\alpha_i\alpha_jK(x_i,x_j) - \frac{1}{2}\sum_{i=1}^{l}\sum_{j=1}^{n}y_iy_j\alpha_i\alpha_j - \frac{1}{2C}\sum_{i=1}^{l}\alpha_i^2 + \sum_{i=1}^{n}\alpha_i$$

$$\text{s.t.} \quad \alpha_i \geq 0, \quad i = 1,\ldots,l, \tag{4.45}$$

where $K(x,x') = (\phi(x) \cdot \phi(x'))$ is the kernel function.

For unsupervised two-class classification problem with the training set (4.14), we use the method in [20, 134] to get the optimization problem based on LSVMs

$$\min_{y_i \in \{-1,+1\}^n} \ \min_{w,b,\xi} \ \frac{1}{2}(\|w\|^2 + b^2) + \frac{C}{2}\sum_{i=1}^{n}\xi_i^2, \tag{4.46}$$

$$\text{s.t.} \quad y_i((w \cdot \phi(x_i)) - b) + \xi_i \geq 1, \tag{4.47}$$

$$-\varepsilon \leq \sum_{i=1}^{n}y_i \leq \varepsilon. \tag{4.48}$$

Let $y = (y_1,\ldots,y_l)^T$, $M = yy^T$, $A \circ B$ denotes component-wise matrix multiplication, following the same procedure in [134], we construct the unsupervised classification algorithm.

Algorithm 4.4 (L-SDP for unsupervised Classification problems)

(1) Given the training set $T = \{x_1,\ldots,x_l\}$, where $x_i \in R^n$, $i = 1,\ldots,l$;
(2) Select an appropriate kernel $K(x,x')$, and parameters v, ε, construct and solve the following problem

$$\min_{M,\delta,u} \ \frac{1}{2}\delta, \tag{4.49}$$

$$\text{s.t.} \quad \begin{pmatrix} K \circ M + M + \frac{1}{C}I & u + e \\ (u+e)^T & \delta \end{pmatrix} \succeq 0, \tag{4.50}$$

$$-\varepsilon e \leq Me \leq \varepsilon e, \tag{4.51}$$

$$M \succeq 0, \qquad \text{diag}(M) = e, \tag{4.52}$$

$$u \geq 0, \tag{4.53}$$

get the solution M^*, δ^* and u^*;
(3) Construct label $y^* = \text{sgn}(t_1)$, where t_1 is eigenvector corresponding to the maximal eigenvalue of M^*.

Similar with Algorithm 4.3, it is easy to extend Algorithm 4.4 to semi-supervised classification algorithm, and is omitted here.

In order to evaluate the performance of L-SDP, we will compared Algorithm 4.4 with Algorithm 4.2 and C-SDP [227]. Firstly we still consider four synthetic data

Table 4.7 Compared results on synthetic data sets

Algorithm	AI	Gaussian	Circles	Joined-circles
L-SDP	9.84	0	0	8.19
C-SDP	9.84	1.67	11.67	28.33
v-SDP	9.84	1.67	1.67	11.48

Table 4.8 Computation time on synthetic data sets

Algorithm	AI	Gaussian	Circles	Joined-circles
L-SDP	1425	1328	1087.6	1261.8
C-SDP	2408.9	1954.9	2080.2	2284.8
v-SDP	2621.8	1891	1837.1	2017.2

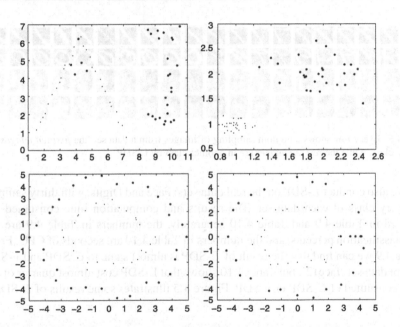

Fig. 4.4 Results of L-SDP on data set 'AI', 'Gaussian', 'circles' and 'joined-circles'

sets including data set AI, Gaussian, circles and joined-circles, which every data set has sixty points. $\varepsilon = 2$, $C = 100$ and Gaussian kernel with appropriate parameter $\sigma = 1$ are selected. Results are showed in Table 4.7, in which the number is the misclassification percent. And Fig. 4.4 illustrates some results on the data sets.

From Table 4.7 we can find that the result of L-SDP is better than that of C-SDP and v-SDP, moreover the time consumed is showed in Table 4.8, in which the numbers are seconds of CPU. From Table 4.8 we can find that L-SDP cost almost half of the time consumed of C-SDP or v-SDP.

Table 4.9 Compared results on Face and Digits data sets

Algorithm	Digits32	Digits65	Digits71	Digits90	Face12	Face34	Face56	Face78
L-SDP	0	0	0	0	8.33	0	0	0
C-SDP	0	0	0	0	1.67	0	0	0
v-SDP	0	0	0	0	1.67	0	0	0

Table 4.10 Computation time on Face and Digits data sets

Algorithm	Digits32	Digits65	Digits71	Digits90	Face12	Face34	Face56	Face78
L-SDP	445.1	446.2	446.4	446.5	519.8	446.3	446.1	446
C-SDP	1951.8	1950.7	1951.6	1953.4	1954.5	1952.1	1950.3	1951
v-SDP	1721.6	1721.5	1722.2	1722.8	1721.4	1722.1	1721	1719.7

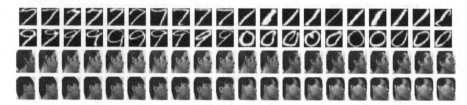

Fig. 4.5 Every *row* shows a random sampling of images from a data set, the *first ten images* are in one class, while the *rest ten images* are in another class by L-SDP

We also conduct L-SDP on the real data sets: Face and Digits, with thirty samples of every class of each data set. The results and computation time consumed are showed in Table 4.9 and Table 4.10 separately, the numbers in Table 4.9 are the misclassification percents, and the numbers in Table 4.10 are seconds of CPU. From Table 4.9 we can find that the result of L-SDP is almost same to C-SDP and v-SDP except data set 'face12', but Table 4.10 shows that L-SDP cost almost quarter of the time consumed of C-SDP or v-SDP. Figure 4.5 illustrates some results of L-SDP.

4.4 Unconstrained Transductive Support Vector Machine

Most of the learning models and systems in artificial intelligence apply inductive inference where a model (a function) is derived from data and this model is further applied on new data. The model is created without taking into account any information about a particular new data. The new data would fit into the model to certain degree so that an error is estimated. The model is in most cases a global model, covering the whole problem space. Creating a global model (function) that would be valid for the whole problem space is a difficult task and in most cases—it is not necessary. The inductive learning and inference approach is useful when a global model of the problem is needed even in its very approximate form, when incremental, on-line learning is applied to adjust this model on new data and trace its evolution.

Generally speaking, inductive inference is concerned with the estimation of a function (a model) based on data from the whole problem space and using this model to predict output values for a new input, which can be any point in this space.

In contrast to the inductive inference, transductive inference methods estimate the value of a potential model (function) only for a single point of the space (the new data vector) utilizing additional information related to this point [206]. This approach seems to be more appropriate for clinical and medical applications of learning systems, where the focus is not on the model, but on the individual patient data. And it is not so important what the global error of a global model over the whole problem space is, but rather—the accuracy of prediction for any individual patient. Each individual data may need an individual, local model rather than a global model which new data tried to be matched into it without taking into account any specific information on where this new data point is located in the space. Transductive inference methods are efficient when the size of the available training set is relatively small [206].

4.4.1 Transductive Support Vector Machine

In contrast to inductive SVM learning described in Chap. 1, the classification problem for transductive SVM to be solved not only includes the training set T (1.1), but also an i.i.d test set S from the same distribution,

$$S = \{x_1^*, \ldots, x_m^*\}, \tag{4.54}$$

where $x_i^* \in R^n$.

Transductive Support Vector Machine (TSVM) take into account this test set S and try to minimize misclassification of just those particular examples. Therefore, in a linearly separable data case, TSVM try to find a labeling $y_1^*, y_2^*, \ldots, y_m^*$ of the test inputs, and the hyperplane $(w \cdot x) + b = 0$ so that it can separate both training and test input with maximum margin, which leads to the following problem:

$$\min_{w,b,y^*} \frac{1}{2} \|w\|^2 \tag{4.55}$$

$$\text{s.t.} \quad y_i((w \cdot x_i) + b) \geq 1, \quad i = 1, \ldots, l, \tag{4.56}$$

$$y_j^*((w \cdot x_j^*) + b) \geq 1, \quad j = 1, \ldots, m, \tag{4.57}$$

where $y^* = (y_1^*, \ldots, y_m^*)$.

Figure 4.6 illustrates this. In Fig. 4.6, positive and negative training points are marked as "+" and "−", test points as "●". The dashed line is the solution of the inductive SVM. The solid line shows the transductive classification.

Fig. 4.6 Transductive classification for linearly separable problem

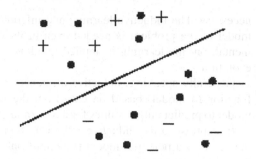

To be able to handle linearly non-separable case, similar to the way in above inductive SVM, the learning process of transductive SVM is formulated as the following optimization problem,

$$\min_{w,b,y^*,\xi,\xi^*} \quad \frac{1}{2}\|w\|^2 + C\sum_{i=1}^{l}\xi_i + C^*\sum_{j=1}^{m}\xi_j^*, \tag{4.58}$$

$$\text{s.t.} \quad y_i((w \cdot x_i) + b) \geq 1 - \xi_i, \quad i = 1,\ldots,l, \tag{4.59}$$

$$y_j^*((w \cdot x_j^*) + b) \geq 1 - \xi_j^*, \quad j = 1,\ldots,m, \tag{4.60}$$

$$\xi_i \geq 0, \quad i = 1,\ldots,l, \tag{4.61}$$

$$\xi_j^* \geq 0, \quad j = 1,\ldots,m, \tag{4.62}$$

where C and C^* are parameters set by the user. They allow trading of margin size against misclassifying training examples or excluding test examples. Some algorithms have been proposed to solve this problem [117].

4.4.2 Unconstrained Transductive Support Vector Machine

Training a transductive SVM means solving the (partly) combinatorial optimization problem (4.58)–(4.62). For a small number of test examples, this problem can be solved optimally simply by trying all possible assignments of $y_1^*,\ldots,y_m^*$ to the two classes. However, this approach become intractable for test sets with more than 10 examples. Previous approaches using branch-and-bound search push the limit to some extent [212], and an efficient algorithm was designed to handle the large test sets common in text classification with 10,000 test examples and more in [117]. It finds an approximate solution to problem (4.58)–(4.62) using a form of local search. Next we will propose a new algorithm to solve problem (4.58)–(4.62) by its special construction [203].

Unconstrained Optimization Problem

First we will transform problem (4.58)–(4.62) to an unconstrained problem.

Theorem 4.5 *Considering the solution* (w, b, y^*, ξ, ξ^*) *of problem* (4.58)–(4.62). *For any* x_i^* *of test set* S, *it must satisfy*

$$y_j^*((w \cdot x_j^*) + b) \geq 0. \tag{4.63}$$

Proof Suppose $(w, b, \xi, \xi^*, y_1^*, \ldots, y_m^*)$ is the solution of problem (4.58)–(4.62). First we will show that for any $x_j^* \in S$, if $(w \cdot x_j^*) + b > 0$, then there must be $y_j^* = 1$.

In fact, if $y_j^* = 1$, then $y_j^*((w \cdot x_j^*) + b) > 0$, and by the constraint (4.60), we have $\xi_j^* \geq 1 - y_j^*((w \cdot x_j^*) + b)$. Because the objective function (4.58) is minimized, so there must be

$$0 \leq \xi_j^* = \min\{0, 1 - y_j^*((w \cdot x_j^*) + b)\} < 1. \tag{4.64}$$

If $y_j^* = -1$, then $y_j^*((w \cdot x_j^*) + b) < 0$, and by the constraint (4.60), we have

$$\xi_j^* \geq 1 - y_j^*((w \cdot x_j^*) + b) > 1, \tag{4.65}$$

the value of objective function is larger than the case $y_j^* = 1$. So if $(w \cdot x_j^*) + b \geq 0$, there must be $y_j^* = 1$ and $y_j^*((w \cdot x_j^*) + b) > 0$.

Following the above logic, for any $x_j^* \in S$, if $(w \cdot x_j^*) + b < 0$, then there must be $y_j^* = -1$ and $y_j^*((w \cdot x_j^*) + b) > 0$.

Obviously, if $(w \cdot x_j^*) + b = 0$, then $y_j^*((w \cdot x_j^*) + b) \geq 0$. Therefore, for any $x_i^* \in S$, there must be $y_j^*((w \cdot x_j^*) + b) \geq 0$. $\qquad\square$

Based on Theorem 4.5, we can rewrite the constraints (4.60) and (4.62) of problem (4.58)–(4.62) as

$$\xi_j^* = (1 - y_j^*(w \cdot x_j^*) + b))_+ = (1 - |(w \cdot x_j^*) + b|)_+, \quad j = 1, \ldots, m. \tag{4.66}$$

Furthermore, rewrite the constraints (4.59) and (4.61) as

$$\xi_i = (1 - y_i((w \cdot x_i) + b)_+, \quad i = 1, 2, \ldots, l, \tag{4.67}$$

where function $(\cdot)_+$ is plus function:

$$(\Delta)_+ = \begin{cases} \Delta, & \Delta \geq 0; \\ 0, & \Delta < 0. \end{cases} \tag{4.68}$$

Therefore problem (4.58)–(4.62) can be transformed as unconstrained optimization problem

$$\min_{w,b} \frac{1}{2}\|w\|^2 + C \sum_{i=1}^{l}(1 - y_i((w \cdot x_i) + b))_+$$

$$+ C^* \sum_{j=1}^{m}(1 - |(w \cdot x_j^*) + b|)_+. \tag{4.69}$$

Smooth Unconstrained Optimization Problem

However, the objective function in problem (4.69) is not differentiable which precludes the usual optimization method, we thus apply the smoothing technique and replace the plus function $(\cdot)_+$ by a very accurate smooth approximation to obtain a smooth Transductive Support Vector Machine.

Therefore, we introduce the approximation function of $(\Delta)_+$

$$P(\Delta, \lambda) = \Delta + \frac{1}{\lambda} \ln(1 + e^{-\lambda \Delta}), \tag{4.70}$$

where $\lambda > 0$ is the smoothing parameter. Obviously function (4.70) is smooth, and it can shown that function $P(\Delta, \lambda)$ converges to $(\Delta)_+$ with $\lambda \to \infty$, so the second part of objective function in problem (4.69) can be approximately formulated as

$$C \sum_{i=1}^{l} P(1 - y_i((\mathbf{w} \cdot \mathbf{x}_i) + b), \lambda). \tag{4.71}$$

And the third part can be approximately formulated as

$$C^* \sum_{j=1}^{m} P(1 - |(\mathbf{w} \cdot \mathbf{x}_j^*) + b|, \lambda). \tag{4.72}$$

However, there still exist a term $|(\mathbf{w} \cdot \mathbf{x}_j^*) + b|$ which is not smooth. So we introduce another approximation function to replace function $|\Delta'|$,

$$P'(\Delta', \mu) = \Delta' + \frac{1}{\mu} \ln(1 + e^{-2\mu \Delta'}) \tag{4.73}$$

It is easy to prove the following theorem.

Theorem 4.6 *Function $P'(\Delta', \mu)$ is smooth and satisfies:*

(i) $P'(\Delta', \mu)$ *is continuously differentiable;*
(ii) $P'(\Delta', \mu)$ *is strictly convex on R;*
(iii) *For any $\mu \in R$, there is $P'(\Delta', \mu) > |\Delta'|$;*
(iv) $\lim_{\mu \to \infty} P'(\Delta', \mu) = |\Delta'|, \forall \Delta' \in R$.

Figure 4.7 describes the function $P'(\Delta', \mu)$ with $\mu = 5$ approximating function $|\Delta'|$. So (4.72) can also be approximately formulated as

$$C^* \sum_{j=1}^{m} P(1 - P'((\mathbf{w} \cdot \mathbf{x}_j^*) + b, \mu), \lambda). \tag{4.74}$$

Fig. 4.7 Smoothing function for $|\Delta|$

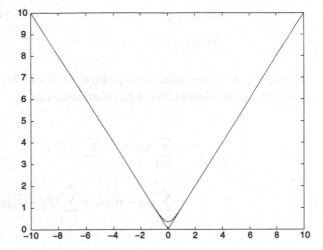

Now when λ, μ is large enough, problem (4.69) can be replaced by approximation smooth problem

$$\min_{w,b} \frac{1}{2}\|w\|^2 + C\sum_{i=1}^{l} P(1 - y_i((w \cdot x_i) + b), \lambda)$$

$$+ C^*\sum_{j=1}^{m} P(1 - P'((w \cdot x_j^*) + b, \mu), \lambda). \qquad (4.75)$$

4.4.3 Unconstrained Transductive Support Vector Machine with Kernels

Solving the smooth problem (4.75) will construct a linear separating surface $(w \cdot x) + b = 0$ for our classification problem. We now describe how to construct a nonlinear separating surface which is implicitly defined by a kernel function $K(\cdot, \cdot)$.

For any fixed values of $y_1^*, \ldots, y_m^*$, the dual problem of problem (4.58)–(4.62) is [206]

$$\min_{\alpha, \beta} \frac{1}{2}\left[\sum_{i,r=1}^{l} \alpha_i \alpha_r y_i y_r K(x_i, x_r) + \sum_{j,r=1}^{m} \beta_j \beta_r y_j^* y_r^* K(x_j^*, x_r^*)\right.$$

$$\left. - 2\sum_{j=1}^{l}\sum_{r=1}^{m} y_j y_r^* \alpha_j \beta_r K(x_j, x_r^*)\right] - \sum_{i=1}^{l} \alpha_i - \sum_{j=1}^{m} \beta_j \qquad (4.76)$$

$$\text{s.t.} \quad \sum_{i=1}^{l} y_i \alpha_i + \sum_{j=1}^{m} y_j^* \beta_j = 0, \qquad (4.77)$$

$$0 \le \alpha_i \le C, \quad i = 1, \ldots, l, \tag{4.78}$$

$$0 \le \beta_j \le C^*, \quad j = 1, \ldots, m. \tag{4.79}$$

Suppose α^*, β^* is the solution of problem (4.76)–(4.79), then the solution with respect to w of problem (4.58)–(4.62) is expressed as

$$\begin{aligned} \mathrm{w} &= \sum_{i=1}^{l} y_i \alpha_i^* x_i + \sum_{j=1}^{m} y_j^* \beta_j^* x_j^* \\ &= \sum_{i=1}^{l} (\tilde{\alpha}_i - \bar{\alpha}_i) x_i + \sum_{j=1}^{m} (\tilde{\beta}_j - \bar{\beta}_j) x_j, \end{aligned} \tag{4.80}$$

where $y_i \alpha_i^*$ is replace by $\tilde{\alpha}_i - \bar{\alpha}_i$, and $y_j^* \beta_j^*$ by $\tilde{\beta}_j - \bar{\beta}_j$.

If we imply 1-norm of $\|\mathrm{w}\|_1$ in problem (4.58)–(4.62), and based on (4.80), approximately replace it by

$$\sum_{i=1}^{l} |\tilde{\alpha}_i - \bar{\alpha}_i| + \sum_{j=1}^{m} |\tilde{\beta}_j - \bar{\beta}_j| \tag{4.81}$$

problem (4.58)–(4.62) is transformed as

$$\min_{\tilde{\alpha}, \bar{\alpha}, \tilde{\beta}, \bar{\beta}, b, \xi, \xi^*} \sum_{i=1}^{l} (\tilde{\alpha}_i + \bar{\alpha}_i) + \sum_{j=1}^{m} (\tilde{\beta}_j^* + \bar{\beta}_j^*) + C \sum_{i=1}^{l} \xi_i + C^* \sum_{j=1}^{m} \xi_j^*, \tag{4.82}$$

$$\text{s.t.} \quad y_i \left(\sum_{k=1}^{l} (\tilde{\alpha}_k - \bar{\alpha}_k) K(x_k, x_i) + \sum_{k=1}^{m} (\tilde{\beta}_k - \bar{\beta}_k) K(x_k^*, x_i) + b \right) \ge 1 - \xi_i,$$

$$i = 1, \ldots, l, \tag{4.83}$$

$$y_j^* \left(\sum_{k=1}^{l} (\tilde{\alpha}_k - \bar{\alpha}_k) K(x_k, x_j^*) + \sum_{k=1}^{m} (\tilde{\beta}_k - \bar{\beta}_k) K(x_k^*, x_j^*) + b \right) \ge 1 - \xi_j^*,$$

$$j = 1, \ldots, m, \tag{4.84}$$

$$\xi_i \ge 0, \quad \tilde{\alpha}_i, \bar{\alpha}_i \ge 0, \quad i = 1, \ldots, l, \tag{4.85}$$

$$\xi_j^* \ge 0, \quad \tilde{\beta}_j, \bar{\beta}_j \ge 0, \quad j = 1, \ldots, m. \tag{4.86}$$

Like the smooth method in the above section, by introducing smooth functions $P(\Delta, \lambda)$ and $P'(\Delta', \mu)$, we can transform problem (4.82)–(4.86) to a smooth problem with kernel

$$\min_{\tilde{\alpha},\bar{\alpha},\tilde{\beta},\bar{\beta},b,\xi,\xi^*} \sum_{i=1}^{l}(\tilde{\alpha}_i + \bar{\alpha}_i) + \sum_{j=1}^{m}(\tilde{\beta}_j + \bar{\beta}_j)$$

$$+ C\sum_{i=1}^{l} P\left(1 - y_i\left(\sum_{k=1}^{l}(\tilde{\alpha}_k - \bar{\alpha}_k)K(x_k, x_i)\right.\right.$$

$$+ \sum_{k=1}^{m}(\tilde{\beta}_k - \bar{\beta}_k)K(x_k^*, x_i) + b\bigg), \lambda\bigg)$$

$$+ C^*\sum_{j=1}^{m} P\left(1 - P'\left(\sum_{k=1}^{l}(\tilde{\alpha}_k - \bar{\alpha}_k)K(x_k, x_j^*)\right.\right.$$

$$+ \sum_{k=1}^{m}(\tilde{\beta}_k - \bar{\beta}_k)K(x_k^*, x_j^*) + b, \mu\bigg), \lambda\bigg).$$

$$(4.87)$$

If λ, μ is chosen large enough, we hope that the solution of problem (4.87) will approximates the solution of problem (4.82)–(4.86). After getting the solution $(\tilde{\alpha}, \bar{\alpha}, \tilde{\beta}, \bar{\beta}, b, \xi, \xi^*)$ of problem (4.87), the decision function is constructed as

$$f(x) = \text{sgn}\left(\sum_{k=1}^{l}(\tilde{\alpha}_k - \bar{\alpha}_k)K(x_k, x) + \sum_{k=1}^{m}(\tilde{\beta}_k - \bar{\beta}_k)K(x_k^*, x) + b\right), \quad (4.88)$$

so the test points in test set S are assigned to two classes by this function, i.e., the values of $y_1^*, \ldots, y_m^*$ are given. Now the new algorithm Unconstrained Transductive Support Vector Machine (UTSVM) are described as follows:

Algorithm 4.7 (UTSVM)

(1) Given a training set $T = \{(x_1, y_1), \ldots, (x_l, y_l)\} \in (R^n \times \{-1, 1\})^l$, and a test set $S = \{x_1^*, \ldots, x_m^*\}$;
(2) Select a kernel $K(\cdot, \cdot)$, and parameters $C > 0$, $C^* > 0$, $\lambda > 0$, $\mu > 0$;
(3) Solve problem (4.87) and get its solution $(\tilde{\alpha}, \bar{\alpha}, \tilde{\beta}, \bar{\beta}, b, \xi, \xi^*)$;
(4) Construct the decision function as (4.88) and get the values of $y_1^*, \ldots, y_m^*$.

Chapter 5
Robust Support Vector Machines

We have introduced ordinal regression problem, multi-class classification problem, unsupervised and semi-supervised problems, which all can be solved efficiently follow the scheme of SVMs, see, e.g. [8, 9, 42, 44, 50, 104, 129, 130, 215, 232]. But in the above approaches, the training data are implicitly assumed to be known exactly. However, in real world applications, the situation is not always the case because the training data subject to measurement and statistical errors [89]. Since the solutions to optimization problems are typically sensitive to training data perturbations, errors in the input data tend to get amplified in the decision function, often resulting in far from optimal solutions [15, 39, 250]. So it will be useful to explore formulations that can yield robust discriminants to such estimation errors.

In this chapter, we first establish robust versions of SVORM, which are represented as a second order cone programming (SOCP) [2]. And as the theoretical foundation, we study the relationship between the solutions of the SOCP and its dual problem. Here the second order cone in Hilbert space is involved. Then, we also establish a multi-class algorithm based on the above robust SVORM for general multi-class classification problem with perturbations. Furthermore, we construct a robust unsupervised and semi-supervised SVC for the problems with uncertainty information.

5.1 Robust Support Vector Ordinal Regression Machine

Ordinal regression problem is presented here as in Sect. 2.3.1:

Suppose a training set is given by

$$T = \{x_i^j\}_{i=1,\dots,l^j}^{j=1,\dots,k}, \tag{5.1}$$

where $x_i^j \in R^n$ is the input the superscript, $j = 1, \dots, k$ denotes the class number and $i = 1, \dots, l^j$ is the index with each class. The task is to find a real value function

Y. Shi et al., *Optimization Based Data Mining: Theory and Applications*,
Advanced Information and Knowledge Processing,
DOI 10.1007/978-0-85729-504-0_5, © Springer-Verlag London Limited 2011

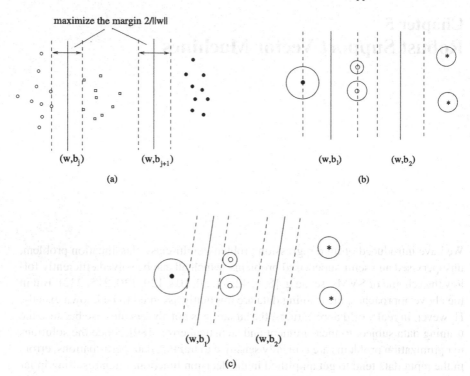

Fig. 5.1 (a) The fixed margin version; (b) the effect of measurement noises; (c) the result of Robust version

$g(x)$ and an orderly real sequence $b_1 \leq \cdots \leq b_{k-1}$ and construct a decision

$$f(x) = \min_{s \in \{1,\ldots,k\}} \{s : g(x) - b_s < 0\}, \tag{5.2}$$

where $b_k = +\infty$.

Consider the linear SVORM in two dimensional space with $k = 3$ as illustrated in Fig. 5.1. The training data x_i^j, $j = 1, \ldots, k$, $i = 1, \ldots, l^j$ used are implicitly assumed to be known exactly and two separation lines are obtained (see Fig. 5.1(a)). However, in real world applications, the data are corrupted by measurements and statistical errors [89]. Errors in the input space tend to get amplified in the decision function, which often results in misclassification. So a robust version of SVORM with sphere perturbations is considered in this paper. Each training point in Fig. 5.1(a) is allowed to move in a circle. In this case, we should modify the original separation lines since they cannot separate the training set in this case (see Fig. 5.1(b)) properly. It seems reasonable to yield separation lines as show in Fig. 5.1(c) which separate three classes of perturbed circles.

Now let us establish robust SVORM. Ordinal regression problem with sphere perturbations is presented as follows:

Suppose a training set is given by

$$T = \{\mathcal{X}_i^j\}_{i=1,\ldots,l^j}^{j=1,\ldots,k}, \tag{5.3}$$

where $\mathcal{X}_i^j$ is the input set: a sphere with the x_i^j and radius r_i^j, namely,

$$\mathcal{X}_i^j = \{x_i^j + r_i^j u_i^j : \|u_i^j\| \leq 1\}, \quad j = 1,\ldots,k, \ i = 1,\ldots,l^j, \tag{5.4}$$

where $u_i^j \in R^n$, $r_i^j \geq 0$ is a constant. The output is the superscript $j \in \{1, 2, \ldots, k\}$ denoting the class number. $i = 1, \ldots, l^j$ is the index within each class. We want to find a decision function that minimizes the misclassification in the worst case, i.e. one that minimizes the maximum misclassification when samples are allowed to move within their corresponding confidence balls.

We consider both the linear separation and the nonlinear separation and, therefore, introduce the transformation

$$\Phi : x \rightarrow \mathbf{x} = \Phi(x), \tag{5.5}$$

$$R^n \rightarrow \mathcal{H}. \tag{5.6}$$

Denote the corresponding kernel as $K(x, x') = (\Phi(x) \cdot \Phi(x'))$. We restrict the kernel to be one of the following two cases:

(Case I)   linear kernel: $K(x, x') = K_1(x, x') = (x \cdot x')$, $\tag{5.7}$

(Case II)   Gaussian kernel: $K(x, x') = K_2(x, x') = \exp\left(\dfrac{-\|x - x'\|^2}{2\sigma^2}\right). \tag{5.8}$

Denote the two corresponding transformations as $\Phi_1(x)$ and $\Phi_2(x)$ respectively. For both cases, the training set (5.3) is transformed to

$$T = \{X_i^j\}_{i=1,\ldots,l^j}^{j=1,\ldots,k}, \tag{5.9}$$

where

$$X_i^j = \{\Phi(\tilde{x}_i^j) \mid \tilde{x}_i^j = x_i^j + r_i^j u_i^j, \ \|u_i^j\| \leq 1\}. \tag{5.10}$$

Notice that for case II, when $\|\tilde{x}_i^j - x_i^j\| \leq r_i^j$, there is

$$\begin{aligned}
\|\Phi(\tilde{x}_i^j) - \Phi(x_i^j)\|^2 &= ((\Phi(\tilde{x}_i^j) - \Phi(x_i^j)) \cdot (\Phi(\tilde{x}_i^j) - \Phi(x_i^j))) \\
&= K(\tilde{x}_i^j, \tilde{x}_i^j) - 2K(\tilde{x}_i^j, x_i^j) + K(x_i^j, x_i^j) \\
&= 2 - \exp(-\|\tilde{x}_i^j - x_i^j\|^2/\sigma^2) \\
&= 2 - 2\exp(-(r_i^j)^2/\sigma^2).
\end{aligned} \tag{5.11}$$

Therefore, for both case I and case II, X_i^j defined by (5.10) is a sphere in Hilbert space $\mathcal{H}$ and its spherical center is $\Phi(x_i^j)$, its radius is r_i^j, where

For case I: $\Phi(x_i^j) = \Phi_1(x_i^j) = x_i^j$, $r_i^j = r_i^j$;   (5.12)

For case II: $\Phi(x_i^j) = \Phi_2(x_i^j)$, $r_i^j = (2 - 2\exp(-(r_i^j)^2/\sigma^2))^{\frac{1}{2}}$.   (5.13)

Thus, as an extended of SVORM, for the training set (5.9) we can establish the following primal optimization problem

$$\min_{w,b,\xi^{(*)}} \frac{1}{2}\|w\|^2 + C\sum_{j=1}^{k}\sum_{i=1}^{l^j}(\xi_i^j + \xi_i^{*j}),$$ (5.14)

$$\text{s.t.}\quad (w \cdot (\Phi(x_i^j) + r_i^j \tilde{u}_i^j)) - b_j \leq -1 + \xi_i^j,$$

$$\forall \tilde{u}_i^j \in \tilde{\mathcal{U}},\ j = 1,\ldots,k,\ i = 1,\ldots,l^j,$$ (5.15)

$$(w \cdot (\Phi(x_i^j) + r_i^j \tilde{u}_i^j)) - b_{j-1} \geq 1 - \xi_i^{*j},$$

$$\forall \tilde{u}_i^j \in \tilde{\mathcal{U}},\ j = 1,\ldots,k,\ i = 1,\ldots,l^j,$$ (5.16)

$$\xi_i^j \geq 0,\quad \xi_i^{*j} \geq 0,\quad j = 1,\ldots,k,\ i = 1,\ldots,l^j,$$ (5.17)

where $\tilde{\mathcal{U}}$ is a unit sphere in Hilbert space, $\Phi(x_i^j)$ and r_i^j are given by (5.12) and (5.13), $b = (b_1,\ldots,b_{k-1})^T$, $b_0 = -\infty$, $b_k = +\infty$, $\xi^{(*)} = (\xi_1^1,\ldots,\xi_{l^1}^1,\ldots,\xi_1^k,\ldots,$ $\xi_{l^k}^k,\xi_1^{*1},\ldots,\xi_{l^1}^{*1},\ldots,\xi_1^{*k},\ldots,\xi_{l^k}^{*k})$.

The above problem is equivalent to

$$\min_{w,b,\xi^{(*)}} \frac{1}{2}t^2 + C\sum_{j=1}^{k}\sum_{i=1}^{l^j}(\xi_i^j + \xi_i^{*j}),$$ (5.18)

$$\text{s.t.}\quad -(w \cdot \Phi(x_i^j)) - r_i^j t + b_j + \xi_i^j \geq 1,\quad j = 1,\ldots,k,\ i = 1,\ldots,l^j,$$ (5.19)

$$(w \cdot \Phi(x_i^j)) - r_i^j t - b_{j-1} + \xi_i^{*j} \geq 1,\quad j = 1,\ldots,k,\ i = 1,\ldots,l^j,$$ (5.20)

$$\xi_i^j \geq 0,\quad \xi_i^{*j} \geq 0,\quad j = 1,\ldots,k,\ i = 1,\ldots,l^j,$$ (5.21)

$$\|w\| \leq t.$$ (5.22)

Next we rewrite the above problem as a second order cone programming (SOCP): introduce the new scalar variables u and v, replace t^2 in the objective by $u - v$ and require that u and v satisfy the linear and second order cone constraints $u + v = 1$ and $\sqrt{t^2 + v^2} \leq u$. Since the latter imply that $t^2 \leq u - v$, problem (5.18)–(5.22) can be reformulated as

$$\min_{w,b,\xi^{(*)},u,v,t} \frac{1}{2}(u-v) + C\sum_{j=1}^{k}\sum_{i=1}^{l^j}(\xi_i^j + \xi_i^{*j}), \tag{5.23}$$

$$\text{s.t.} \quad -(w \cdot \Phi(x_i^j)) - r_i^j t + b_j + \xi_i^j \geq 1, \quad j=1,\ldots,k,\ i=1,\ldots,l^j,$$

$$(w \cdot \Phi(x_i^j)) - r_i^j t - b_{j-1} + \xi_i^{*j} \geq 1, \quad j=1,\ldots,k,\ i=1,\ldots,l^j,$$

$$\xi_i^j \geq 0, \quad \xi_i^{*j} \geq 0, \quad j=1,\ldots,k,\ i=1,\ldots,l^j, \tag{5.24}$$

$$\|w\| \leq t, \tag{5.25}$$

$$u + v = 1, \tag{5.26}$$

$$\sqrt{t^2 + v^2} \leq u, \tag{5.27}$$

where $\Phi(x_i^j)$ and r_i^j are given by (5.12) and (5.13), $b = (b_1, \ldots, b_{k-1})^{\mathrm{T}}$, $b_0 = -\infty$, $b_k = +\infty$, $\xi^{(*)} = (\xi_1^1, \ldots, \xi_{l^1}^1, \ldots, \xi_1^k, \ldots, \xi_{l^k}^k, \xi_1^{*1}, \ldots, \xi_{l^1}^{*1}, \ldots, \xi_1^{*k}, \ldots, \xi_{l^k}^{*k})$.

This problem is the robust counterpart of the primal problem of robust SVORM.

The solution of the above problem is obtained by solving its dual problem. The dual problem is

$$\max_{\alpha^{(*)},\beta,\gamma,z_u,z_v} \sum_{j=1}^{k}\sum_{i=1}^{l^j}(\alpha_i^j + \alpha_i^{*j}) + \beta, \tag{5.28}$$

$$\text{s.t.} \quad \gamma \leq \sum_{j,i} r_i^j(\alpha_i^j + \alpha_i^{*j}) - \sqrt{\sum_{j,i}\sum_{j',i'}(\alpha_i^{*j} - \alpha_i^j)(\alpha_{i'}^{*j'} - \alpha_{i'}^{j'})K(x_i^j, x_{i'}^{j'})},$$
$$\tag{5.29}$$

$$\sqrt{\gamma^2 + z_v^2} \leq z_u, \tag{5.30}$$

$$\beta + z_v = -\frac{1}{2}, \quad \beta + z_u = \frac{1}{2}, \tag{5.31}$$

$$\sum_{i=1}^{l^j}\alpha_i^j = \sum_{i=1}^{l^{j+1}}\alpha_i^{*j+1}, \quad j=1,2,\ldots,k-1, \tag{5.32}$$

$$0 \leq \alpha_i^j, \alpha_i^{*j} \leq C, \quad j=1,2,\ldots,k,\ i=1,2,\ldots,l^j, \tag{5.33}$$

where $K(x,x')$ and r_i^j are given by (5.7)–(5.8) and (5.12)–(5.13), $\alpha^{(*)} = (\alpha_1^1, \ldots, \alpha_{l^1}^1, \ldots, \alpha_1^k, \ldots, \alpha_{l^k}^k, \alpha_1^{*1}, \ldots, \alpha_{l^1}^{*1}, \ldots, \alpha_1^{*k}, \ldots, \alpha_{l^k}^{*k})^{\mathrm{T}}$, $\alpha_i^{*1} = 0$, $i=1,2,\ldots,n^1$, $\alpha_i^k = 0$, $i=1,2,\ldots,n^k$.

Next we show the relation between the solutions of primal problem and dual problem.

Definition 5.1 ([71]) In Hilbert space $\mathcal{H}$ second order cone is defined as follows

$$Q = \{z = (z_0, \bar{z}^T)^T \in \mathcal{H} : z_0 \in R, \|\bar{z}\| \leq z_0\}. \tag{5.34}$$

Lemma 5.2 ([71]) *In Definition 5.1 the second order cone is self dual, i.e., $Q^* = Q$, where Q^* is the dual cone of the second order cone Q.*

From Definition 5.1 and Lemma 5.2, Lemma 15 in [2] can be generalized to Hilbert space as follows:

Lemma 5.3 (Complementary conditions) *Suppose that $z = (z_1^T, \ldots, z_m^T)^T \in Q$ and $y = (y_1^T, \ldots, y_m^T)^T \in Q$, where $Q = Q_1 \times Q_2 \times \cdots \times Q_m$ is a direct product of m second order cones in Hilbert space, (i.e., $z_i \succeq_{Q_i} 0$, $y_i \succeq_{Q_i} 0$, $i = 1, 2, \ldots, m$). Then $z^T y = 0$ (i.e., $z_i^T y_i = 0, i = 1, \ldots, m$) is equivalent to*

$$\text{(i)} \quad z_i^T y_i = z_{i0} y_{i0} + \bar{z}_i^T \bar{y}_i = 0, \quad i = 1, 2, \ldots, m, \tag{5.35}$$

$$\text{(ii)} \quad z_{i0} \bar{y}_i + y_{i0} \bar{z}_i = 0, \quad i = 1, 2, \ldots, m, \tag{5.36}$$

where $z_i = (z_{i0}, \bar{z}_i^T)^T$, $y_i = (y_{i0}, \bar{y}_i^T)^T$.

Proof The proof is similar to that of Lemma 15 in [2], and is omitted here. $\square$

Now we give two theorems about the relation between the solutions of primal problem (5.23)–(5.27) and dual problem (5.28)–(5.33) for robust nonlinear SVORM in Hilbert space.

Theorem 5.4 *Suppose that $((\bar{\alpha}^{(*)})^T, \bar{\beta}, \bar{\gamma}, \bar{z}_u, \bar{z}_v)^T$ is any optimal solution of the dual problem (5.28)–(5.33), where $\bar{\alpha}^{(*)} = (\bar{\alpha}_1^1, \ldots, \bar{\alpha}_{l1}^1, \ldots, \bar{\alpha}_1^k, \ldots, \bar{\alpha}_{lk}^k, \bar{\alpha}_1^{*1}, \ldots, \bar{\alpha}_{l1}^{*1}, \ldots, \bar{\alpha}_1^{*k}, \ldots, \bar{\alpha}_{lk}^{*k})^T$. If for any $j = 1, \ldots, k - 1$, there exists a subscript i such that either the component $\bar{\alpha}_i^j \in (0, C)$ or $\bar{\alpha}_i^{*j+1} \in (0, C)$, then for the primal problem (5.23)–(5.27), there is a solution $(\bar{w}, \bar{b}, \bar{\xi}^{(*)}, \bar{u}, \bar{v}, \bar{t})$ such that*

$$\bar{w} = \frac{\bar{\gamma}}{\bar{\gamma} - \sum_{j=1}^k \sum_{i=1}^{l^j} r_i^j (\bar{\alpha}_i^j + \bar{\alpha}_i^{*j})} \sum_{j=1}^k \sum_{i=1}^{l^j} (\bar{\alpha}_i^{*j} - \bar{\alpha}_i^j) \Phi(x_i^j) \tag{5.37}$$

and $\bar{b} = (\bar{b}_1, \ldots, \bar{b}_{k-1})^T$, where $\bar{b}_j$ is defined by, if there exists $\bar{\alpha}_i^j \in (0, C)$,

$$\bar{b}_j = 1 + \sum_{j'=1}^k \sum_{i'=1}^{l^{j'}} (\bar{\alpha}_{i'}^{*j'} - \bar{\alpha}_{i'}^{j'}) K(x_{i'}^{j'}, x_i^j) - \bar{\gamma} r_i^j; \tag{5.38}$$

if there exists $\bar{\alpha}_i^{*j+1} \in (0, C)$,

$$\bar{b}_j = -1 + \sum_{j'=1}^{k} \sum_{i'=1}^{l^{j'}} (\bar{\alpha}_{i'}^{*j'} - \bar{\alpha}_{i'}^{j'}) K(x_{i'}^{j'}, x_i^{j+1}) + \bar{\gamma} r_i^{j+1}, \qquad (5.39)$$

where $K(x, x')$ and r_i^j are given by (5.7)–(5.8) and (5.12)–(5.13).

Proof Firstly, we prove that the problem (5.28)–(5.33) is strictly feasible. Because the problem (5.28)–(5.33) is a convex programming problem in finite dimensional space, we can construct a strictly feasible solution by the following way. Let $\hat{\beta} = -\frac{3}{10}, \hat{\gamma} = \frac{3}{10}, \hat{z}_u = \frac{4}{5}, \hat{z}_v = -\frac{1}{5}$, Easily verified that $\hat{\beta}, \hat{\gamma}, \hat{z}_u, \hat{z}_v$ satisfy the constrains (5.30)–(5.31) and the strictly inequality of (5.30). For given the penalty parameter $C > 0$, we must be able to find an $\hat{\alpha}^{(*)}$ so that $((\hat{\alpha}^{(*)})^T, \hat{\beta}, \hat{\gamma}, \hat{z}_u, \hat{z}_v)^T$ satisfies the constrains (5.29), (5.32) and (5.33) and the strictly inequality (5.29) and (5.33). So the problem is strictly feasible.

The Lagrange function of the dual problem (5.28)–(5.33) is

$$L(\cdot) = -\sum_{j=1}^{k} \sum_{i=1}^{l^j} (\alpha_i^{*j} + \alpha_i^j) - \beta - t \sum_{j=1}^{k} \sum_{i=1}^{l^j} [r_i^j (\alpha_i^{*j} + \alpha_i^j) - \gamma]$$

$$- w^T \left(\sum_{j=1}^{k} \sum_{i=1}^{l^j} (\alpha_i^j - \alpha_i^{*j}) \Phi(x_i^j) \right) - z_1 z_u - z_2 z_v - z_\gamma \gamma$$

$$+ \sum_{j=1}^{k} \sum_{i=1}^{l^j} \xi_i^j (\alpha_i^j - C)$$

$$+ \sum_{j=1}^{k} \sum_{i=1}^{l^j} \xi_i^{*j} (\alpha_i^{*j} - C) - \sum_{j=1}^{k} \sum_{i=1}^{l^j} \eta_i^j \alpha_i^j - \sum_{j=1}^{k} \sum_{i=1}^{l^j} \eta_i^{*j} \alpha_i^{*j}$$

$$+ \sum_{j=1}^{k-1} b_j \left(\sum_{i=1}^{l^j} \alpha_i^j - \sum_{i=1}^{l^{j+1}} \alpha_i^{*j} \right) + u \left(\beta + z_u - \frac{1}{2} \right)$$

$$+ v \left(\beta + z_v + \frac{1}{2} \right), \qquad (5.40)$$

where $t, z_1, z_2, z_\gamma, u, v \in R, w \in \mathcal{H}, b \in R^{k-1}, \xi^{(*)}$ and $\eta^{(*)} \in R^{2l}$ are the Lagrange multiplies and satisfy $\|w\| \le t, \sqrt{z_2^2 + z_\gamma^2} \le z_1, \xi^{(*)} \ge 0$ and $\eta^{(*)} \ge 0$. For the solution $(\bar{\alpha}^{(*)})^T, \bar{\beta}, \bar{\gamma}, \bar{z}_u, \bar{z}_v)^T$, according to the result in [22], there exist Lagrange multiply vectors $(\bar{t}, \bar{w})^T, (\bar{u}, \bar{v}, \bar{z}_\gamma)^T, \bar{\xi}^{(*)}, \bar{\eta}^{(*)}$, which satisfy the following KKT-conditions

$$-1 - \bar{t} r_i^j - (\bar{\mathbf{w}} \cdot \Phi(x_i^j)) + \bar{b}_j + \bar{\xi}_i^j - \bar{\eta}_i^j = 0, \quad j = 1, \ldots, k-1, \ i = 1, \ldots, l^j,$$

$$(5.41)$$

$$\bar{\xi}_i^j \geq 0, \quad \bar{\eta}_i^j \geq 0, \quad j = 1, \ldots, k-1, \ i = 1, \ldots, l^j, \tag{5.42}$$

$$-1 - \bar{t} r_i^j + (\bar{\mathbf{w}} \cdot \Phi(x_i^j)) - \bar{b}_{j-1} + \bar{\xi}_i^{*j} - \bar{\eta}_i^{*j} = 0, \quad j = 2, \ldots, k, i = 1, \ldots, l^j,$$

$$(5.43)$$

$$\bar{\xi}_i^{*j} \geq 0, \quad \bar{\eta}_i^{*j} \geq 0, \quad j = 1, \ldots, k-1, \ i = 1, \ldots, l^j, \tag{5.44}$$

$$0 \leq \bar{\alpha}_i^j, \bar{\alpha}_i^{*j} \leq C, \quad j = 1, \ldots, k-1, \ i = 1, \ldots, l^j, \tag{5.45}$$

$$\sum_{i=1}^{l^j} \bar{\alpha}_i^j = \sum_{i=1}^{l^{j+1}} \bar{\alpha}_i^{j+1}, \quad j = 1, \ldots, k-1, \tag{5.46}$$

$$\bar{\xi}_i^j (\bar{\alpha}_i^j - C) = 0, \quad \bar{\xi}_i^{*j} (\bar{\alpha}_i^{*j} - C) = 0, \quad j = 1, \ldots, k-1, \ i = 1, \ldots, l^j, \quad (5.47)$$

$$\bar{\eta}_i^j \bar{\alpha}_i^j = 0, \quad \bar{\eta}_i^{*j} \bar{\alpha}_i^{*j} = 0, \quad j = 1, \ldots, k-1, \ i = 1, \ldots, l^j, \tag{5.48}$$

$$\begin{pmatrix} \bar{t} \\ \bar{\mathbf{w}} \end{pmatrix}^{\mathrm{T}} \begin{pmatrix} \sum_{j=1}^{k} \sum_{i=1}^{l^j} r_i^j (\bar{\alpha}_i^j + \bar{\alpha}_i^{*j}) - \bar{\gamma} \\ \sum_{j=1}^{k} \sum_{i=1}^{l^j} (\bar{\alpha}_i^j - \bar{\alpha}_i^{*j}) \Phi(x_i^j) \end{pmatrix} = 0, \quad \begin{pmatrix} \bar{t} \\ \bar{\mathbf{w}} \end{pmatrix} \in \mathcal{Q}, \ \bar{t} \in R, \ \bar{\mathbf{w}} \in \mathcal{H},$$

$$(5.49)$$

$$\begin{pmatrix} \bar{u} \\ \bar{v} \\ \bar{z}_\gamma \end{pmatrix}^{\mathrm{T}} \begin{pmatrix} \bar{z}_u \\ \bar{z}_v \\ \bar{\gamma} \end{pmatrix} = 0, \quad \begin{pmatrix} \bar{u} \\ \bar{v} \\ \bar{z}_\gamma \end{pmatrix} \in \mathcal{Q}_3, \ \bar{z}_\gamma = \bar{t}. \tag{5.50}$$

However (5.41)–(5.42) imply

$$-1 - \bar{t} r_i^j - (\bar{\mathbf{w}} \cdot \Phi(x_i^j)) + \bar{b}_j + \bar{\xi}_i^j \geq 0, \quad j = 1, \ldots, k-1, \ i = 1, \ldots, l^j, \quad (5.51)$$

$$\bar{\xi}_i^j \geq 0, \quad j = 1, \ldots, k, \ i = 1, \ldots, l^j. \tag{5.52}$$

Similarly, (5.43)–(5.44) imply

$$-1 - \bar{t} r_i^j + (\bar{\mathbf{w}} \cdot \Phi(x_i^j)) - \bar{b}_{j-1} + \bar{\xi}_i^{*j} \geq 0, \quad j = 2, \ldots, k, \ i = 1, \ldots, l^j, \quad (5.53)$$

$$\bar{\xi}_i^{*j} \geq 0, \quad j = 1, \ldots, k, \ i = 1, \ldots, l^j. \tag{5.54}$$

Therefore, $(\bar{\mathbf{w}}, \bar{b}, \bar{\xi}^{(*)}, \bar{u}, \bar{v}, \bar{t})$ satisfies the constraint conditions of the primal problem (5.23)–(5.27), i.e., it is a feasible solution.

Since $\begin{pmatrix} \bar{t} \\ \bar{\mathbf{w}} \end{pmatrix} \in \mathcal{Q}$,

$$\begin{pmatrix} \sum_{j=1}^{k} \sum_{i=1}^{l^j} r_i^j (\bar{\alpha}_i^j + \bar{\alpha}_i^{*j}) - \bar{\gamma} \\ \sum_{j=1}^{k} \sum_{i=1}^{l^j} (\bar{\alpha}_i^j - \bar{\alpha}_i^{*j}) \Phi(x_i^j) \end{pmatrix} \in \mathcal{Q},$$

where Q is second order cone in Hilbert space. According to Lemma 5.3, (5.49) is equivalent to

$$\bar{t}\left(\sum_{j=1}^{k}\sum_{i=1}^{l^j}r_i^j(\bar{\alpha}_i^j+\bar{\alpha}_i^{*j})-\bar{\gamma}\right)+\bar{w}^T\left(\sum_{j=1}^{k}\sum_{i=1}^{l^j}(\bar{\alpha}_i^j-\bar{\alpha}_i^{*j})\Phi(x_i^j)\right)=0, \quad (5.55)$$

$$\bar{t}\left(\sum_{j=1}^{k}\sum_{i=1}^{l^j}(\bar{\alpha}_i^j-\bar{\alpha}_i^{*j})\Phi(x_i^j)\right)+\left(\sum_{j=1}^{k}\sum_{i=1}^{l^j}r_i^j(\bar{\alpha}_i^j+\bar{\alpha}_i^{*j})-\bar{\gamma}\right)\bar{w}=0. \quad (5.56)$$

According to Lemma 15 in [2], (5.50) is equivalent to

$$\bar{u}\bar{z}_u+\bar{v}\bar{z}_v+\bar{t}\bar{\gamma}=0, \quad (5.57)$$

$$\bar{u}\begin{pmatrix}\bar{v}\\\bar{t}\end{pmatrix}+\bar{z}_u\begin{pmatrix}\bar{z}_v\\\bar{\gamma}\end{pmatrix}=0. \quad (5.58)$$

So considering (5.55)–(5.56), (5.57)–(5.58) and $\bar{z}_u-\bar{z}_v=1$, we get

$$\bar{w}=\frac{\bar{\gamma}}{(\bar{\gamma}-\sum_{j=1}^{k}\sum_{i=1}^{l^j}r_i^j(\bar{\alpha}_i^j+\bar{\alpha}_i^{*j}))}\sum_{j=1}^{k}\sum_{i=1}^{l^j}(\bar{\alpha}_i^{*j}-\bar{\alpha}_i^j)\Phi(x_i^j), \quad (5.59)$$

$$\bar{u}=\bar{z}_u, \qquad \bar{v}=\bar{z}_v, \qquad \bar{t}=-\bar{\gamma}. \quad (5.60)$$

Furthermore, from the equations (5.41)–(5.50), (5.59)–(5.60), and let $\alpha_i^{*1}=0$, $i=1,2,\ldots,n^1$, $\alpha_i^k=0$, $i=1,2,\ldots,n^k$, we have

$$-\frac{1}{2}(\bar{u}-\bar{v})-C\sum_{j=1}^{k}\sum_{i=1}^{l^j}(\bar{\xi}_i^j+\bar{\xi}_i^{*j})$$

$$=-\frac{1}{2}(\bar{z}_u+\bar{z}_v)-C\sum_{j=1}^{k}\sum_{i=1}^{l^j}(\bar{\xi}_i^j+\bar{\xi}_i^{*j})$$

$$+\sum_{j=1}^{k-1}\sum_{i=1}^{l^j}\bar{\alpha}_i^j(-1-\bar{t}r_i^j-(\bar{w}\cdot\Phi(x_i^j))+\bar{b}_j+\bar{\xi}_i^j-\bar{\eta}_i^j)$$

$$+\sum_{j=2}^{k}\sum_{i=1}^{l^j}\bar{\alpha}_i^{*j}(-1-\bar{t}r_i^j+(\bar{w}\cdot\Phi(x_i^j))-\bar{b}_{j-1}+\bar{\xi}_i^{*j}-\bar{\eta}_i^{*j})$$

$$=\bar{\beta}-\sum_{j=1}^{k}\sum_{i=1}^{l^j}\bar{\xi}_i^j(C-\bar{\alpha}_i^j)-\sum_{j=1}^{k}\sum_{i=1}^{l^j}\bar{\xi}_i^{*j}(C-\alpha_i^{*j})$$

$$+ \sum_{j=1}^{k-1}\sum_{i=1}^{l^j} \bar{\alpha}_i^j \bar{\eta}_i^j + \sum_{j=2}^{k}\sum_{i=1}^{l^j} \bar{\alpha}_i^{*j} \bar{\eta}_i^{*j} + \sum_{j=1}^{k} \bar{b}_j \left(\sum_{i=1}^{l^j} \bar{\alpha}_i^j - \sum_{i=1}^{l^{j+1}} \bar{\alpha}_i^{*j+1} \right)$$

$$- \bar{t} \sum_{j=1}^{k}\sum_{i=1}^{l^j} r_i^j (\bar{\alpha}_i^j + \bar{\alpha}_i^{*j}) - \sum_{j=1}^{k}\sum_{i=1}^{l^j} (\bar{\alpha}_i^j - \bar{\alpha}_i^{*j})(\bar{w} \cdot \Phi(x_i^j))$$

$$- \sum_{j=1}^{k}\sum_{i=1}^{l^j} (\bar{\alpha}_i^j + \bar{\alpha}_i^{*j})$$

$$= -\bar{\beta} - \sum_{j=1}^{k}\sum_{i=1}^{l^j} (\bar{\alpha}_i^j + \bar{\alpha}_i^{*j}).$$

The equation above shows that the objective function value of the dual problem (5.28)–(5.33) at $((\bar{\alpha}^{(*)})^T, \bar{\beta}, \bar{\gamma}, \bar{z}_u, \bar{z}_v)^T$ is equal to that of the primal problem (5.23)–(5.27) at $(\bar{w}, \bar{b}, \bar{\xi}^{(*)}, \bar{u}, \bar{v}, \bar{t})$. According to Theorem 4 in [71], we conclude that $(\bar{w}, \bar{b}, \bar{\xi}^{(*)}, \bar{u}, \bar{v}, \bar{t})$ is a solution of the primal problem.

In order to get the expressions (5.38) and (5.39), we consider KKT-condition (5.41)–(5.50) and reach that

$$\bar{\alpha}_i^j (-1 - \bar{t} r_i^j - (\bar{w} \cdot \Phi(x_i^j)) + \bar{b}_j + \bar{\xi}_i^j) = 0, \quad j = 1, \ldots, k-1, \ i = 1, \ldots, l^j,$$

$$(5.61)$$

$$\bar{\alpha}_i^{*j} (-1 - \bar{t} r_i^j + (\bar{w} \cdot \Phi(x_i^j)) - \bar{b}_{j-1} + \bar{\xi}_i^{*j}) = 0, \quad j = 2, \ldots, k, \ i = 1, \ldots, l^j,$$

$$(5.62)$$

$$\bar{\xi}_i^j (\bar{\alpha}_i^j - C) = 0, \quad \bar{\xi}_i^{*j}(\bar{\alpha}_i^{*j} - C) = 0, \quad j = 1, \ldots, k-1, \ i = 1, \ldots, l^j.$$

$$(5.63)$$

So the expressions (5.38) and (5.39) follow.   □

Above the theorem shows that how to compute w and b under some conditions on $\bar{\alpha}^{(*)}$. The following theorem gives that how to compute w and b in general case.

Theorem 5.5 *Suppose that $((\bar{\alpha}^{(*)})^T, \bar{\beta}, \bar{\gamma}, \bar{z}_u, \bar{z}_v)^T$ is any optimal solution of the dual problem (5.28)–(5.33), where $\bar{\alpha}^{(*)} = (\bar{\alpha}_1^1, \ldots, \bar{\alpha}_{l^1}^1, \ldots, \bar{\alpha}_1^k, \ldots, \bar{\alpha}_{lk}^k, \bar{\alpha}_1^{*1}, \ldots,$ $\bar{\alpha}_{l^1}^{*1}, \ldots, \bar{\alpha}_1^{*k}, \ldots, \bar{\alpha}_{lk}^{*k})^T$. Then the primal problem (5.23)–(5.27) has a solution $(\bar{w}, \bar{b}, \bar{\xi}^{(*)}, \bar{u}, \bar{v}, \bar{t})$ such that*

$$\{(w, b) = (w, b_1, \ldots, b_{k-1}) \mid w = \bar{w}, \bar{b}_j \in [b_j^{\text{dn}}, b_j^{\text{up}}]\}, \quad j = 1, \ldots, k-1, \quad (5.64)$$

where

$$\bar{w} = \frac{\bar{\gamma}}{\bar{\gamma} - \sum_{j=1}^{k}\sum_{i=1}^{l^j} r_i^j (\bar{\alpha}_i^j + \bar{\alpha}_i^{*j})} \sum_{j=1}^{k}\sum_{i=1}^{l^j} (\bar{\alpha}_i^{*j} - \bar{\alpha}_i^j)\Phi(x)_i^j, \quad (5.65)$$

$$b_j^{\mathrm{dn}} = \max\left\{ \max_{i \in I_1^j}((\bar{w} \cdot \Phi(x)_i^j) - \bar{\gamma} r_i^j + 1), \max_{i \in I_4^j}((\bar{w} \cdot \Phi(x)_i^{j+1}) + \bar{\gamma} r_i^{j+1} - 1) \right\},$$

$$(5.66)$$

$$b_j^{\mathrm{up}} = \min\left\{ \min_{i \in I_3^j}((\bar{w} \cdot \Phi(x)_i^j) - \bar{\gamma} r_i^j + 1), \min_{i \in I_2^j}((\bar{w} \cdot \Phi(x)_i^{j+1}) + \bar{\gamma} r_i^{j+1} - 1) \right\},$$

$$(5.67)$$

and

$$I_1^j = \{i \in \{1, \ldots, l^j\} | \alpha_i^j = 0\}, \qquad I_2^j = \{i \in \{1, \ldots, l^{j+1}\} | \alpha_i^{*j+1} = 0\},$$

$$I_3^j = \{i \in \{1, \ldots, l^j\} | \alpha_i^j = C\}, \qquad I_4^j = \{i \in \{1, \ldots, l^{j+1}\} | \alpha_i^{*j+1} = C\},$$

where $\Phi(x_i^j)$ and r_i^j are given by (5.12) and (5.13).

Proof Reminded that the proof of Theorem 5.4, we see that what we need to prove is only the $\bar{b}$'s expressions (5.64), (5.66) and (5.67). In fact, following (5.61)–(5.63), we have

$$b_j \geq 1 + (w \cdot \Phi(x_i^j)) - \bar{\gamma} r_i^j, \quad i \in I_1^j = \{i \in \{1, \ldots, l^j\} \mid \alpha_i^j = 0\},$$

$$b_j \leq -1 + (w \cdot \Phi(x_i^{j+1})) + \bar{\gamma} r_i^{j+1}, \quad i \in I_2^j = \{i \in \{1, \ldots, l^{j+1}\} \mid \alpha_i^{*j+1} = 0\},$$

$$b_j \leq 1 + (w \cdot \Phi(x_i^j)) - \bar{\gamma} r_i^j, \quad i \in I_3^j = \{i \in \{1, \ldots, l^j\} \mid \alpha_i^j = C\},$$

$$b_j \geq -1 + (w \cdot \Phi(x_i^{j+1})) + \bar{\gamma} r_i^{j+1}, \quad i \in I_4^j = \{i \in \{1, \ldots, l^{j+1}\} \mid \alpha_i^{*j+1} = C\}.$$

Combining the above equations, we get $\bar{b}_j \in [b_j^{\mathrm{dn}}, b_j^{\mathrm{up}}]$, where b_j^{up} and b_j^{dn} are defined by (5.66) and (5.67) respectively.   □

This leads to the following algorithm

Algorithm 5.6 (Robust SVORM (R-SVORM))

(1) Given a training set (5.9);
(2) Select $C > 0$ and a kernel. There are two choices:
    Case I: $K(x, x') = K_1(x, x') = (x \cdot x')$;
    Case II: $K(x, x') = K_2(x, x') = \exp(\frac{-\|x - x'\|^2}{2\sigma^2})$;
(3) Solve the dual problem (5.28)–(5.33) and get its solution $((\alpha^{(*)})^{\mathrm{T}}, \beta, \gamma, z_u, z_v)^{\mathrm{T}}$;
(4) Compute

$$g(x) = \frac{\gamma}{(\gamma - \sum_{j=1}^{k} \sum_{i=1}^{l^j} r_i^j (\alpha_i^j + \alpha_i^{*j}))} \sum_{j=1}^{k} \sum_{i=1}^{l^j} (\alpha_i^{*j} - \alpha_i^j) K(x_i^j, x), \quad (5.68)$$

where $r_i^j = r_i^j$ for case I, $r_i^j = (2 - 2\exp(-(r_i^j)^2/2\sigma^2))^{\frac{1}{2}}$ for case II;

(5) For $j = 1, \ldots, k - 1$, execute the following steps:

(5.1) Choose some component $\alpha_i^j \in (0, C)$ in $\alpha^{(*)}$. If we get such i, let

$$b_j = 1 + \sum_{j'=1}^{k} \sum_{i'=1}^{l^{j'}} (\alpha_{i'}^{*j'} - \alpha_{i'}^{j'}) K(x_{i'}^{j'}, x_i^j) - \gamma r_i^j, \qquad (5.69)$$

otherwise go to step (5.2);

(5.2) Choose some component $\alpha_i^{*j+1} \in (0, C)$ in $\alpha^{(*)}$. If we get such i, let

$$b_j = \sum_{j'=1}^{k} \sum_{i'=1}^{l^{j'}} (\alpha_{i'}^{*j'} - \alpha_{i'}^{j'}) K(x_{i'}^{j'}, x_i^{j+1}) + \gamma r_i^{j+1} - 1, \qquad (5.70)$$

otherwise go to step (5.3);

(5.3) Let

$$b_j = \frac{1}{2}(b_j^{dn} + b_j^{up}), \qquad (5.71)$$

where

$$b_j^{dn} = \max\left\{\max_{i \in I_1^j}(g(x_i^j) - \gamma r_i^j + 1), \max_{i \in I_4^j}(g(x_i^{j+1}) + \gamma r_i^{j+1} - 1)\right\},$$

$$b_j^{up} = \min\left\{\min_{i \in I_3^j}(g(x_i^j) + -\gamma r_i^j + 1), \min_{i \in I_2^j}(g(x_i^{j+1}) + \gamma r_i^{j+1} - 1)\right\},$$

and

$$I_1^j = \{i \in \{1, \ldots, l^j\} \mid \alpha_i^j = 0\},$$

$$I_2^j = \{i \in \{1, \ldots, l^{j+1}\} \mid \alpha_i^{*j+1} = 0\},$$

$$I_3^j = \{i \in \{1, \ldots, l^j\} \mid \alpha_i^j = C\},$$

$$I_4^j = \{i \in \{1, \ldots, l^{j+1}\} \mid \alpha_i^{*j+1} = C\};$$

(6) If there exists $j \in \{1, \ldots, k\}$ such that $b_j \leq b_{j-1}$, the algorithm stop or go to step (2);

(7) Define $b_k = +\infty$ and construct the decision function

$$f(x) = \min_{s \in \{1,\ldots,k\}} \{s : g(x) - b_s < 0\}. \qquad (5.72)$$

Obviously, the above algorithm is reduced to the original SVORM algorithm by choosing $r_i^j = 0$, $j = 1, \ldots, k$, $i = 1, \ldots, l^j$.

5.2   Robust Multi-class Algorithm

Now we turn to our robust multi-class algorithm solving the multi-class classification problem with perturbation. The multi-class classification problem with perturbation is presented as follows:

Supposing that given a training set

$$T = \{(\mathcal{X}_1, y_1), \ldots, (\mathcal{X}_l, y_l)\}, \tag{5.73}$$

where $\mathcal{X}_i$ is the input set: a sphere with x_i and radius r_i, namely,

$$\mathcal{X}_i = \{x_i + r_i u_i : \|u_i\| \le 1\}, \quad i = 1, \ldots, l, \tag{5.74}$$

the output is $y_i \in \mathcal{Y} = \{1, 2, \ldots, k\}$, l is the number of all training points. The task is to find a decision function $f(x) : R^n \to \mathcal{Y}$.

For standard multi-class problem without perturbation, a series of algorithms based on SVORM is proposed in [230]; Particularly a multi-class algorithm based on 3-class SVORM is developed in detail [232]. Essentially speaking, our robust multi-class algorithm is obtained from the latter one (Algorithm 2 in [232]) with replacing the SVORM by the above robust SVORM. More exactly, the robust multi-algorithm is described as follows:

Algorithm 5.7 (Robust Multi-class Algorithm)

(1) Given the training set (5.73). Construct a set P containing $k(k-1)/2$ class pair:

$$P = \{(s_{11}, s_{21}), \ldots, (s_{1,k(k-1)/2}, s_{2,k(k-1)/2})\}$$

$$= \{(s_{1i}, s_{2i}) \mid s_{1i} < s_{2i}, \ (s_{1i}, s_{2i}) \in (\mathcal{Y} \times \mathcal{Y}), \ i = 1, \ldots, k(k-1)/2\}, \tag{5.75}$$

where $\mathcal{Y} = \{1, \ldots, k\}$. Set $m = 1$;

(2) Let $(t_1, t_2) = (s_{1m}, s_{2m})$ and transform the training set (5.73) into the following form:

$$\tilde{T} = \{\tilde{\mathcal{X}}_i^j\}_{i=1,\ldots,l^j}^{j=1,2,3}, \tag{5.76}$$

where

$$\{\tilde{\mathcal{X}}_i^1, \ i = 1, \ldots, l^1\} = \{\mathcal{X}_i \mid y_i = t_1\},$$

$$\{\tilde{\mathcal{X}}_i^2, \ i = 1, \ldots, l^2\} = \{\mathcal{X}_i \mid y_i \in \mathcal{Y}\setminus\{t_1, t_2\}\},$$

$$\{\tilde{\mathcal{X}}_i^3, \ i = 1, \ldots, l^3\} = \{\mathcal{X}_i \mid y_i = t_2\};$$

(3) For the training set (5.76), execute Algorithm 5.6, with Gaussian kernel and $k = 3$ and get the decision function $f_{t_1,t_2}(x)$;

(4) Suppose that an input $\bar{x}$ is given, where $\|u\| \le 1$. When $f_{t_1,t_2}(\bar{x}) = 1$, a positive vote is added on class t_1, and no votes are added on the other classes; when $f_{t_1,t_2}(\bar{x}) = 3$, a positive vote is added on class t_2, and no votes are added on the other classes; when $f_{t_1,t_2}(\bar{x}) = 2$, a negative vote is added on both class t_1 and class t_2, and no votes are added on the other classes;

(5) If $m \neq k(k-1)/2$, set $m = m + 1$, return to the step (3);
(6) Calculate the total votes of each class by adding the positive and negative votes on this class. The input $\bar{x}$ is assigned to the class that gets the most votes.

5.3 Numerical Experiments

The numerical experiments are implemented on a PC (256 RAM, CPU 2.66 GHz) using SeDuMi1.0 [191] as a solver is developed by J.Sturm for optimization problems over symmetric cones including SOCP. Our experiments are concerned with two algorithms proposed above.

5.3.1 Numerical Experiments of Algorithm 5.6

Algorithm 5.6 with both linear kernel and Gaussian kernel are implemented. Our numerical experiments follow the approach in [250]. In fact, following [42, 74], the ordinal regression problems are obtained from the regression problems in [74] by discretizing their output values. Due to the time consuming, only 4 regression problems are selected among the 29 ones in [74]. These 4 problems are the smallest ones according to the number of the training points.

In order to test our algorithms, the measurement errors in the inputs of the training points are considered. Assume that x_i^j given by the dataset is the measurement value and the real value $\bar{x}_i^j$ is given by

$$\bar{x}_i^j = x_i^j + r_i^j u_i^j, \tag{5.77}$$

where $\|u_i^j\| = 1$. For simplicity, r_i^j is assumed to be a constant independent of i and j, i.e. $r_i^j = r$. On the other hand, for the test point, there is also a perturbation around its attribute x_i^j generated by the same way. More precisely, the input $\bar{x}_i^j$ of the test point is obtained by $\bar{x}_i^j = x_i^j + r_i^j u_i^j$ with the same constant r and the noise u_i^j is generated randomly from the normal distribution and scaled on the unit sphere.

The parameters in Algorithm 5.6 with both linear kernel and Gaussian kernel are chosen by ten-fold cross validation: in Algorithm 5.6 with linear kernel, for data sets "Diabetes" and "Triazines" $C = 1000$, for data sets "Pyrimidines" $C = 10$, for data sets "Wisconsin" $C = 10000$; in Algorithm 1 with Gaussian kernel, for four data sets, the penalty parameters are the same $C = 1000$; the kernel parameters, for data sets "Diabetes" and "Triazines" $\sigma = 1$, for data sets " Pyrimidines" $\sigma = 4$, for data sets "Wisconsin" $\sigma = 0.25$. The numerical results for Algorithm 5.6 with linear kernel (R-LSVORM) and Algorithm 5.6 with Gaussian kernel (R-SVORM) are given in Table 5.1 and Table 5.2 respectively. What we are concerned here is

Table 5.1 For Algorithm 5.6 with linear kernel the percentage of tenfold testing error for datasets with noise

Dataset	Instances	Dimension	Class	Algorithm	r				
					0.1	0.2	0.3	0.4	0.5
Diabetes	43	2	5	R-LSVORM	0.4884	0.4651	0.4651	0.4651	0.4651
				LSVORM	0.4884	0.4884	0.4884	0.4884	0.4884
Pyrimidines	74	27	5	R-LSVORM	0.5811	0.5811	0.5811	0.5811	0.5811
				LSVORM	0.5946	0.5811	0.6081	0.6351	0.7027
Triazines	186	60	5	R-LSVORM	0.5215	0.5215	0.5215	0.5215	0.5215
				LSVORM	0.5376	0.6183	0.7366	0.6882	0.7258
Wisconsin	194	32	5	R-LSVORM	0.7320	0.7113	0.6495	0.6804	0.6959
				LSVORM	0.7320	0.7165	0.7423	0.7629	0.7216

Table 5.2 For Algorithm 5.6 with Gaussian kernel the percentage of tenfold testing error for datasets with noise

Dataset	Algorithm	r				
		0.1	0.2	0.3	0.4	0.5
Diabetes	R-SVORM	0.4651	0.4651	0.4651	0.4651	0.4651
	SVORM	0.5349	0.5349	0.5581	0.4884	0.5116
Pyrimidines	R-SVORM	0.4865	0.5811	0.5811	0.5811	0.5946
	SVORM	0.4865	0.5811	0.6351	0.6892	0.6081
Triazines	R-SVORM	0.5215	0.5215	0.5215	0.5215	0.5215
	SVORM	0.5323	0.6613	0.7581	0.8280	0.8333
Wisconsin	R-SVORM	0.7216	0.7474	0.7216	0.7216	0.7216
	SVORM	0.7990	0.7784	0.7784	0.7784	0.7784

the percentage of tenfold testing error, which are shown with different noise level r. In addition, for comparison, the results corresponding to the SVORM without perturbations are also listed in these tables. It can be seen that the performance of our Robust SVORM is better. Since all of these problems are 5-class classification, the results on the percentages of tenfold testing error meaningful, e.g. they are comparable to that in [74].

5.3.2 Numerical Experiments of Algorithm 5.7

Algorithm 5.7 is also implemented in our numerical experiments. The datasets are taken from the UCI machine learning repository [21]. They are Iris, Wine and Glass.

Table 5.3 For Algorithm 5.7 the percentage of tenfold testing error for datasets with noise

Dataset	Instances	Dimension	Class	Algorithm	r			
					0.1	0.2	0.3	0.4
Iris	150	4	3	I	0.0467	0.0467	0.0533	0.0667
				II	0.0733	0.0800	0.0733	0.0867
Wine	178	13	3	I	0.0674	0.0449	0.0506	0.0618
				II	0.0730	0.0506	0.0674	0.1236
Glass	214	9	6	I	0.3692	0.5374	0.5514	0.5514
				II	0.5561	0.6355	0.5981	0.5935

I: Algorithm 5.7 (Robust Multi-class Algorithm). II: Algorithm 5.7 in [232] (multi-class algorithm based on 3-class SVORM)

The characteristics of three datasets are also shown in Table 5.3. The perturbations is imposed on these datasets by the way similar to that in the above subsection. In our numerical experiment, the parameters are selected as follows: the penalty $C = 10^5$ for three datasets, the kernel parameters $\sigma = 2$ for Iris and Glass, $\sigma = 10$ for Wine. The result of experiment is listed in Table 5.3.

Table 5.3 shows that the percentage of tenfold testing correctness Algorithm 5.7 on three datasets with various noise levels r. In addition, for comparison, we also list the corresponding results obtained by the algorithm without considering the perturbations, i.e. Algorithm 5.7 in our paper [232]. It can be observed that the performance of the robust model is consistently better than of the original model.

In this section, we have established the robust SVORM with both linear kernel and Gaussian kernel for ordinal regression problem with noise. And based on the latter one, we have also established a multi-class algorithm for multi-class classification problem with noise. Preliminary numerical experiments show that the performance of our robust models are better than that of the corresponding original models without perturbation.

It should be noted that the robust multi-class algorithm proposed here is able to be extended to a class of algorithms. In fact, for a k-class classification problem, an algorithm based on p-class robust SVORM with Gaussian kernel can be established. The efficiency of this class of algorithms is an interesting topic to be studied further.

5.4 Robust Unsupervised and Semi-supervised Bounded C-Support Vector Machine

In this section, we will propose robust model for unsupervised and semi-supervised C-Support Vector Machine.

5.4.1 Robust Linear Optimization

The general optimization problem under parameter uncertainty is as follows:

$$\max\ f_0(x, \widetilde{D}_0), \tag{5.78}$$

$$\text{s.t.}\quad f_i(x, \widetilde{D}_i) \geq 0, \quad i \in I, \tag{5.79}$$

$$x \in X, \tag{5.80}$$

where $f_i(x, \widetilde{D}_i), i \in \{0\} \cup I$ are given functions, X is a given set and $\widetilde{D}_i, i \in \{0\} \cup I$ is the vector of uncertain coefficients. When the uncertain coefficients $\widetilde{D}_i$ take values equal to their expected values $\widetilde{D}_i^0$, it is the normal formulation of problem (5.78)–(5.80).

In order to address parameter uncertainty problem (5.78)–(5.80), Ben-Tal and Nemirovski [15, 16] and independently EI-Ghaoui *et al.* [65, 66] proposed to solve the following robust optimization problem

$$\max\ \min_{D_0 \in \mathcal{U}_0}\ f_0(x, D_0), \tag{5.81}$$

$$\text{s.t.}\quad \min_{D_i \in \mathcal{U}_i}\ f_i(x, D_i) \geq 0, \quad i \in I, \tag{5.82}$$

$$x \in X, \tag{5.83}$$

where $\mathcal{U}_i, i \in \{0\} \cup I$ are given uncertainty sets. In the robust optimization framework (5.81)–(5.83), Melvyn Sim [186] consider the uncertainty set $\mathcal{U}$ as follows

$$\mathcal{U} = \left\{ D \mid \exists u \in \Re^{|N|} : D = D^0 + \sum_{j \in N} \Delta D^j u_j, \ \|u\| \leq \Omega \right\},$$

where D^0 is the nominal value of the data, ΔD^j ($j \in N$) is a direction of data perturbation, and Ω is a parameter controlling the trade-off between robustness and optimality (robustness increases as Ω increases). We restrict the vector norm $\| \cdot \|$ as $\|u\| = \|u^+\|$, such as l_1 and l_2. When we select the norm as l_1, then the corresponding perturbation region is a polyhedron and the robust counterpart is also an Linear Programming. When the norm is l_2, then the corresponding uncertainty set $\mathcal{U}$ is an ellipsoid and the robust counterpart becomes an Second Order Cone Programming (SOCP), detailed robust counterpart can be seen in [186].

5.4.2 Robust Algorithms with Polyhedron

In practice, the training data have perturbations since they are usually corrupted by measurement noise. When considering the measurement noise, we assume training data $x_i \in R^n, i = 1, \ldots, l$, which has perturbed as $\widetilde{x}_i$, concretely, $\widetilde{x}_{ij} = x_{ij} + \Delta x_{ij} z_{ij}$,

$i = 1, \ldots, l$, $j = 1, \ldots, n$, $\|z_i\|_p \leq \Omega$. z_i is a random variable, when select its norm as l_1 norm, that, is $\|z_i\|_1 \leq \Omega$, while it is equivalent with $\sum_{j=1}^{n} |z_{ij}| \leq \Omega$, $i = 1, \ldots, l$. Considering $\tilde{x}_{ij} = x_{ij} + \Delta x_{ij} z_{ij}$,

$$\sum_{j=1}^{n} \left| \frac{\tilde{x}_{ij} - x_{ij}}{\Delta x_{ij}} \right| \leq \Omega, \quad i = 1, \ldots, l, \tag{5.84}$$

so the perturbation region of x_i is a polyhedron.

Considering Bounded C-Support Vector Machine [225] for the training data having perturbations as mentioned above, we get the optimization problem

$$\min_{w,b,\xi} \frac{1}{2}(\|w\|^2 + b^2) + C\sum_{i=1}^{l} \xi_i, \tag{5.85}$$

$$\text{s.t.} \quad iy_i((w \cdot \tilde{x}_i) + b) \geq 1 - \xi_i, \quad i = 1, \ldots, l, \tag{5.86}$$

$$\xi_i \geq 0, \quad i = 1, \ldots, l. \tag{5.87}$$

Because constraint (5.86) is infinite and problem (5.85)–(5.87) is semi-infinite optimization problem, there seems no good method to resolve it directly. Due to robust linear optimization, we tend to find its robust counterpart. In Sim's proposed robust framework [186], constraint $y_i((w \cdot \tilde{x}_i) + b) \geq 1 - \xi_i$ is equivalent to

$$y_i((w \cdot x_i) + b) - 1 + \xi_i \geq \Omega t_i, \quad t_i \geq 0, \tag{5.88}$$

$$|\Delta x_{ij}| w_j y_i \leq t_i, \quad j = 1, \ldots, n, \tag{5.89}$$

$$-|\Delta x_{ij}| w_j y_i \leq t_i, \quad j = 1, \ldots, n. \tag{5.90}$$

Then problem (5.85)–(5.87) turns to be

$$\min_{w,b,\xi,t} \frac{1}{2}(\|w\|^2 + b^2) + C\sum_{i=1}^{l} \xi_i, \tag{5.91}$$

$$\text{s.t.} \quad y_i((w \cdot x_i) + b) - 1 + \xi_i \geq \Omega t_i, \quad i = 1, \ldots, l, \tag{5.92}$$

$$t_i \geq 0, \quad i = 1, \ldots, l, \tag{5.93}$$

$$|\Delta x_{ij}| w_j y_i \leq t_i, \quad j = 1, \ldots, n, \, i = 1, \ldots, l, \tag{5.94}$$

$$-|\Delta x_{ij}| w_j y_i \leq t_i, \quad j = 1, \ldots, n, \, i = 1, \ldots, l, \tag{5.95}$$

$$\xi_i \geq 0, \quad i = 1, \ldots, l. \tag{5.96}$$

Therefore we get the programming based on (5.85)–(5.87) for unsupervised classification problem

$$\min_{y_i \in \{-1,1\}^l} \min_{w,b,\xi,t} \frac{1}{2}(\|w\|^2 + b^2) + C\sum_{i=1}^{l} \xi_i, \tag{5.97}$$

$$\text{s.t.} \quad y_i((w \cdot x_i) + b) - 1 + \xi_i \geq \Omega t_i, \quad i = 1, \ldots, l, \tag{5.98}$$

$$|\Delta x_{ij}|w_j y_i \le t_i, \quad j=1,\ldots,n, \ i=1,\ldots,l, \tag{5.99}$$

$$-|\Delta x_{ij}|w_j y_i \le t_i, \quad j=1,\ldots,n, \ i=1,\ldots,l, \tag{5.100}$$

$$t_i \ge 0, \quad i=1,\ldots,l, \tag{5.101}$$

$$\xi_i \ge 0, \quad i=1,\ldots,l, \tag{5.102}$$

$$-\varepsilon \le \sum_{i=1}^{l} y_i \le \varepsilon. \tag{5.103}$$

Next, we will relax problem (5.97)–(5.103) to SDP to get its approximate solutions. Follows the idea described in Chap. 4, we get the optimization problem [244]

$$\min_{\widetilde{M},\theta,\overline{h}_i,\overline{\overline{h}}_i,\kappa_i,\varphi_i,\overline{\varphi}_i, i=1,\ldots,l} \frac{1}{2}\theta, \tag{5.104}$$

$$\text{s.t.} \quad \begin{pmatrix} G \circ \widetilde{M} & \varsigma^{\mathrm{T}} \\ \varsigma & \theta - 2Ce^{\mathrm{T}}\overline{\overline{h}} \end{pmatrix} \succeq 0, \tag{5.105}$$

$$\kappa \ge 0, \quad \varphi_i \ge 0, \quad \overline{\varphi}_i \ge 0, \quad i=1,\ldots,l, \tag{5.106}$$

$$-(2n+1)\varepsilon e \le \widetilde{M}e \le (2n+1)\varepsilon e, \tag{5.107}$$

$$\overline{h} \ge 0, \quad \overline{\overline{h}} \ge 0, \tag{5.108}$$

$$\widetilde{M} \succeq 0, \quad \operatorname{diag}(\widetilde{M}) = e, \tag{5.109}$$

where $\varsigma = (1 + \overline{h}_1 - \overline{\overline{h}}_1 + \Omega\kappa_1, \ldots, 1 + \overline{h}_l - \overline{\overline{h}}_l + \Omega\kappa_l, -\kappa_1 e^{\mathrm{T}} + \varphi_1^{\mathrm{T}}, \ldots, -\kappa_l e^{\mathrm{T}} + \varphi_l^{\mathrm{T}}, -\kappa_1 e^{\mathrm{T}} + \overline{\varphi}_1^{\mathrm{T}}, \ldots, -\kappa_l e^{\mathrm{T}} + \overline{\varphi}_l^{\mathrm{T}})^{\mathrm{T}}$, and

$$\widetilde{M}_{(2nl+l)\times(2nl+l)} = \widetilde{Y}\widetilde{Y}^{\mathrm{T}}, \tag{5.110}$$

$$\widetilde{Y} = (y_1,\ldots,y_l, y_1,\ldots,y_1,\ldots,y_l,\ldots,y_l, y_1,\ldots,y_1,\ldots,y_l,\ldots,y_l)^{\mathrm{T}}, \tag{5.111}$$

$$G_{(2nl+l)\times(2nl+l)} = \begin{pmatrix} G1 & G2^{\mathrm{T}} & -G2^{\mathrm{T}} \\ G2 & G3 & -G3 \\ -G2 & -G3 & G3 \end{pmatrix}, \tag{5.112}$$

$$G1 = \begin{pmatrix} x_1^{\mathrm{T}}x_1 + 1 & \cdots & x_1^{\mathrm{T}}x_l + 1 \\ \vdots & \vdots & \vdots \\ x_l^{\mathrm{T}}x_1 + 1 & \cdots & x_l^{\mathrm{T}}x_l + 1 \end{pmatrix}, \tag{5.113}$$

$$G3_{ln\times ln} = \begin{pmatrix} G3_{11} & G3_{12} & \cdots & G3_{1l} \\ G3_{21} & G3_{22} & \cdots & G3_{2l} \\ \vdots & \vdots & \ddots & \vdots \\ G3_{l1} & G3_{l2} & \cdots & G3_{ll} \end{pmatrix}, \tag{5.114}$$

in which

$$G3_{11} = \text{diag}(\Delta x_{11}^2, \ldots, \Delta x_{1n}^2),$$

$$G3_{12} = \text{diag}(|\Delta x_{11} \Delta x_{21}|, \ldots, \Delta|x_{1n} \Delta x_{2n}|),$$

$$G3_{1l} = \text{diag}(|\Delta x_{11} \Delta x_{l1}|, \ldots, |\Delta x_{1n} \Delta x_{ln}|),$$

$$G3_{21} = \text{diag}(|\Delta x_{11} \Delta x_{21}|, \ldots, |\Delta x_{1n} \Delta x_{2n}|),$$

$$G3_{22} = \text{diag}(\Delta x_{21}^2, \ldots, \Delta x_{2n}^2),$$

$$G3_{2l} = \text{diag}(|\Delta x_{l1} \Delta x_{21}|, \ldots, |\Delta x_{ln} \Delta x_{2n}|),$$

$$G3_{l1} = \text{diag}(|\Delta x_{11} \Delta x_{l1}|, \ldots, |\Delta x_{1n} \Delta x_{ln}|),$$

$$G3_{l2} = \text{diag}(|\Delta x_{l1} \Delta x_{21}|, \ldots, |\Delta x_{ln} \Delta x_{2n}|),$$

$$G3_{ll} = \text{diag}(\Delta x_{l1}^2, \ldots, \Delta x_{ln}^2), \tag{5.115}$$

$$G2_{ln \times l} = \begin{pmatrix}
x_{11}|\Delta x_{11}| & x_{21}|\Delta x_{11}| & \ldots & x_{l1}|\Delta x_{11}| \\
x_{12}|\Delta x_{12}| & x_{22}|\Delta x_{12}| & \ldots & x_{l2}|\Delta x_{12}| \\
\vdots & \vdots & \ddots & \vdots \\
x_{1n}|\Delta x_{1n}| & x_{2n}|\Delta x_{1n}| & \ldots & x_{ln}|\Delta x_{1n}| \\
x_{11}|\Delta x_{21}| & x_{21}|\Delta x_{21}| & \ldots & x_{l1}|\Delta x_{21}| \\
x_{12}|\Delta x_{22}| & x_{22}|\Delta x_{22}| & \ldots & x_{l2}|\Delta x_{22}| \\
\vdots & \vdots & \ddots & \vdots \\
x_{1n}|\Delta x_{2n}| & x_{2n}|\Delta x_{2n}| & \ldots & x_{ln}|\Delta x_{2n}| \\
\vdots & \vdots & \vdots & \vdots \\
x_{11}|\Delta x_{l1}| & x_{21}|\Delta x_{l1}| & \ldots & x_{l1}|\Delta x_{l1}| \\
x_{12}|\Delta x_{l2}| & x_{22}|\Delta x_{l2}| & \ldots & x_{l2}|\Delta x_{l2}| \\
\vdots & \vdots & \ddots & \vdots \\
x_{1n}|\Delta x_{ln}| & x_{2n}|\Delta x_{ln}| & \ldots & x_{ln}|\Delta x_{ln}|
\end{pmatrix}. \tag{5.116}$$

Because $\widetilde{M}$ has complicated structure and there is not an efficient rounding method for it. In order to use of rounding method that eigenvector corresponding to maximal eigenvalue of optimal solution M^*, which $M = (y_1, \ldots, y_l)^{\text{T}}(y_1, \ldots, y_l)$, we will find the relationship of $\widetilde{M}$ and M. Set

$$\widetilde{I} = (I_l; I1_{n \times l}^{\text{T}}; I2_{n \times l}^{\text{T}}; \ldots; Il_{n \times l}^{\text{T}}; I1_{n \times l}^{\text{T}}; I2_{n \times l}^{\text{T}}; \ldots; Il_{n \times l}^{\text{T}})^{\text{T}}, \tag{5.117}$$

where I_l is identity matrix with $l \times l$, and $Ii_{n \times l}$ $(i = 1, \ldots, l)$ is the matrix which the elements of ith column are all ones and the rest elements are zeros. So that

$$\widetilde{M} = \widetilde{I} M \widetilde{I}^{\text{T}}. \tag{5.118}$$

Clearly, $\text{diag}(\widetilde{M}) = e$ if and only if $\text{diag}(M) = e$; $-(2n+1)\varepsilon e \leq \widetilde{M} e \leq (2n+1)\varepsilon e$ if and only if $-\varepsilon e \leq Me \leq \varepsilon e$; $\widetilde{M} \succeq 0$ if and only if $M \succeq 0$. Finally the semi-definite programming is

$$\min_{M,\theta,\overline{h}_i,\overline{\overline{h}}_i,\kappa_i,\varphi_i,\overline{\varphi}_i,i=1,\dots,l} \frac{1}{2}\theta, \tag{5.119}$$

$$\text{s.t.} \quad \kappa \geq 0, \qquad \varphi_i \geq 0, \quad \overline{\varphi}_i \geq 0, \quad i = 1,\dots,l, \qquad M \succeq 0, \tag{5.120}$$

$$-\varepsilon e \leq Me \leq \varepsilon e, \qquad \overline{h} \geq 0, \qquad \overline{\overline{h}} \geq 0, \qquad \text{diag}(M) = e, \tag{5.121}$$

$$\begin{pmatrix} G \circ (\widetilde{I}M\widetilde{I}^{\mathrm{T}}) & \varsigma \\ \varsigma^{\mathrm{T}} & \theta - 2Ce^{\mathrm{T}}\overline{\overline{h}} \end{pmatrix} \succeq 0. \tag{5.122}$$

Algorithm 5.8 (PRC-SDP)

(1) Given training set training data

$$T = \{\widetilde{x}_1,\dots,\widetilde{x}_l\}, \tag{5.123}$$

where $\widetilde{x}_{ij} = x_{ij} + \Delta x_{ij}z_{ij}$, $i = 1,\dots,l$, $j = 1,\dots,n$, $\|z_i\|_1 \leq \Omega$, where $\Omega > 0$ is chosen prior;
(2) Select appropriate kernel $K(x,x')$ and parameter $C > 0$, $\varepsilon > 0$;
(3) Solve problem (5.119)–(5.122), get the solution M^*, set $y^* = \text{sgn}(t_1)$, where t_1 is eigenvector corresponding to the maximal eigenvalue of M^*.

It is easy to extend the unsupervised classification algorithm to semi-supervised classification algorithm follow the above method, and it omitted here.

5.4.3 Robust Algorithm with Ellipsoid

Considering the measurement noise, we also assume training data $x_i \in R^n$, $i = 1,\dots,l$, which has perturbed as $\widetilde{x}_i$, $\widetilde{x}_{ij} = x_{ij} + \Delta x_{ij}z_{ij}$, $i = 1,\dots,l$, $j = 1,\dots,n$, $\|z_i\|_p \leq \Omega$. z_i is a random variable, when its norm equals to l_2 norm, that is $\|z_i\|_2 \leq \Omega$, which is equivalent with $\sum_{j=1}^{n} z_{ij}^2 \leq \Omega^2$, $i = 1,\dots,l$. Considering $\widetilde{x}_{ij} = x_{ij} + \Delta x_{ij}z_{ij}$, then

$$\sum_{j=1}^{n} \left(\frac{\widetilde{x}_{ij} - x_{ij}}{\Delta x_{ij}}\right)^2 \leq \Omega^2, \quad i = 1,\dots,l, \tag{5.124}$$

so the perturbation region of x_i is an ellipsoid.

In the Sim's proposed robust framework [186], constraint $y_i((w \cdot \widetilde{x}_i) + b) \geq 1 - \xi_i$ is equivalent to

$$y_i((w \cdot x_i) + b) - 1 + \xi_i \geq \Omega t_i, \tag{5.125}$$

$$(t_i, y_i\Delta x_{i1}w_1, \dots, y_i\Delta x_{in}w_n)^{\mathrm{T}} \in L_i^{n+1}. \tag{5.126}$$

Same with the above section, we can get the semi-definite programming

$$\min_{M,\theta,\kappa_i,\overline{\kappa}_i,h_i,i=1,\dots,l} \frac{1}{2}\theta, \tag{5.127}$$

$$\text{s.t.} \quad y_i((w \cdot x_i)+b)-1+\xi_i \geq \Omega t_i, \tag{5.128}$$

$$(t_i, y_i\Delta x_{i1}w_1,\dots,y_i\Delta x_{in}w_n)^{\mathrm{T}} \in L_i^{n+1}, \tag{5.129}$$

$$-\varepsilon e \leq Me \leq \varepsilon e, \qquad \text{diag}(M)=e, \tag{5.130}$$

$$M \succeq 0, \qquad \overline{\kappa}_i \geq 0, \quad i=1,\dots,l, \tag{5.131}$$

$$\begin{pmatrix} G \circ (\widetilde{I}M\widetilde{I}^{\mathrm{T}}) & \eta \\ \eta^{\mathrm{T}} & \theta - 2\Omega Ce^{\mathrm{T}}\overline{\kappa} \end{pmatrix} \succeq 0, \tag{5.132}$$

where

$$\eta = \left(\kappa_1 - \overline{\kappa}_1 + \frac{1}{\Omega},\dots,\kappa_l - \overline{\kappa}_l + \frac{1}{\Omega}, h_1,\dots,h_l\right)^{\mathrm{T}}, \tag{5.133}$$

$$\widetilde{I} = (I_l; I1_{n\times l}^{\mathrm{T}}; I2_{n\times l}^{\mathrm{T}}; \cdots; Il_{n\times l}^{\mathrm{T}})^{\mathrm{T}}, \tag{5.134}$$

$$G_{(nl+l)\times(nl+l)} = \begin{pmatrix} G1 & G2^{\mathrm{T}} \\ G2 & G3 \end{pmatrix}, \tag{5.135}$$

$$G1 = \begin{pmatrix} \frac{1}{\Omega^2}x_1^{\mathrm{T}}x_1 + 1 & \cdots & \frac{1}{\Omega^2}x_1^{\mathrm{T}}x_l + 1 \\ \vdots & \vdots & \vdots \\ \frac{1}{\Omega^2}x_l^{\mathrm{T}}x_1 + 1 & \cdots & \frac{1}{\Omega^2}x_l^{\mathrm{T}}x_l + 1 \end{pmatrix}, \tag{5.136}$$

$$G3_{ln\times ln} = \begin{pmatrix} G3_{11} & G3_{12} & \cdots & G3_{1l} \\ G3_{21} & G3_{22} & \cdots & G3_{2l} \\ \vdots & \vdots & \ddots & \vdots \\ G3_{l1} & G3_{l2} & \cdots & G3_{ll} \end{pmatrix}, \tag{5.137}$$

$$G3_{11} = \text{diag}(\Delta x_{11}^2,\dots,\Delta x_{1n}^2),$$

$$G3_{12} = \text{diag}(\Delta x_{11}\Delta x_{21},\dots,\Delta x_{1n}\Delta x_{2n}),$$

$$G3_{1l} = \text{diag}(\Delta x_{11}\Delta x_{l1},\dots,\Delta x_{1n}\Delta x_{ln}),$$

$$G3_{21} = \text{diag}(\Delta x_{11}\Delta x_{21},\dots,\Delta x_{1n}\Delta x_{2n}),$$

$$G3_{22} = \text{diag}(\Delta x_{21}^2,\dots,\Delta x_{2n}^2), \tag{5.138}$$

$$G3_{2l} = \text{diag}(\Delta x_{l1}\Delta x_{21},\dots,\Delta x_{ln}\Delta x_{2n}),$$

$$G3_{l1} = \text{diag}(\Delta x_{11}\Delta x_{l1},\dots,\Delta x_{1n}\Delta x_{ln}),$$

$$G3_{l2} = \text{diag}(\Delta x_{l1}\Delta x_{21},\dots,\Delta x_{ln}\Delta x_{2n}),$$

$$G3_{ll} = \text{diag}(\Delta x_{l1}^2,\dots,\Delta x_{ln}^2),$$

$$
G2_{ln \times l} =
\begin{pmatrix}
\frac{1}{\Omega}x_{11}\Delta x_{11} & \frac{1}{\Omega}x_{21}\Delta x_{11} & \cdots & \frac{1}{\Omega}x_{l1}\Delta x_{11} \\
\frac{1}{\Omega}x_{12}\Delta x_{12} & \frac{1}{\Omega}x_{22}\Delta x_{12} & \cdots & \frac{1}{\Omega}x_{l2}\Delta x_{12} \\
\vdots & \vdots & \ddots & \vdots \\
\frac{1}{\Omega}x_{1n}\Delta x_{1n} & \frac{1}{\Omega}x_{2n}\Delta x_{1n} & \cdots & \frac{1}{\Omega}x_{ln}\Delta x_{1n} \\
\frac{1}{\Omega}x_{11}\Delta x_{21} & \frac{1}{\Omega}x_{21}\Delta x_{21} & \cdots & \frac{1}{\Omega}x_{l1}\Delta x_{21} \\
\frac{1}{\Omega}x_{12}\Delta x_{22} & \frac{1}{\Omega}x_{22}\Delta x_{22} & \cdots & \frac{1}{\Omega}x_{l2}\Delta x_{22} \\
\vdots & \vdots & \ddots & \vdots \\
\frac{1}{\Omega}x_{1n}\Delta x_{2n} & \frac{1}{\Omega}x_{2n}\Delta x_{2n} & \cdots & \frac{1}{\Omega}x_{ln}\Delta x_{2n} \\
\vdots & \vdots & \vdots & \vdots \\
\frac{1}{\Omega}x_{11}\Delta x_{l1} & \frac{1}{\Omega}x_{21}\Delta x_{l1} & \cdots & \frac{1}{\Omega}x_{l1}\Delta x_{l1} \\
\frac{1}{\Omega}x_{12}\Delta x_{l2} & \frac{1}{\Omega}x_{22}\Delta x_{l2} & \cdots & \frac{1}{\Omega}x_{l2}\Delta x_{l2} \\
\vdots & \vdots & \ddots & \vdots \\
\frac{1}{\Omega}x_{1n}\Delta x_{ln} & \frac{1}{\Omega}x_{2n}\Delta x_{ln} & \cdots & \frac{1}{\Omega}x_{ln}\Delta x_{ln}
\end{pmatrix}.
\tag{5.139}
$$

Algorithm 5.9 (ERC-SDP)

(1) Given training set training data

$$
T = \{\tilde{x}_1, \ldots, \tilde{x}_l\},
\tag{5.140}
$$

where $\tilde{x}_{ij} = x_{ij} + \Delta x_{ij} z_{ij}$, $i = 1, \ldots, l$, $j = 1, \ldots, n$, $\|z_i\|_2 \leq \Omega > 0$, where Ω is chosen prior;

(2) Select appropriate kernel $K(x, x')$ and parameter $C > 0$, $\varepsilon > 0$;

(3) Solve problem (5.127)–(5.132), get the solution M^*, set $y^* = \text{sgn}(t_1)$, where t_1 is eigenvector corresponding to the maximal eigenvalue of M^*.

It is also easy to extend the unsupervised classification algorithm to semi-supervised classification algorithm.

5.4.4 Numerical Results

In this section, through numerical experiments, we will test two algorithm (PRC-SDP and ERC-SDP) on various data sets using SeDuMi library. In order to evaluate the influence of the robust trade-off parameter Ω, we will set value of Ω from 0.25 to 1.25 with increment 0.25 on synthetic data set 'AI', which have 19 data points in R^2. Let parameters $\varepsilon = 1$, $C = 100$ and directions of data perturbations are produced randomly. Results are showed in Table 5.4 (in which the number is the misclassification percent) and illustrated in Fig. 5.2.

We also conduct two algorithms on Digits data sets which can be obtained from http://www.cs.toronto.edu/~roweis/data.html. Selecting number '3' and '2', number

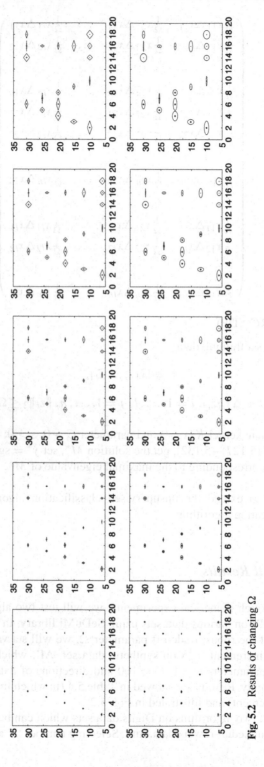

Fig. 5.2 Results of changing Ω

Table 5.4 Results of changing Ω

Ω	0.25	0.5	0.75	1	1.25
PRC-SDP	6/19	4/19	2/19	2/19	2/19
ERC-SDP	4/19	4/19	4/19	4/19	4/19

Table 5.5 Results of changing Ω on data set '3-2'

Ω	0.25	0.5	0.75	1
PRC-SDP	1/20	1/20	1/20	1/20
ERC-SDP	1/20	2/20	0	3/20

Ω	1.5	2	2.5	3	3.5	4
PRC-SDP	1/20	1/20	1/20	1/20	1/20	1/20
ERC-SDP	2/20	2/20	4/20	3/20	5/20	3/20

Table 5.6 Results of changing Ω on data set '7-1'

Ω	0.25	0.5	0.75	1
PRC-SDP	2/20	2/20	2/20	2/20
ERC-SDP	3/20	1/20	5/20	3/20

Ω	1.5	2	2.5	3	3.5	4
PRC-SDP	2/20	2/20	2/20	2/20	2/20	2/20
ERC-SDP	5/20	4/20	4/20	4/20	3/20	2/20

'7' and '1' as training sets respectively. Every number has ten samples of 256 dimensions. Because the problems in PRC-SDP and ERC-SDP have $2ln + l^2 + 3l + 1$ and $ln + l^2 + 2l + 1$ variables respectively (l is the number and n dimension of training data respectively), it seems difficult for software SeDuMi to solve, therefore we use principal component analysis to reduce the dimension n from 256 to 19. Here we also set value of Ω from 0.25 to 1 with increment 0.25 and from 1.5 to 4 with increment 0.5, and directions of data perturbations are produced randomly. To evaluate the robust classification performance, a labeled data set was taken and the labels are removed, then run robust unsupervised classification algorithms, and labeled each of the resulting class with the majority class according to the original training labels, then measured the number of misclassification. The results are showed in Table 5.5 and Table 5.6, in which the number is the misclassification percent.

Chapter 6
Feature Selection via l_p-Norm Support Vector Machines

Though support vector machine has been a promising tool in machine learning, but it does not directly obtain the feature importance. Identifying a subset of features which contribute most to classification is also an important task in classification. The benefit of feature selection is twofold. It leads to parsimonious models that are often preferred in many scientific problems, and it is also crucial for achieving good classification accuracy in the presence of redundant features [77, 252]. We can combine SVM with various feature selection strategies, Some of them are "filters": general feature selection methods independent of SVM. That is, these methods select important features first and then SVM is applied for classification. On the other hand, some are wrapper-type methods: modifications of SVM which choose important features as well as conduct training/testing. In the machine learning literature, there are several proposals for feature selection to accomplish the goal of automatic feature selection in the SVM [26, 99, 216, 252, 254], in some of which they applied the l_0-norm, l_1-norm or l_∞-norm SVM and got competitive performance. The interesting one is l_1-norm SVM, where the 2-norm vector w of the objective function is replaced by 1-norm in the standard SVM model. Naturally, we observe that it leads to more sparse solution when p norm is reduced from 2-norm to 1-norm and the more spare solutions when p $(0 < p < 1)$ is decreased further.

We will propose two models in this chapter, l_p-norm C-support vector classification (l_p-SVC) and l_p-norm proximal support vector machine (l_p-PSVM), which separately combines C-SVC and PSVM [79] with feature selection strategy by introducing the l_p-norm $(0 < p < 1)$.

6.1 l_p-Norm Support Vector Classification

For a two-class classification problem, the training set is given by

$$T = \{(x_1, y_1), \ldots, (x_l, y_l)\} \in (R^n \times \{-1, 1\})^l, \tag{6.1}$$

where $x_i = ([x_i]_1, \ldots, [x_i]_n)^T \in R^n$ and $y_i \in \{-1, 1\}, i = 1, \ldots, l$. Standard C-support vector machine aims to build a decision function by solving the following primal problem:

Y. Shi et al., *Optimization Based Data Mining: Theory and Applications*,
Advanced Information and Knowledge Processing,
DOI 10.1007/978-0-85729-504-0_6, © Springer-Verlag London Limited 2011

$$\min_{w,b,\xi} \frac{1}{2}\|w\|^2 + C\sum_{i=1}^{l}\xi_i, \qquad\qquad (6.2)$$

$$\text{s.t.}\quad y_i((w\cdot x_i)+b) \geq 1-\xi_i, \quad i=1,\ldots,l, \qquad (6.3)$$

$$\xi_i \geq 0, \quad i=1,\ldots,l, \qquad\qquad (6.4)$$

where $\|w\|^2$ is the l_2-norm of w.

6.1.1 l_p-SVC

Naturally, we expect that using the l_p $(0 < p < 1)$-norm to substitute the l_2-norm or l_1-norm to find more sparse solution, which means w has more zero components thus less features could be chosen, where

$$\|w\|_p = \left(\sum_{i=1}^{n}|w_i|^p\right)^{\frac{1}{p}}, \qquad\qquad (6.5)$$

in the sense

$$\|w\|_0 = \lim_{p\to 0}\|w\|_p^p = \lim_{p\to 0}\left(\sum_{i=1}^{n}|w_i|^p\right) = \sharp\{i \mid w_i \neq 0\}, \qquad (6.6)$$

and

$$\|w\|_1 = \lim_{p\to 1}\|w\|_p^p = \lim_{p\to 1}\left(\sum_{i=1}^{n}|w_i|^p\right). \qquad\qquad (6.7)$$

Therefore we introduce l_p-norm $(0 < 1 < p)$ in problem (6.2)–(6.4) and it turns to be

$$\min_{w,b,\xi} \|w\|_p^p + C\sum_{i=1}^{l}\xi_i, \qquad\qquad (6.8)$$

$$\text{s.t.}\quad y_i((w\cdot x_i)+b) \geq 1-\xi_i, \quad i=1,\ldots,l, \qquad (6.9)$$

$$\xi_i \geq 0, \quad i=1,\ldots,l. \qquad\qquad (6.10)$$

Problem (6.8)–(6.10) is then called l_p-SVC problem. The first part in the objective function is nonconvex and non-Lipschitz continuous in R^n.

Now we can construct algorithm l_p-SVC as follows:

Algorithm 6.1 (l_p-SVC)

(1) Given a training set $T = \{(x_1, y_1),\ldots, (x_l, y_l)\} \in (R^n \times \{-1, 1\})^l$;
(2) Select appropriate parameters C and p;
(3) Solve problem (6.8)–(6.10) and get the solution (w^*, b^*);
(4) Construct the decision function as

$$f(x) = \text{sgn}((w^* \cdot x) + b^*). \qquad\qquad (6.11)$$

6.1.2 Lower Bound for Nonzero Entries in Solutions of l_p-SVC

Solving the nonconvex, non-Lipschitz continuous minimization problem (6.8)–(6.10) is very difficult. Most optimization algorithms are only efficient for smooth and convex problems. Of course we can apply some approximation methods to solve (6.8)–(6.10), but numerical solutions generated by the approximation methods have many entries with small values. Can we consider these entries to be zero entries in the solution of the original minimization problem (6.8)–(6.10)? Moreover, how to verify the accuracy of numerical solutions? In this section we will establish a lower bound for the absolute value of nonzero entries in every local optimal solution of the model, which can be used to eliminate zero entries precisely in any numerical solution. Furthermore, this lower bound clearly shows the relationship between the sparsity of the solution and the choice of the regularization parameter and norm, so that our lower bound theory can be used for selecting desired model parameters and norms.

Follow the idea of paper [35], we give out the lower bound for the absolute value of nonzero entries in every local optimal solution of problem (6.8)–(6.10).

Let $\mathcal{Z}_p^*$ denote the set of local solutions of problem (6.8)–(6.10), then for any $z^* = (w^*, b^*, \xi^*) \in \mathcal{Z}_p^*$, we have:

Theorem 6.2 (Lower bound) *Let*

$$L_i = \left(p \Big/ \left(C \sum_{i=1}^{l} |x_i|_j \right) \right)^{\frac{1}{1-p}}, \quad i = 1, \ldots, n, \tag{6.12}$$

then for any $z^* = (w^*, b^*, \xi^*) \in \mathcal{Z}_p^*$, we have

$$w_i^* \in (-L_i, L_i) \quad \Rightarrow \quad w_i^* = 0, \quad i = 1, \ldots, n. \tag{6.13}$$

Proof Suppose $z^* = (w^*, b^*, \xi^*) \in \mathcal{Z}_p^*$, with $\|w^*\|_0 = k$, which means the number of nonzero variables in w^* is k, without loss of generality, we assume

$$w^* = (w_1^*, \ldots, w_k^*, 0, \ldots, 0)^T. \tag{6.14}$$

Let $\tilde{w}^* = (w_1^*, \ldots, w_k^*)^T$, $\tilde{x}_i = ([x_i]_1, \ldots, [x_i]_k)^T \in R^k$, $i = 1, \ldots, l$, and construct a new problem

$$\min_{\tilde{w}, \tilde{b}, \tilde{\xi}} \|\tilde{w}\|_p^p + C \sum_{i=1}^{l} \tilde{\xi}_i, \tag{6.15}$$

$$\text{s.t.} \quad y_i((\tilde{w} \cdot \tilde{x}_i) + \tilde{b}) \geq 1 - \tilde{\xi}_i, \quad i = 1, \ldots, l, \tag{6.16}$$

$$\tilde{\xi}_i \geq 0, \quad i = 1, \ldots, l. \tag{6.17}$$

We denote the feasible set of problem (6.15)–(6.17) as

$$\tilde{D} = \{(\tilde{w}, \tilde{b}, \tilde{\xi}) | y_i((\tilde{w} \cdot \tilde{x}_i) + \tilde{b}) \geq 1 - \tilde{\xi}_i, \ i = 1, \ldots, l; \ \tilde{\xi}_i \geq 0, \ i = 1, \ldots, l\}. \tag{6.18}$$

Because $|\tilde{w}^*| > 0$, then $\|\tilde{w}\|_p^p$ is continuously differentiable at $\tilde{w}^*$, moreover

$$\|\tilde{w}^*\|_p^p + C\sum_{i=1}^{l}\xi_i$$

$$= \|w^*\|_p^p + C\sum_{i=1}^{l}\xi_i$$

$$\leq \min\left\{\|w\|_p^p + C\sum_{i=1}^{l}\xi_i \,\middle|\, (w,b,\xi) \in D;\ [x]_j = 0,\ j = k+1,\ldots,n\right\}$$

$$= \min\left\{\|\tilde{w}\|_p^p + C\sum_{i=1}^{l}\tilde{\xi}_i \,\middle|\, (\tilde{w},\tilde{b},\tilde{\xi}) \in \tilde{D}\right\} \tag{6.19}$$

in a neighborhood of (w^*, b^*, ξ^*). We find that $(\tilde{w}^*, b^*, \xi^*)$ must be a local minimizer of problem (6.15)–(6.17). Hence the first order necessary condition for (6.15)–(6.17) holds at $(\tilde{w}^*, b^*, \xi^*)$, i.e. KKT conditions:

$$y_i((\tilde{w}\cdot\tilde{x}_i)+b^*) \geq 1 - \xi_i^*, \quad i = 1,\ldots,l; \tag{6.20}$$

$$\xi_i^* \geq 0, \quad i = 1,\ldots,l; \tag{6.21}$$

$$\alpha_i^*(y_i((\tilde{w}\cdot\tilde{x}_i)+b^*)-1+\xi_i^*) = 0, \quad i = 1,\ldots,l; \tag{6.22}$$

$$\alpha_i^* \geq 0, \quad i = 1,\ldots,l; \tag{6.23}$$

$$\nabla_{\tilde{w}^*,b^*,\xi^*}L(\alpha^*,w,b,\xi) = 0, \tag{6.24}$$

where

$$L(\alpha^*,w,b,\xi) = \|\tilde{w}\|_p^p + C\sum_{i=1}^{l}\tilde{\xi}_i - \sum_{i=1}^{l}\alpha_i^*(y_i((\tilde{w}\cdot\tilde{x}_i)+b^*)-1+\xi_i^*)$$

is the Lagrange function of problem (6.15)–(6.17), and $\alpha_i^*, i = 1,\ldots,l$ are Lagrange multipliers.

From above conditions, we can deduce that $p(|\tilde{w}^*|^{p-1}\cdot\mathrm{sign}(\tilde{w}^*)) = \sum_{i=1}^{l}\alpha_i^* y_i \tilde{x}_i$ and $0 \leq \alpha_i^* \leq C, i = 1,\ldots,l$. Therefore we obtain

$$p|\tilde{w}^*|^{p-1} = \left|\sum_{i=1}^{l}\alpha_i^* y_i \tilde{x}_i\right| \leq \sum_{i=1}^{l}\alpha_i|\tilde{x}_i| \leq C\sum_{i=1}^{l}|x_i|. \tag{6.25}$$

Note that $0 < p < 1$, we find

$$|\tilde{w}_j^*| \geq L_i = \left(p\middle/\left(C\sum_{i=1}^{l}|x_i|_j\right)\right)^{\frac{1}{1-p}}. \tag{6.26}$$

$\square$

Remark 6.3 Result in Theorem 6.2 clearly shows the relationship between the sparsity of the solution and the choice of parameter C and norm $\|\cdot\|_p$. For sufficiently small C, the number of nonzero entries of w^* in any local minimizer of problem

(6.8)–(6.10) reduces to 0 for $0 < p < 1$. Hence our lower bound theory can be used for selecting model parameters C and p.

6.1.3 Iteratively Reweighted l_q-SVC for l_p-SVC

Following the idea of iteratively reweighted l_q ($q = 1, 2$) minimization proposed in [30], we construct an iteratively reweighted l_1-SVC to approximately solve l_p-SVC problem, which alternates between estimating w_p and redefining the weights. The algorithm is as follows:

Algorithm 6.4 (Iteratively Reweighted l_q-SVC for l_p-SVC)

(1) Set the iteration count K to zero and $\beta_i^{(0)} = 1$, $i = 1, \ldots, n$; set $q \geq 1$;
(2) Solve the weighted l_q-SVM problem

$$\min_{w,b,\xi} \sum_{i=1}^{n} \beta_i^{(K)} |w_i|^q + C \sum_{i=1}^{l} \xi_i, \tag{6.27}$$

$$\text{s.t.} \quad y_i((w \cdot x_i) + b) \geq 1 - \xi_i, \quad i = 1, \ldots, l, \tag{6.28}$$

$$\xi_i \geq 0, \quad i = 1, \ldots, l, \tag{6.29}$$

and get the solution $w^{(K)}$ w.r.t. w;
(3) Update the weights: for each $i = 1, \ldots, n$,

$$\beta_i^{(K+1)} = \frac{p}{q}(|w_i^{(K)}| + \varepsilon)^{p-1} |w_i|^{1-q}, \tag{6.30}$$

here ε is a positive parameter to ensure that the algorithm is well-defined;
(4) Terminate on convergence or where K attains a specified maximum number of iteration $K_{\max}$. Otherwise, increment K and go to step 2.

The typical choices of q are $q = 1$ or $q = 2$. That is, we relax l_p regularization to l_1 or l_2 regularization. If we choose $q = 2$, problem solved in step 2 will be a standard C-SVC except for an iteratively weighted w, which assures that we can apply standard C-SVC to solve large-scale problems with available softwares.

6.2 l_p-Norm Proximal Support Vector Machine

Based on the results of l_p-SVC, we apply l_p norm to the proximal support vector machine (PSVM). For a classification problem as formulation (6.1), PSVM aims to build a decision function by solving the following primal problem:

$$\min_{w,b,\eta} \frac{1}{2}(\|w\|^2 + b^2) + \frac{C}{2} \sum_{i=1}^{l} \eta_i^2, \tag{6.31}$$

$$\text{s.t.} \quad y_i((w \cdot x_i) + b) = 1 - \eta_i, \quad i = 1, \ldots, l, \tag{6.32}$$

Fig. 6.1 Geometric
explanation of PSVM

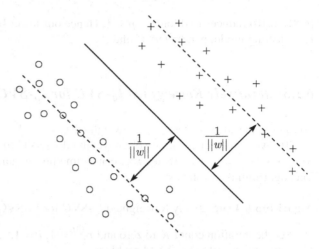

in which $\|w\|^2$ is the l_2-norm of w. Figure 6.1 describes its geometric explanation
in R^2, the planes $(w \cdot x) + b = \pm 1$ around which points of the points "o" and points
"+" cluster and which are pushed apart by the optimization problem (6.31)–(6.32).
PSVM leads to an extremely fast and simple algorithm for generating a linear or
nonlinear classifier that merely requires the solution of a single system of linear
equations, and has been efficiently applied to many fields.

In order to get more sparse solutions, we substitute the l_2-norm of w and b by
l_p-norm $(0 < 1 < p)$ in problem (6.31)–(6.32) and it turns to be

$$\min_{w,b,\eta} \lambda(\|w\|_p^p + |b|^p) + \sum_{i=1}^{l} \eta_i^2, \tag{6.33}$$

$$\text{s.t.} \quad y_i((w \cdot x_i) + b) = 1 - \eta_i, \quad i = 1, \ldots, l. \tag{6.34}$$

Obviously, problem (6.33)–(6.34) is equivalent to the following unconstrained minimization problem

$$\min_{z \in R^{n+1}} \|Az - e\|_2^2 + \lambda \|z\|_p^p, \tag{6.35}$$

where

$$z = (w^T, b)^T \in R^{n+1}, \tag{6.36}$$

$$A = \begin{pmatrix} y_1 x_1^T, y_1 \\ \vdots \\ y_l x_l^T, y_l \end{pmatrix} \in R^{l \times (n+1)}, \tag{6.37}$$

$$e = (1, \ldots, 1)^T \in R^l. \tag{6.38}$$

We call problem (6.35) l_p-norm PSVM problem. Paper [35] established two lower
bounds for the absolute value of nonzero entries in every local optimal solution of
the general model (6.35), which can be used to eliminate zero entries precisely in
any numerical solution. Therefor, we apply them to describe the performance for
feature selection of l_p-norm PSVM for classification.

6.2.1 Lower Bounds for Nonzero Entries in Solutions of l_p-PSVM

Like the proving procedure of the l_p-SVC, let $\mathcal{Q}_p^*$ denote the set of local solutions of problem, then for any $q^* \in \mathcal{Q}_p^*$, we have the corresponding theorems in [35] for model (6.35):

Theorem 6.5 (First bound) *Let* $Lp = (\frac{\lambda p}{2\beta})^{\frac{1}{1-p}}$, *where* $\beta = \|A\|\|e\|$, *then for any* $q^* \in \mathcal{Q}_p^*$, *we have*

$$q_i^* \in (-Lp, Lp) \quad \Rightarrow \quad q_i^* = 0, \quad i = 1, \dots, n+1. \tag{6.39}$$

Theorem 6.6 (Second bound) *Let* $Lp_i = (\frac{\lambda p(1-p)}{2\|a_i\|^2})^{\frac{1}{2-p}}$, $i = 1, \dots, n+1$, *where* a_i *is the ith column of the matrix A (6.37), then for any* $q^* \in \mathcal{Q}_p^*$, *we have*

$$q_i^* \in (-Lq_i, Lq_i) \quad \Rightarrow \quad q_i^* = 0, \quad i = 1, \dots, n+1. \tag{6.40}$$

Just as pointed out by [35], the above two theorems clearly shows the relationship between the sparsity of the solution and the choice of the regularization parameter λ and norm $\| \cdot \|_p$. The lower bounds is not only useful for identification of zero entries in all local optimal solutions from approximation ones, but also for selection of the regularization parameter λ and norm p.

6.2.2 Smoothing l_p-PSVM Problem

Obviously, problem (6.35) is a nonconvex, non-Lipschitz continuous minimization problem, [35] proposed an efficient method to solve it. They first smooth it by choosing appropriate smoothing function and then apply the smoothing conjugate gradient method (SCG) [36] to solve it, which guarantees that any accumulation point of a sequence generated by this method is a Clarke stationary point of problem (6.35). Here we first introduce the smoothing function and then smooth problem (6.35). Let $s_\mu(\cdot)$ be a smoothing function of $|t|$, which takes the formulations as

$$s_\mu(t) = \begin{cases} |t|, & \text{if } |t| > \frac{\mu}{2}, \\ \frac{t^2}{\mu} + \frac{\mu}{4}, & \text{if } |t| \le \frac{\mu}{2}, \end{cases} \tag{6.41}$$

so the smoothed version of problem (6.35) is

$$\min_{z \in R^{n+1}} \|Az - e\|_2^2 + \lambda \sum_{i=1}^{n+1} (s_\mu(z_i))^p. \tag{6.42}$$

Therefore, based on solving (6.42), and determining the nonzero elements of solutions by two bounds, we can establish the l_p-norm PSVM algorithm for both feature selection and classification problem:

Table 6.1 Experiments on Heart disease dataset ($l = 270, n = 13$)

λ	p								
	0.1	0.2	0.3	0.4	0.5	0.6	0.7	0.8	0.9
2^0	(0, 0)	(0, 0)	(0, **1**)	(0, 0)	(0, 0)	(0, 0)	(0, 0)	(0, 0)	(0, 0)
2^1	(0, 0)	(0, **1**)	(0, 0)	(0, 0)	(0, **1**)	(0, **1**)	(0, 0)	(0, 0)	(0, 0)
2^2	(0, 0)	(0, **1**)	(0, **1**)	(0, **1**)	(0, **1**)	(0, **1**)	(0, **1**)	(0, 0)	(0, 0)
2^3	(0, **1**)	(0, **1**)	(0, **1**)	(0, **1**)	(0, **1**)	(0, **1**)	(0, **1**)	(0, **1**)	(0, **1**)
2^4	(0, **1**)	(0, **1**)	(0, **1**)	(0, **1**)	(0, **1**)	(0, **1**)	(0, **1**)	(0, **1**)	(0, **1**)
2^5	(0, 2)	(0, **3**)	(0, **3**)	(0, **3**)	(0, **3**)	(0, **3**)	(0, **3**)	(0, 2)	(0, 1)
2^6	(0, 7)	(0, 7)	(0, **8**)	(0, **8**)	(0, **8**)	(0, 6)	(0, 4)	(0, 4)	(0, 3)
2^7	(0, 9)	(0, 10)	(0, **11**)	(0, **11**)	(0, **11**)	(0, **11**)	(0, 8)	(0, 7)	(0, 5)

Algorithm 6.7 (l_p-PSVM)

(1) Given a training set $T = \{(x_1, y_1), \ldots, (x_l, y_l)\} \in (R^n \times \{-1, 1\})^l$;
(2) Select appropriate parameters λ and p;
(3) Solve problem (6.42) using SCG method and get the solution $z^* = (w^{*\mathrm{T}}, b^*)^\mathrm{T}$;
(4) Set the variables w_i^* to zero if it satisfies either of the two bounds, get the sparse solution w^*;
(5) Select the features corresponding to nonzero elements of w^*;
(5) Construct the decision function as

$$f(x) = \mathrm{sgn}((w^* \cdot x) + b^*). \qquad (6.43)$$

6.2.3 Numerical Experiments

In this section, based on several UCI datasets we apply Algorithm 6.7 to investigate the performance of feature selection by the choice of λ and norm $\| \cdot \|_p$ model. The computational results are conducted on a Dell PC (1.80 GHz, 1.80 GHz, 512 MB of RAM) with using Matlab 7.4.

For every dataset, we scale it with each feature to [0, 1], and then choose parameters $\lambda \in \{2^0, 2^1, \ldots, 2^7\}$, $p \in [0.1, 0.9]$ with step 0.1. Tables 6.1, 6.2, 6.3 and 6.4 describe the sparsity of solutions of problem (6.42) under corresponding (λ, p), $(\sharp_1, \sharp_2)$ in each table means the number of zero variables in solutions w^* determined by bound (6.39) and bound (6.40) separately. Each Bold number denotes the maximum number for a given λ and varying p.

From all the four tables we can see that: bound (6.40) gives out more sparsity than bound (6.39); for bound (6.40), the sparsity value takes its maximum mainly at $p \in (0.2, 0.6)$ for any given λ, which also can be roughly estimated by

$$p^*(\lambda) = \arg \max_{0 < p < 1} (\lambda p (1 - p))^{1/(2-p)} \qquad (6.44)$$

Table 6.2 Experiments on German credit dataset ($l = 1000, n = 24$)

λ	p								
	0.1	0.2	0.3	0.4	0.5	0.6	0.7	0.8	0.9
2^0	(0, 1)	(0, **2**)	(0, 1)	(0, 1)	(0, **2**)	(0, 1)	(0, 1)	(0, 1)	(0, 1)
2^1	(0, **2**)	(0, **2**)	(0, **2**)	(0, 1)	(0, 1)	(0, 1)	(0, 1)	(0, 0)	(0, 0)
2^2	(0, 2)	(0, **3**)	(0, 2)	(0, 2)	(0, 1)	(0, 2)	(0, 1)	(0, 1)	(0, 1)
2^3	(0, 3)	(0, **5**)	(0, 4)	(0, 2)	(0, 2)	(0, 2)	(0, 1)	(0, 1)	(0, 0)
2^4	(0, **5**)	(0, **5**)	(0, **5**)	(0, **5**)	(0, 4)	(0, 3)	(0, 2)	(0, 1)	(0, 1)
2^5	(0, **7**)	(0, **7**)	(0, 6)	(0, **7**)	(0, **7**)	(0, 6)	(0, 4)	(0, 1)	(0, 1)
2^6	(1, 7)	(0, **10**)	(0, **10**)	(0, 9)	(0, 9)	(0, 7)	(0, 7)	(0, 3)	(0, 1)
2^7	(1, 12)	(1, **13**)	(0, **13**)	(0, **13**)	(0, 12)	(0, 10)	(0, 9)	(0, 6)	(0, 2)

Table 6.3 Experiments on Australian credit dataset ($l = 690, n = 14$)

λ	p								
	0.1	0.2	0.3	0.4	0.5	0.6	0.7	0.8	0.9
2^0	(0, **1**)	(0, **1**)	(0, **1**)	(0, **1**)	(0, **1**)	(0, **1**)	(0, 0)	(0, **1**)	(0, 0)
2^1	(0, **1**)	(0, **1**)	(0, **1**)	(0, **1**)	(0, **1**)	(0, 0)	(0, 0)	(0, 0)	(0, 0)
2^2	(0, **1**)	(0, **1**)	(0, **2**)	(0, **1**)	(0, **1**)	(0, **1**)	(0, **1**)	(0, **1**)	(0, 0)
2^3	(0, **1**)	(0, **1**)	(0, **2**)	(0, **1**)	(0, **2**)	(0, **1**)	(0, **1**)	(0, **1**)	(0, 0)
2^4	(0, 3)	(0, 4)	(0, **5**))	(0, 4)	(0, 2)	(0, **1**)	(0, **1**)	(0, **1**)	(0, **1**)
2^5	(1, 5)	(0, **6**)	(1, **6**)	(0, **6**)	(0, **6**)	(0, 5)	(0, 2)	(0, 2)	(0, **1**)
2^6	(0, **6**)	(0, **6**)	(0, **6**)	(0, **6**)	(0, **6**)	(0, **6**)	(0, **6**)	(0, 3)	(0, 2)
2^7	(0, 7)	(1, 9)	(1, 9)	(0, **10**)	(0, **10**)	(0, 8)	(0, 7)	(0, 6)	(0, 3)

Table 6.4 Experiments on Sonar dataset ($l = 208, n = 60$)

λ	p								
	0.1	0.2	0.3	0.4	0.5	0.6	0.7	0.8	0.9
2^0	(0, 1)	(0, **5**)	(0, 2)	(0, 1)	(0, 3)	(0, 1)	(0, 0)	(0, 0)	(0, 0)
2^1	(0, 2)	(0, **4**)	(0, 3)	(0, 2)	(0, 2)	(0, 2)	(0, 0)	(0, 4)	(0, 0)
2^2	(0, 11)	(0, **9**)	(0, 7)	(0, 7)	(0, 7)	(0, 3)	(0, 3)	(0, 0)	(0, 1)
2^3	(0, 12)	(0, **17**)	(0, 13)	(0, 13)	(0, 11)	(0, 8)	(0, 4)	(0, 2)	(0, 1)
2^4	(0, 19)	(0, 23)	(0, **26**)	(0, 23)	(0, 22)	(0, 19)	(0, 14)	(0, 8)	(0, 1)
2^5	(0, 29)	(0, 42)	(0, **44**)	(1, **44**)	(0, **44**)	(0, 40)	(0, 32)	(0, 16)	(0, 5)
2^6	(1, 46)	(1, 58)	(0, **60**)	(1, **60**)	(0, **60**)	(0, 59)	(0, 55)	(0, 44)	(0, 22)
2^7	(1, **60**)	(2, **60**)	(1, **60**)	(1, **60**)	(0, **60**)	(0, **60**)	(1, **60**)	(0, **60**)	(0, 52)

Table 6.5 Numerical results

Dataset	Algorithm	
	l_p-PSVM	l_1-PSVM
Heart	79.63% ($\bar{\lambda} = 2^6$, $\bar{p} = 0.3$, $\bar{\sharp} = 8$)	79.63% ($\tilde{\lambda} = 1$, $\tilde{\sharp} = 3$)
Australian	85.8% ($\bar{\lambda} = 2^5$, $\bar{p} = 0.3$, $\bar{\sharp} = 6$)	85.94% ($\tilde{\lambda} = 128$, $\tilde{\sharp} = 2$)
Sonar	77.51% ($\bar{\lambda} = 2^6$, $\bar{p} = 0.2$, $\bar{\sharp} = 58$)	75.62% ($\tilde{\lambda} = 16$, $\tilde{\sharp} = 36$)
German	75.7% ($\bar{\lambda} = 2^7$, $\bar{p} = 0.3$, $\bar{\sharp} = 13$)	76.1% ($\tilde{\lambda} = 4$, $\tilde{\sharp} = 13$)

for $\lambda \in (0, 2^7)$ if we scale a_i such that $\|a_i\| = 1$; the number of nonzero entries in any local minimizer of (6.42) reduces when λ becomes larger.

Comparison with l_1-PSVM

We compare l_p-PSVM with algorithm l_1-PSVM in this part, and if $p = 1$ the problem (6.42) turns to be a convex problem

$$\min_{z \in R^{n+1}} \|Az - e\|_2^2 + \lambda \|z\|_1. \tag{6.45}$$

For every dataset, we use 5-fold cross-validation to choose the appropriate parameters ($\bar{\lambda}$, $\bar{p}$) for l_p-PSVM and $\tilde{\lambda}$ for algorithm l_1-PSVM, the Table 6.5 gives out the numerical results, where $\bar{\sharp}$ mean the number of zero variables in $\bar{w}$ of algorithm l_p-PSVM determined by bound (6.40), $\tilde{\sharp}$ means the number of zero variables in $\tilde{w}$ of algorithm l_1-PSVM.

From Table 6.5 we find that l_p-PSVM successes in finding more sparse solution with higher accuracy than or almost the same with l_1-PSVM.

Part II
Multiple Criteria Programming:
Theory and Algorithms

Chapter 7
Multiple Criteria Linear Programming

7.1 Comparison of Support Vector Machine and Multiple Criteria Programming

Given a set of n variables about the records $X^{\mathrm{T}} = (x_1, \ldots, x_l)$, let $x_i = (x_{i1}, \ldots, x_{in})^{\mathrm{T}}$ be the development sample of data for the variables, where $i = 1, \ldots, l$ and l is the sample size. Suppose this is a linearly separable case, and we want to determine the coefficients or weights for an appropriate subset of the variables (or attributes), denoted by $w = (w_1, \ldots, w_n)^{\mathrm{T}}$, and a threshold b to separate two classes: say G (Good) and B (Bad); that is

$$(w \cdot x_i) \leqslant b, \quad \text{for } x_i \in G \tag{7.1}$$

and

$$(w \cdot x_i) \geqslant b, \quad \text{for } x_i \in B, \tag{7.2}$$

where x_i are vector values of the subset of the variables and $(w \cdot x_i)$ is the inner product.

In the formulation of Support Vector Machine (SVM), membership of each x_i in class $+1$ (Bad) or -1 (Good) specified by an $l \times l$ diagonal matrix $Y = \{y_{ii}\}$ with $+1$ and -1 entries. Given two bounding planes $(w \cdot x_i) = b \pm 1$, the above separation problem becomes

$$(w \cdot x_i) \leqslant b - 1, \qquad y_{ii} = -1, \tag{7.3}$$

$$(w \cdot x_i) \geqslant b + 1, \qquad y_{ii} = +1. \tag{7.4}$$

It is shown as

$$Y(Xw - eb) \geqslant e, \tag{7.5}$$

where e is a vector of ones.

A standard Support Vector Machine formulation, which can be tackled using quadratic programming, is as follows:

Y. Shi et al., *Optimization Based Data Mining: Theory and Applications*, 119
Advanced Information and Knowledge Processing,
DOI 10.1007/978-0-85729-504-0_7, © Springer-Verlag London Limited 2011

Let ξ_i be defined as a slack variable, then $\xi = (\xi_1, \ldots, \xi_l)^{\mathrm{T}}$ is a slack vector. A SVM is stated as

$$\min \ \frac{1}{2}\|w\|_2^2 + C\|\xi\|_2^2, \tag{7.6}$$

$$\text{s.t.} \quad Y(Xw - eb) \geqslant e - \xi, \tag{7.7}$$

$$\xi \geqslant 0, \tag{7.8}$$

where e is a vector of ones and $C > 0$.

In the formulation of multiple criteria programming, the variable ξ_i is viewed as the overlapping with respect of the training sample x_i. Let β_i be the distance from the training sample x_i to the discriminator $(w \cdot x) = b$ (separating hyperplane). The two-group constraints can be written as

$$(w \cdot x_i) \leqslant b + \xi_i - \beta_i, \qquad y_{ii} = -1 \quad \text{(Good)}, \tag{7.9}$$

and

$$(w \cdot x_i) \geqslant b - \xi_i + \beta_i, \qquad y_{ii} = +1 \quad \text{(Bad)}. \tag{7.10}$$

This can be written as $Y(Xw - eb) \geqslant \beta - \xi$, where e is a vector of ones. Therefore, a quadratic form of multiple criteria programming can be formulated as

$$\min \ \|\xi\|_2^2 - \|\beta\|_2^2, \tag{7.11}$$

$$\text{s.t.} \quad Y(Xw - eb) \geqslant \beta - \xi, \tag{7.12}$$

$$\xi, \beta \geqslant 0. \tag{7.13}$$

Comparing the above two formulations, we can see that the multiple criteria programming model is similar to the Support Vector Machine model in terms of the formation by considering minimization of overlapping of the data. However, the former tries to measure all possible distances β from the training samples x_i to separating hyperplane, while the latter fixes the distance as 1 (through bounding planes $(w \cdot x) = b \pm 1$) from the support vectors. Although the interpretation can vary, the multiple criteria programming model addresses more control parameters than the Support Vector Machine, which may provide more flexibility for better separation of data under the framework of the mathematical programming.

7.2 Multiple Criteria Linear Programming

Now we turn our attention to discussion on the formulation of multiple criteria programming. In linear discriminate analysis, data separation can be achieved by two opposite objectives. The first objective separates the observations by minimizing the sum of the deviations (MSD) among the observations. The second maximizes

Fig. 7.1 Geometric meaning of MCLP

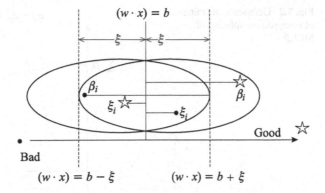

the minimum distances (MMD) of observations from the critical value [76]. As we see from the above section, the overlapping of data ξ should be minimized while the distance β has to be maximized. However, it is difficult for traditional linear programming to optimize MMD and MSD simultaneously. According to the concept of Pareto optimality, we can seek the best trade-off of the two measurements [182, 183]. The first Multiple Criteria Linear Programming (MCLP) model can be described as follows:

$$\min \sum_{i=1}^{l} \xi_i, \tag{7.14}$$

$$\max \sum_{i=1}^{l} \beta_i, \tag{7.15}$$

$$\text{s.t.} \quad (w \cdot x_i) = b + y_i(\xi_i - \beta_i), \quad i = 1, \ldots, l, \tag{7.16}$$

$$\xi, \beta \geqslant 0. \tag{7.17}$$

Here, ξ_i is the overlapping and β_i the distance from the training sample x_i to the discriminator $(w \cdot x_i) = b$ (classification separating hyperplane). If $y_i \in \{1, -1\}$ denotes the label of x_i and l is the number of samples, a training set can be interpreted as pair $\{x_i, y_i\}$, where x_i are the vector values of the variables and $y_i \in \{1, -1\}$ (note, we use y_{ii} in the matrix form of Sect. 7.1) is the label in the classification case. The weights vector w and the bias b are the unknown variables to the optimized for the two objectives. Note that alternatively, the constraint (7.16) can also be written as $y_i((w \cdot x_i) - b) = \xi_i - \beta_i$. The geometric meaning of the model is shown in Fig. 7.1.

The objectives in problem (7.14)–(7.17), the original form of MCLP, are difficult to optimize. In order to facilitate the computation, the compromise solution [184] is employed for reforming the above model so that we can systematically identify the best trade-off between $-\sum \xi_i$ and $\sum \beta_i$ for an optimal solution. Note that because we have to consider the objective space of this two criteria problem, instead of minimizing $\sum \xi_i$, we use maximizing $-\sum \xi_i$. The "ideal value" of $-\sum \xi_i$ and $\sum \beta_i$

Fig. 7.2 Geometric meaning
of compromise solution of
MCLP

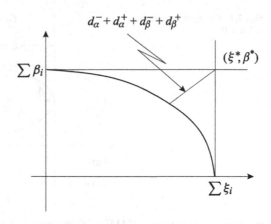

are assumed to be constant $\xi^* > 0$ and $\beta^* > 0$, respectively. Then, if $-\sum \xi_i > \xi^*$, we define the regret measure as $-d_\xi^+ = \sum \xi_i + \xi^*$; otherwise, it is 0. If $-\sum \xi_i < \xi^*$, the regret measure is defined as $d_\xi^- = \xi^* + \sum \xi_i$; otherwise, it is 0. Thus, we have (i) $\xi^* + \sum \xi_i = d_\xi^- - d_\xi^+$, (ii) $|\xi^* + \sum \xi_i| = d_\xi^- + d_\xi^+$, and (iii) $d_\xi^-, d_\xi^+ \geq 0$. Similarly, we derive (i) $\beta^* - \sum \beta_i = d_\beta^- - d_\beta^+$, (ii) $|\beta^* - \sum \beta_i| = d_\beta^- + d_\beta^+$, and (iii) $d_\beta^-, d_\beta^+ \geq 0$. The two-class MCLP model has evolved to the following model:

$$\min\ d_\xi^+ + d_\xi^- + d_\beta^+ + d_\beta^-, \tag{7.18}$$

$$\text{s.t.}\quad \xi^* + \sum_{i=1}^l \xi_i = d_\xi^- - d_\xi^+, \tag{7.19}$$

$$\beta^* - \sum_{i=1}^l \beta_i = d_\beta^- - d_\beta^+, \tag{7.20}$$

$$(w \cdot x_i) = b + y_i(\xi_i - \beta_i), \quad i = 1, \dots, l, \tag{7.21}$$

$$\xi, \beta \geq 0, \qquad d_\xi^+, d_\xi^-, d_\beta^+, d_\beta^- \geq 0. \tag{7.22}$$

Here ξ^* and β^* are given, w and b are unrestricted. The geometric meaning of model (7.18)–(7.22) is shown as in Fig. 7.2.

Note that in problem (7.18)–(7.22), to avoid a trivial solution, one can use $w > 1$. The value of b will also affect the solution. Although b is a variable in problem (7.18)–(7.22), for some applications, the user can choose a fixed value of b to get a solution as the classifier. [162] studied the mechanism of choosing b to improve the accuracy of the performance in classification of credit card holder's behaviors. In order to analyze a large size dataset, the Linux-based MCLP classification algorithm was developed to implement the above model (7.18)–(7.22) [128].

7.3 Multiple Criteria Linear Programming for Multiple Classes

A multiple-class problem of data classification by using multiple criteria linear programming can be described as:

Given a set of n variables or attributes, let $x_i = (x_{i1}, \ldots, x_{in})^T \in R^n$ be the sample observations of data for the variables, where $i = 1, \ldots, l$ and l is the sample size. If a given problem can be predefined as s different classes, $G_1, \ldots, G_s$, then the boundary between the jth and $(j+1)$th classes can be b_j, $j = 1, \ldots, s - 1$. We want to determine the coefficients for an appropriate subset of the variables, denoted by $w = (w_1, \ldots, w_n)^T$ and scalars b_j such that the separation of these classes can be described as follows:

$$(w \cdot x_i) \leqslant b_1, \quad \forall x_i \in G_1, \tag{7.23}$$

$$b_{k-1} \leqslant (w \cdot x_i) \leqslant b_k, \quad \forall x_i \in G_k, \ k = 2, \ldots, s - 1, \tag{7.24}$$

$$(w \cdot x_i) \geqslant b_{s-1}, \quad \forall x_i \in G_s, \tag{7.25}$$

where $\forall x_j \in G_k, k = 1, \ldots, s$, means that the data case x_j belongs to the class G_k.

In the data separation, $(w \cdot x_i)$ is called the score of data case i, which is a linear combination of the weighted values of attribute variables w. For example, in the case of credit card portfolio analysis, $(w \cdot x_i)$ may represent the aggregated value of the ith card holder's score for his or her attributes of age, salary, eduction, and residency under consideration. Even though the boundary b_j is defined as a *scalar* in the above data separation, generally, b_j may be treated as "variable" in the formulation. However, if there is no feasible solution about "variable" b_j in the real data analysis, it should be predetermined as a control parameter according to the experience of the analyst.

The quality of classification is measured by minimizing the total overlapping of data and maximizing the distances of every data to its classes boundary simultaneously. Let ξ_i^j be the overlapping degree with respect of data case x_i within G_j and G_{j+1}, and β_i^j be the distance from x_i within G_j and G_{j+1} to its adjusted boundaries.

By incorporating both ξ_i^j and β_i^j into the separation inequalities, a multiple criteria linear programming (MCLP) classification model can be defined as:

$$\min \sum_i \sum_j \xi_i^j \quad \text{and} \quad \max \sum_i \sum_j \beta_i^j, \tag{7.26}$$

$$\text{s.t.} \quad (w \cdot x_i) = b_1 + \xi_i^1 - \beta_i^1, \quad \forall x_i \in G_1, \tag{7.27}$$

$$b_{k-1} - \xi_i^{k-1} + \beta_i^{k-1} = (w \cdot x_i) = b_k + \xi_i^k - \beta_i^k,$$

$$\forall x_i \in G_k, \ k = 2, \ldots, s - 1, \tag{7.28}$$

$$(w \cdot x_i) = b_{s-1} - \xi_i^{s-1} + \beta_i^{s-1}, \quad \forall x_i \in G_s, \tag{7.29}$$

$$b_{k-1} + \xi_i^{k-1} \leqslant b_k - \xi_i^k, \quad k = 2, \ldots, s - 1, \ i = 1, \ldots, l, \tag{7.30}$$

$$\xi_i^j, \beta_i^j \geqslant 0, \quad j = 1, \ldots, s - 1, \ i = 1, \ldots, l, \tag{7.31}$$

where x_i are given, w and b_j are unrestricted. Note that the constraints (7.30) ensure the existence of the boundaries.

If minimizing the total overlapping of data, maximizing the distances of every data to its class boundary, or a given combination of both criteria is considered separately, model (7.26)–(7.31) is reduced to linear programming (LP) classification (known as linear discriminant analysis), which is initiated by Freed and Glover. However, the single criterion LP could not determine the "best tradeoff" of two misclassification measurements. Therefore, the above model is potentially better than LP classification in identifying the best trade-off of the misclassifications for data separation.

Although model (7.26)–(7.31) can be theoretically solved by the MC-simplex method for all possible trade-offs of both criteria functions, the available software still cannot handle the real-life database or data warehouse with a terabyte of data. To facilitate the computation on the real-life data, a compromise solution approach is employed to reform model (7.26)–(7.31) for the "best tradeoff" between $\sum_i \sum_j \xi_i^j$ and $\sum_i \sum_j \beta_i^j$. Let us assume the "ideal values" for $s-1$ classes overlapping $(-\sum_i \xi_i^1, \ldots, -\sum_i \xi_i^{s-1})$ be $(\xi_*^1, \ldots, \xi_*^{s-1}) > 0$, and the "ideal values" of $(\sum_i \beta_i^1, \ldots, \sum_i \beta_i^{s-1})$ be $(\beta_*^1, \ldots, \beta_*^{s-1}) > 0$. Selection of the ideal values depends on the nature and data format of the problem.

When $-\sum_i \xi_i^j > \xi_*^j$, we define the regret measure as $-d_{\xi j}^+ = \xi_*^j + \sum_i \xi_i^j$; otherwise, it is 0, where $j = 1, \ldots, s-1$. When $-\sum_i \xi_i^j < \xi_*^j$, we define the regret measure as $d_{\xi j}^- = \xi_*^j + \sum_i \xi_i^j$; otherwise, it is 0, where $j = 1, \ldots, s-1$. Thus, we have:

Theorem 7.1

$$\text{(i)}\quad \xi_*^j + \sum_i \xi_i^j = d_{\xi j}^- - d_{\xi j}^+; \tag{7.32}$$

$$\text{(ii)}\quad \left| \xi_*^j + \sum_i \xi_i^j \right| = d_{\xi j}^- + d_{\xi j}^+; \tag{7.33}$$

$$\text{(iii)}\quad d_{\xi j}^-, d_{\xi j}^+ \geqslant 0, \quad j = 1, \ldots, s-1. \tag{7.34}$$

Similarly, we can derive:

Corollary 7.2

$$\text{(i)}\quad \beta_*^j - \sum_i \beta_i^j = d_{\beta j}^- - d_{\beta j}^+; \tag{7.35}$$

$$\text{(ii)}\quad \left| \beta_*^j + \sum_i \beta_i^j \right| = d_{\beta j}^- + d_{\beta j}^+; \tag{7.36}$$

$$\text{(iii)}\quad d_{\beta j}^-, d_{\beta j}^+ \geqslant 0, \quad j = 1, \ldots, s-1. \tag{7.37}$$

Applying the above results into model (7.26)–(7.31), it is reformulated as:

$$\min \sum_{j=1}^{s-1}(d_{\xi j}^- + d_{\xi j}^+ + d_{\beta j}^- + d_{\beta j}^+) \tag{7.38}$$

$$\text{s.t.} \quad \xi_*^j + \sum_i \xi_i^j = d_{\xi j}^- - d_{\xi j}^+, \quad j = 1, \dots, s-1, \tag{7.39}$$

$$\beta_*^j - \sum_i \beta_i^j = d_{\beta j}^- - d_{\beta j}^+, \quad j = 1, \dots, s-1, \tag{7.40}$$

$$(w \cdot x_i) = b_1 + \xi_i^1 - \beta_i^1, \quad \forall x_i \in G_1, \tag{7.41}$$

$$b_{k-1} - \xi_i^{k-1} + \beta_i^{k-1} = (w \cdot x_i) = b_k + \xi_i^k - \beta_i^k,$$
$$\forall x_i \in G_k, \; k = 2, \dots, s-1, \tag{7.42}$$

$$(w \cdot x_i) = b_{s-1} - \xi_i^{s-1} + \beta_i^{s-1}, \quad \forall x_i \in G_s, \tag{7.43}$$

$$b_{k-1} + \xi_i^{k-1} \leqslant b_k - \xi_i^k, \quad k = 2, \dots, s-1, \; i = 1, \dots, l, \tag{7.44}$$

$$\xi_i^j, \beta_i^j \geqslant 0, \quad j = 1, \dots, s-1, \; i = 1, \dots, l, \tag{7.45}$$

$$d_{\xi j}^-, d_{\xi j}^+, d_{\beta j}^-, d_{\beta j}^+ \geqslant 0, \quad j = 1, \dots, s-1, \tag{7.46}$$

where x_i, ξ_*^j and β_*^j are given, w and b_j are unrestricted.

Once the adjusted boundaries $b_{k-1} + \xi_i^{k-1} \leqslant b_k - \xi_i^k$, $k = 2, \dots, s-1$, $i = 1, \dots, l$, are properly chosen, model (7.38)–(7.46) relaxes the conditions of data separation so that it can consider as many overlapping data as possible in the classification process. We can call model (7.38)–(7.46) is a "weak separation formula", With this motivation, we can build a "medium separation formula" on the absolute class boundaries in the following model (7.47)–(7.55) and a "strong separation formula" which contains as few overlapping data as possible in the following model (7.56)–(7.64).

$$\min \sum_{j=1}^{s-1}(d_{\xi j}^- + d_{\xi j}^+ + d_{\beta j}^- + d_{\beta j}^+) \tag{7.47}$$

$$\text{s.t.} \quad \xi_*^j + \sum_i \xi_i^j = d_{\xi j}^- - d_{\xi j}^+, \quad j = 1, \dots, s-1, \tag{7.48}$$

$$\beta_*^j - \sum_i \beta_i^j = d_{\beta j}^- - d_{\beta j}^+, \quad j = 1, \dots, s-1, \tag{7.49}$$

$$(w \cdot x_i) = b_1 - \beta_i^1, \quad \forall x_i \in G_1, \tag{7.50}$$

$$b_{k-1} + \beta_i^{k-1} = (w \cdot x_i) = b_k - \beta_i^k, \quad \forall x_i \in G_k, \; k = 2, \dots, s-1, \tag{7.51}$$

$$(w \cdot x_i) = b_{s-1} + \beta_i^{s-1}, \quad \forall x_i \in G_s, \tag{7.52}$$

$$b_{k-1} + \varepsilon \leqslant b_k - \xi_i^k, \quad k = 2, \ldots, s-1, \ i = 1, \ldots, l, \tag{7.53}$$

$$\xi_i^j, \beta_i^j \geqslant 0, \quad j = 1, \ldots, s-1, \ i = 1, \ldots, l, \tag{7.54}$$

$$d_{\xi j}^-, d_{\xi j}^+, d_{\beta j}^-, d_{\beta j}^+ \geqslant 0, \quad j = 1, \ldots, s-1, \tag{7.55}$$

where x_i, ε, ξ_*^j and β_*^j are given, w and b_j are unrestricted.

$$\min \sum_{j=1}^{s-1} (d_{\xi j}^- + d_{\xi j}^+ + d_{\beta j}^- + d_{\beta j}^+) \tag{7.56}$$

$$\text{s.t.} \quad \xi_*^j + \sum_i \xi_i^j = d_{\xi j}^- - d_{\xi j}^+, \quad j = 1, \ldots, s-1, \tag{7.57}$$

$$\beta_*^j - \sum_i \beta_i^j = d_{\beta j}^- - d_{\beta j}^+, \quad j = 1, \ldots, s-1, \tag{7.58}$$

$$(w \cdot x_i) = b_1 - \xi_i^1 - \beta_i^1, \quad \forall x_i \in G_1, \tag{7.59}$$

$$b_{k-1} + \xi_i^{k-1} + \beta_i^{k-1} = (w \cdot x_i) = b_k - \xi_i^k - \beta_i^k,$$
$$\forall x_i \in G_k, \ k = 2, \ldots, s-1, \tag{7.60}$$

$$(w \cdot x_i) = b_{s-1} + \xi_i^{s-1} + \beta_i^{s-1}, \quad \forall x_i \in G_s, \tag{7.61}$$

$$b_{k-1} + \xi_i^{k-1} \leqslant b_k - \xi_i^k, \quad k = 2, \ldots, s-1, \ i = 1, \ldots, l, \tag{7.62}$$

$$\xi_i^j, \beta_i^j \geqslant 0, \quad j = 1, \ldots, s-1, \ i = 1, \ldots, l, \tag{7.63}$$

$$d_{\xi j}^-, d_{\xi j}^+, d_{\beta j}^-, d_{\beta j}^+ \geqslant 0, \quad j = 1, \ldots, s-1, \tag{7.64}$$

where x_i, ξ_*^j and β_*^j are given, w and b_j are unrestricted.

A loosing relationship of above three models is given as:

Theorem 7.3

(i) *If a data case x_i is classified in a given G_j by model (7.56)–(7.64), then it may be in G_j by using models (7.47)–(7.55) and (7.38)–(7.46);*

(ii) *If a data case x_i is classified in a given G_j by model (7.47)–(7.55), then it may be in G_j by using model (7.38)–(7.46).*

Proof It follows the facts that for a certain value of $\varepsilon > 0$, the feasible solutions of model (7.56)–(7.64) is the feasible solutions of models (7.47)–(7.55) and (7.38)–(7.46), and the feasible solutions of model (7.47)–(7.55) is these of model (7.38)–(7.46). □

Remark 7.4 Conceptually, the usefulness of these formulas should depend on the nature of a given database. If the database contains a few overlapping data, model (7.56)–(7.64) may be used. Otherwise, model (7.47)–(7.55) or (7.38)–(7.46) should

Table 7.1 A two-class data set of customer status

Case	Age	Income	Student	Credit_rating	Class: buys_computer	Training results
x_1	31..40	High	No	Fair	Yes	Success
x_2	>40	Medium	No	Fair	Yes	Success
x_3	>40	Low	Yes	Fair	Yes	Success
x_4	31..40	Low	Yes	Excellent	Yes	Success
x_5	≤30	Low	Yes	Fair	Yes	Success
x_6	>40	Medium	Yes	Fair	Yes	Success
x_7	≤30	Medium	Yes	Excellent	Yes	Success
x_8	31..40	Medium	No	Excellent	Yes	Failure
x_9	31..40	High	Yes	Fair	Yes	Success
x_{10}	≤30	High	No	Fair	No	Success
x_{11}	≤30	High	No	Excellent	No	Success
x_{12}	>40	Low	Yes	Excellent	No	Failure
x_{13}	≤30	Medium	No	Fair	No	Success
x_{14}	>40	Medium	No	Excellent	No	Success

be applied. In many real data analysis, we can always find a feasible solution for model (7.38)–(7.46) if proper values of boundaries b_j are chosen as control parameters. Comparing with conditions of data separation, it is not easier to find the feasible solutions for model (7.47)–(7.55) and/or model (7.38)–(7.46). However, the precise theoretical relationship between three models deserves a further ad careful study.

Example 7.5 As an illustration, we use a small training data set in Table 7.1 to show how two-class model works.

Suppose whether or not a customer buys computer relates to the attribute set {Age, Income, Student, Credit_rating}, we first define the variables Age, Income, Student and Credit_rating by numeric numbers as follows:

For Age: "≤ 30" assigned to be "3", "31... 40" to be "2", and "≥ 40" to be "1". For Income: "high" assigned to be "3", "medium" to be "2", and "low" to be "1". For Student: "yes" assigned to be "2", and "no" to be "1". For Credit_rating: "excellent" assigned to be "2" and "fair" to be "1". G_1 = {yes to buys_computer} and G_2 = {no to buys_computer}.

Then, let $j = 1, 2$ and $i = 1, \ldots, 14$, model (7.38)–(7.46) for this problem to classify the customer's status for {buy_computers} is formulated by

$$\min \; d_\xi^- + d_\xi^+ + d_\beta^- + d_\beta^+ \tag{7.65}$$

$$\text{s.t.} \quad \xi^* + \sum_i \xi_i = d_\xi^- - d_\xi^+, \tag{7.66}$$

$$\beta^* - \sum_i \beta_i = d_\beta^- - d_\beta^+, \tag{7.67}$$

$$2w_1 + 3w_2 + w_3 + w_4 = b + \xi_1 - \beta_1, \tag{7.68}$$

$$w_1 + 2w_2 + w_3 + w_4 = b + \xi_2 - \beta_2, \tag{7.69}$$

$$w_1 + w_2 + 2w_3 + w_4 = b + \xi_3 - \beta_3, \tag{7.70}$$

$$2w_1 + w_2 + 2w_3 + 2w_4 = b + \xi_4 - \beta_4, \tag{7.71}$$

$$3w_1 + w_2 + 2w_3 + w_4 = b + \xi_5 - \beta_5, \tag{7.72}$$

$$w_1 + 2w_2 + 2w_3 + w_4 = b + \xi_6 - \beta_6, \tag{7.73}$$

$$3w_1 + 2w_2 + 2w_3 + 2w_4 = b + \xi_7 - \beta_7, \tag{7.74}$$

$$2w_1 + 2w_2 + w_3 + 2w_4 = b + \xi_8 - \beta_8, \tag{7.75}$$

$$2w_1 + 3w_2 + 2w_3 + w_4 = b + \xi_9 - \beta_9, \tag{7.76}$$

$$3w_1 + 3w_2 + w_3 + w_4 = b + \xi_{10} - \beta_{10}, \tag{7.77}$$

$$3w_1 + 3w_2 + w_3 + 2w_4 = b + \xi_{11} - \beta_{11}, \tag{7.78}$$

$$w_1 + w_2 + 2w_3 + 2w_4 = b + \xi_{12} - \beta_{12}, \tag{7.79}$$

$$3w_1 + 2w_2 + w_3 + w_4 = b + \xi_{13} - \beta_{13}, \tag{7.80}$$

$$w_1 + 2w_2 + w_3 + 2w_4 = b + \xi_{14} - \beta_{14}, \tag{7.81}$$

$$\xi_i, \beta_i \geq 0, \quad i = 1, \ldots, 14, \tag{7.82}$$

$$d_\xi^-, d_\xi^+, d_\beta^-, d_\beta^+ \geq 0, \tag{7.83}$$

where ξ^*, β^* are given, w_i, $i = 1, \ldots, 4$, and b are unrestricted.

Before solving the above problem for data separation, we have to choose the values for the control parameters ξ^*, β^* and b. Suppose we use $\xi^* = 0.1$, $\beta^* = 30000$ and $b = 1$. Then, the optimal solution of this linear programming for the classifier is obtained as column 7 of Table 7.1, where only case x_8 and x_{12} are misclassified. In other words, cases $\{x_1, x_2, x_3, x_4, x_5, x_6, x_7, x_9\}$ are correctly classified in G_1, while cases $\{x_{10}, x_{11}, x_{13}, x_{14}\}$ are found in G_2.

Similarly, when we apply model (7.56)–(7.64) and model (7.47)–(7.55) with $\varepsilon = 0$, one of learning processes provides the same results where cases $\{x_1, x_2, x_3, x_5, x_8\}$ are correctly classified in G_1, while cases $\{x_{10}, x_{11}, x_{12}, x_{14}\}$ are correctly found in G_2. Then, we see that cases $\{x_1, x_2, x_3, x_5\}$ classified in G_1 by model (7.56)–(7.64) are also in G_1 by models (7.47)–(7.55) and (7.38)–(7.46), and cases $\{x_{10}, x_{11}, x_{14}\}$ classified in G_2 by model (7.56)–(7.64) are in G_2 by model (7.47)–(7.55) and (7.38)–(7.46). This is consistent to Theorem 7.3.

7.4 Penalized Multiple Criteria Linear Programming

For many real-life data mining applications, the sample sizes of different classes vary; namely, the training set is unbalanced. Normally, given a dataset of binary cases (Good vs. Bad), there are many more Good records than Bad records. In the training process, the better classifier would be difficult to find if we use model (7.18)–(7.22). To overcome the difficulty with MCLP approach, we proposed the following penalized MCLP method (7.84)–(7.88) in dealing with the real-life credit scoring problem.

$$\min \ d_\xi^- + d_\xi^+ + d_\beta^- + d_\beta^+, \tag{7.84}$$

$$\text{s.t.} \quad \xi^* + p \frac{n_2}{n_1} \sum_{i \in B} \xi_i + \sum_{i \in G} \xi_i = d_\xi^- - d_\xi^+, \tag{7.85}$$

$$\beta^* - p \frac{n_2}{n_1} \sum_{i \in B} \beta_i - \sum_{i \in G} \beta_i = d_\beta^- - d_\beta^+, \tag{7.86}$$

$$(w \cdot x_i) = b + y_i(\xi_i - \beta_i), \quad i = 1, \dots, l, \tag{7.87}$$

$$\xi^*, \beta^* \geqslant 0, \qquad d_\xi^-, d_\xi^+, d_\beta^-, d_\beta^+ \geqslant 0, \tag{7.88}$$

where n_1 and n_2 are the numbers of samples corresponding to the two classes, and $p \geqslant 1$ is the penalized parameter.

The distance is balanced on the two sides of b with the parameters n_1/n_2, even when there are less "Bad" ($+1$) class records to the right of the credit score separating hyperplane b. The value of p enhances the effect of "Bad" distance and penalized much more if we wish more "Bad" to the right of the separating hyperplane.

If $n_1 = n_2$, $p = 1$, the model above degenerates to the original MCLP model (7.14)–(7.17). If $n_1 < n_2$, then $p \geqslant 1$ is used to make the "Bad" catching rate of PMCLP higher than that of MCLP with the same n_1, n_2.

7.5 Regularized Multiple Criteria Linear Programs for Classification

Given an matrix $X \in R^{l \times n}$ and vectors $d, c \in R_+^l$, the multiple criteria linear programming (MCLP) has the following version

$$\min_{u,v} \ d^T u - c^T v,$$

$$\text{s.t.} \quad (w \cdot x_i) + u_i - v_i = b, \quad i = 1, 2, \dots, l_1,$$

$$(w \cdot x_i) - u_i + v_i = b, \quad i = l_1 + 1, l_1 + 2, \dots, l, \tag{7.89}$$

$$u, v \geqslant 0,$$

where x_i is the ith row of X which contains all given data.

However, we cannot ensure this model always has a solution. Obviously the feasible set of MCLP is nonempty, as the zero vector is a feasible point. For $c \geqslant 0$, the objective function may not have a lower bound on the feasible set. To ensure the existence of solution, we add regularization terms in the objective function, and consider the following regularized MCLP

$$\min_z \frac{1}{2} w^T H w + \frac{1}{2} u^T Q u + d^T u - c^T v, \tag{7.90}$$

$$\text{s.t.} \quad (w \cdot x_i) + u_i - v_i = b, \quad i = 1, 2, \ldots, l_1, \tag{7.91}$$

$$(w \cdot x_i) - u_i + v_i = b, \quad i = l_1 + 1, l_1 + 2, \ldots, l, \tag{7.92}$$

$$u, v \geqslant 0, \tag{7.93}$$

where $z = (w, u, v, b) \in R^{n+m+m+1}$, $H \in R^{n \times n}$ and $Q \in R^{l \times l}$ are symmetric positive definite matrices. The regularized MCLP is a convex quadratic program. Although the objective function

$$f(z) := \frac{1}{2} w^T H w + \frac{1}{2} u^T Q u + d^T u - c^T v$$

is not a strictly convex function, we can show that (7.90)–(7.93) always has a solution. Moreover, the solution set of (7.90)–(7.93) is bounded if H, Q, d, c are chosen appropriately.

Let $I_1 \in R^{l_1 \times l_1}$, $I_2 \in R^{(l-l_1) \times (l-l_1)}$ be identity matrices,

$$X_1 = \begin{pmatrix} x_1 \\ \vdots \\ x_{l_1} \end{pmatrix}, \qquad X_2 = \begin{pmatrix} x_{l_1+1} \\ \vdots \\ x_l \end{pmatrix}, \tag{7.94}$$

$$X = \begin{pmatrix} X_1 \\ X_2 \end{pmatrix}, \qquad E = \begin{pmatrix} I_1 & 0 \\ 0 & -I_2 \end{pmatrix}, \tag{7.95}$$

and $e \in R^m$ be the vector whose all elements are 1. Let

$$B = (X \ E \ -E \ -e). \tag{7.96}$$

The feasible set of (7.90)–(7.93) is given by

$$\mathcal{F} = \{z \mid Bz = 0, u \geqslant 0, v \geqslant 0\}. \tag{7.97}$$

Since (7.90)–(7.93) is a convex program with linear constraints, the KKT condition is a necessary and sufficient condition for optimality. To show that $f(z)$ is bounded on $\mathcal{F}$, we will consider the KKT system of (7.90)–(7.93).

Without loss of generality, we assume that $l_1 > 0$ and $l - l_1 > 0$.

Theorem 7.6 *The RMCLP (7.90)–(7.93) is solvable.*

Proof We show that under the assumption that $l_1 > 0$, $l - l_1 > 0$, the objective function has a lower bound. Note that the first terms in the objective function are nonnegative. If there is sequence z^k in $\mathcal{F}$ such that $f(z^k) \rightarrow -\infty$, then there is i such that $v_i^k \rightarrow \infty$, which, together with the constraints of (7.90)–(7.93), implies that there must be j such that $|w_j^k| \rightarrow \infty$ or $u_j^k \rightarrow \infty$. However, the objective function has quadratic terms in x and u which are larger than the linear terms when $k \rightarrow \infty$. This contradicts $f(z^k) \rightarrow -\infty$. Therefore, by Frank-Wolfe Theorem, the regularized MCLP (7.90)–(7.93) always has a solution. We complete the proof. $\square$

Now we show that the solution set of problem (7.90)–(7.93) is bounded if parameters H, Q, d, c are chosen appropriately.

Theorem 7.7 *Suppose that* $XH^{-1}X^{\mathrm{T}}$ *is nonsingular. Let* $G = (XH^{-1}X^{\mathrm{T}})^{-1}$, $\mu = 1/e^{\mathrm{T}}Ge$ *and*

$$M = \begin{pmatrix} Q + EGE - \mu EGee^{\mathrm{T}}GE & -EGE + \mu EGee^{\mathrm{T}}GE \\ -EGE + \mu EGee^{\mathrm{T}}GE & EGE - \mu EGee^{\mathrm{T}}GE \end{pmatrix}, \quad (7.98)$$

$$q = \begin{pmatrix} d \\ -c \end{pmatrix}, \quad y = \begin{pmatrix} u \\ v \end{pmatrix}. \quad (7.99)$$

Then problem (7.90)–(7.93) *is equivalent to the linear complementarity problem*

$$My + q \geq 0, \qquad y \geq 0, \qquad y^{\mathrm{T}}(My + q) = 0. \quad (7.100)$$

If we choose Q *and* H *such that* M *is a positive semidefinite matrix and* c, d *satisfy*

$$d + 2Qe > (\mu EGee^{\mathrm{T}}GE - EGE)e > c, \quad (7.101)$$

then problem (7.90)–(7.93) *has a nonempty and bounded solution set.*

Proof Let us consider the KKT condition of (7.90)–(7.93)

$$Hw + X^{\mathrm{T}}\lambda = 0, \quad (7.102)$$

$$-c - E\lambda - \beta = 0, \quad (7.103)$$

$$Qu + E\lambda + d - \alpha = 0, \quad (7.104)$$

$$Bz = 0, \quad (7.105)$$

$$e^{\mathrm{T}}\lambda = 0, \quad (7.106)$$

$$u \geq 0, \qquad \alpha \geq 0, \qquad \alpha^{\mathrm{T}}u = 0, \quad (7.107)$$

$$v \geq 0, \qquad \beta \geq 0, \qquad \beta^{\mathrm{T}}v = 0. \quad (7.108)$$

From the first three equalities in the KKT condition, we have

$$w = -H^{-1}X^{\mathrm{T}}\lambda, \quad (7.109)$$

$$\beta = -c - E\lambda, \tag{7.110}$$

$$\alpha = Qu + E\lambda + d. \tag{7.111}$$

Substituting w in the 4th equality in the KKT condition gives

$$\lambda = G(Eu - Ev - eb). \tag{7.112}$$

Furthermore, from the 5th equality in the KKT condition, we obtain

$$b = \mu e^{T} GE(u - v). \tag{7.113}$$

Therefore, β and α can be defined by u, v as

$$\beta = -c - EG(Eu - Ev - b) = -c - EG(Eu - Ev - \mu ee^{T} GE(u - v)) \tag{7.114}$$

and

$$\alpha = Qu + EG(Eu - Ev - eb) = Qu + EG(Eu - Ev - \mu ee^{T} GE(u - v)). \tag{7.115}$$

This implies that the KKT condition can be written as the linear complementarity problem (7.100). Since problem (7.90)–(7.93) is a convex problem, it is equivalent to the linear complementarity problem (7.100).

Let $u = 2e$, $v = e$ and $y_0 = (2e, e)$. Then from (7.101), we have

$$My_0 + q = \begin{pmatrix} 2Qe + EGEe - \mu EGee^{T} GHe + d \\ \mu EGee^{T} GEe - EGEe - c \end{pmatrix} > 0, \tag{7.116}$$

which implies that y_0 is a strictly feasible point of (7.100). Therefore, when M is a positive semidefinite matrix, the solution set of (7.100) is nonempty and bounded.

Let $y^{*} = (u^{*}, v^{*})$ be a solution of (7.100), then $z^{*} = (x^{*}, u^{*}, v^{*}, b^{*})$ with

$$b^{*} = \mu e^{T} GE(u^{*} - v^{*}) \quad \text{and}$$

$$w^{*} = -HX^{T} G(Eu^{*} - Ev^{*} - \mu ee^{T} GE(u^{*} - v^{*}))$$

is a solution of (7.90)–(7.93). Moreover, from the KKT condition, it is easy to verify that the boundedness of the solution set of (7.100) implies the boundedness of the solution set of (7.90)–(7.93). □

Chapter 8
MCLP Extensions

8.1 Fuzzy MCLP

As described in Chap. 7, research of linear programming (LP) approaches to classification problems was initiated by Freed and Glover [75]. A simple version seeks MSD can be written as:

$$\min \sum_i \xi_i \tag{8.1}$$

$$\text{s.t.} \quad (w \cdot x_i) \leq b + \xi_i, \quad x_i \in B, \tag{8.2}$$

$$(w \cdot x_i) \geq b - \xi_i, \quad x_i \in G, \tag{8.3}$$

where x_i are given, w and b are unrestricted, and $\xi_i \geq 0$.

The alternative of the above model is to find MMD:

$$\min \sum_i \beta_i \tag{8.4}$$

$$\text{s.t.} \quad (w \cdot x_i) \geq b - \beta_i, \quad x_i \in B, \tag{8.5}$$

$$(w \cdot x_i) \leq b + \beta_i, \quad x_i \in G, \tag{8.6}$$

where x_i are given, w and b are unrestricted, and $\xi_i \geq 0$.

A graphical representation of these models in terms of ξ is shown as Fig. 8.1. We note that the key of the two-class linear classification models is to use a linear combination of the minimization of the sum of ξ_i or/and maximization of the sum of β_i to reduce the two criteria problem into a single criterion. The advantage of this conversion is to easily utilize all techniques of LP for separation, while the disadvantage is that it may miss the scenario of trade-offs between these two separation criteria.

Shi *et al.* [183] applied the compromise solution of multiple criteria linear programming (MCLP) to minimize the sum of ξ_i and maximize the sum of β_i simulta-

Y. Shi et al., *Optimization Based Data Mining: Theory and Applications*,
Advanced Information and Knowledge Processing,
DOI 10.1007/978-0-85729-504-0_8, © Springer-Verlag London Limited 2011

Fig. 8.1 Overlapping case in two-class separation

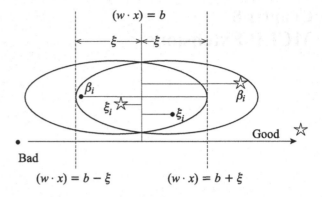

neously. A two-criteria linear programming model is stated as

$$\min \sum_i \xi_i \quad \text{and} \quad \min \sum_i \beta_i \tag{8.7}$$

$$\text{s.t.} \quad (w \cdot x_i) = b + \xi_i - \beta_i, \quad x_i \in G, \tag{8.8}$$

$$(w \cdot x_i) = b - \xi_i + \beta_i, \quad x_i \in B, \tag{8.9}$$

where x_i are given, w and b are unrestricted, ξ_i and $\beta_i \geq 0$.

Suppose a threshold τ is set up as priori, the FLP approach on classifications is to first train (e.g., solve repeatedly) model (8.1)–(8.3) for MSD and model (8.4)–(8.6) for MMD, respectively. If the accuracy rate of either (8.1)–(8.3) or/and (8.4)–(8.6) exceeds threshold τ, the approach terminates. This is the result of known LP classification methods [75, 76]. Generally, if both (8.1)–(8.3) and (8.4)–(8.6) cannot offer the results to meet threshold τ, then the following fuzzy approach is applied.

According to Zimmermann [253], in formulating a FLP problem, the objectives ($\min \sum_i \xi_i$ and $\max \sum_i \beta_i$) and constraints (($w \cdot x_i$) $= b + \xi_i - \beta_i, x_i \in G$; ($w \cdot x_i$) $= b - \xi_i + \beta_i, x_i \in B$) of model (8.7)–(8.9) are redefined as fuzzy sets F and W with corresponding membership functions $\mu_F(w)$ and $\mu_W(w)$ respectively. In this case the fuzzy decision set D is defined as $D = F \cap W$, and the membership function is defined as $\mu_D(w) = \{\mu_F(w), \mu_W(w)\}$. In a maximization problem, w_1 is a "better" decision than w_2 if $\mu_D(w_1) \geq \mu_D(w_2)$. Thus, it can be considered appropriately to select w^* such that

$$\max_w \mu_D(w) = \max_w \min\{\mu_F(w), \mu_W(w)\} = \min\{\mu_F(w^*), \mu_W(w^*)\} \tag{8.10}$$

is the maximized solution.

Let y_{1L} be MSD and y_{2U} be MMD, then one can assume that the value of $\min \sum_i \xi_i$ to be y_{1U} and $\max \sum_i \beta_i$ to be y_{2L}. Note that normally, the "upper bound" y_{1U} related to (8.1)–(8.3) and the "lower bound" y_{2L} related to (8.1)–(8.3) do not exist for the formulations. Let $F_1 = \{w : y_{1L} \leq \sum_i \xi_i \leq y_{1U}\}$ and $F_2 = \{w : y_{2L} \leq \sum_i \beta_i \leq y_{2U}\}$ and their membership functions can be expressed respectively by:

$$\mu_{F_1}(w) = \begin{cases} 1, & \text{if } \sum_i \xi_i \geq y_{1U}, \\ \dfrac{\sum_i \xi_i - y_{1L}}{y_{1U} - y_{1L}}, & \text{if } y_{1L} < \sum_i \xi_i < y_{1U}, \\ 0, & \text{if } \sum_i \xi_i \leq y_{1L}, \end{cases} \tag{8.11}$$

$$\mu_{F_2}(w) = \begin{cases} 1, & \text{if } \sum_i \beta_i \geq y_{2U}, \\ \dfrac{\sum_i \beta_i - y_{2L}}{y_{2U} - y_{2L}}, & \text{if } y_{2L} < \sum_i \beta_i < y_{2U}, \\ 0, & \text{if } \sum_i \beta_i \leq y_{2L}. \end{cases} \tag{8.12}$$

Then the fuzzy set of the objective functions is $F = F_1 \cap F_2$ and its membership function is $\mu_F(w) = \min\{\mu_{F_1}(w), \mu_{F_2}(w)\}$. Using the crisp constraint set $W = \{w : (w \cdot x_i) = b + \xi_i - \beta_i, x_i \in G; (w \cdot x_i) = b - \xi_i + \beta_i, x_i \in B\}$, the fuzzy set of the decision problem is $F = F_1 \cap F_2 \cap W$, and its membership function is $\mu_D(w) = \mu_{F_1 \cap F_2 \cap W}(w)$. Zimmermann [253] has shown that the "optimal solution" of $\max_w \mu_D(w) = \max_w \min\{\mu_{F_1}(w), \mu_{F_2}(w), \mu_W(w)\}$ is an efficient solution of model (8.7)–(8.9), and this problem is equivalent to the following linear program (see Fig. 7.2):

$$\max \eta \tag{8.13}$$

$$\text{s.t.} \quad \eta \leq \frac{\sum_i \xi_i - y_{1L}}{y_{1U} - y_{1L}}, \tag{8.14}$$

$$\eta \leq \frac{\sum_i \beta_i - y_{2L}}{y_{2U} - y_{2L}}, \tag{8.15}$$

$$(w \cdot x_i) = b + \xi_i - \beta_i, \quad x_i \in G, \tag{8.16}$$

$$(w \cdot x_i) = b - \xi_i + \beta_i, \quad x_i \in B, \tag{8.17}$$

where x_i, y_{1L}, y_{1U}, y_{2L} and y_{2U} are known, w and b are unrestricted, and ξ_i, β_i, $\eta \geq 0$.

Note that model (8.13)–(8.17) will produce a value of η with $1 > \eta \geq 0$. To avoid the trivial solution, one can set up $\eta > \varepsilon \geq 0$, for a given ε. Therefore, seeking Maximum η in the FLP approach becomes the standard of determining the classifications between Good and Bad records in the database. Any point of hyperplane $0 < \eta < 1$ over the shadow area represents the possible determination of classifications by the FLP method. Whenever model (8.13)–(8.17) has been trained to meet the given threshold τ, it is said that the better classifier has been identified.

A procedure of using the FLP method for data classifications can be captured by the flowchart of Fig. 8.2. Note that although the boundary of two classes b is the unrestricted variable in model (8.13)–(8.17), it can be presumed by the analyst according to the structure of a particular database. First, choosing a proper value of b can speed up solving (8.13)–(8.17). Second, given a threshold τ, the best data separation can be selected from a number of results determined by different b values.

Fig. 8.2 A flowchart of
fuzzy linear programming
classification method

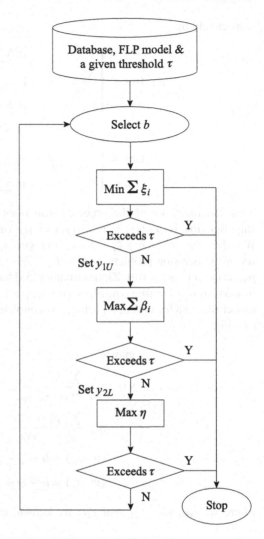

Therefore, the parameter b plays a key role in this paper to achieve and guarantee
the desired accuracy rate τ. For this reason, the FLP classification method uses b as
an important control parameter as shown in Fig. 8.2.

8.2 FMCLP with Soft Constraints

Optimization models in operations research assume that the data are precisely
known, that constrains delimit a crisp set of feasible decisions, and that criteria are
well defined and easy to be formalized [63].

For such a problem, we employ the concept of the following model [63]:

$$\tilde{\max} \; f = cw, \tag{8.18}$$

$$\text{s.t.} \quad (w \cdot x_i) \lesssim b, \tag{8.19}$$

$$w \geq 0. \tag{8.20}$$

Here the symbol $\lesssim$ denotes a relaxed version of $\leq$ and assumes the existence of a vector μ of membership function μ_{D_i}, $i = 1, \ldots, m$. m is the number of the condition function. The symbol $\tilde{\max}$ denotes a relaxed version of max and assumes the existence of a vector μ of membership function μ_F. Then, the problem of model (8.18)–(8.20) can be transferred to the fuzzy linear programming problem with membership function.

$$B(w^*) = \max_{w \in W}(D(W) \wedge F(W)) = \max\{\lambda \mid D(w) \geq \lambda, \; F(w) \geq \lambda, \; \lambda \geq 0\}$$

$$= \max\{\lambda \mid D_1(w) \geq \lambda, \ldots, D_m(w) \geq \lambda, \; F(w) \geq \lambda, \; \lambda \geq 0\}, \tag{8.21}$$

where fuzzy sets $D(w)$, $F(w)$ are transferred from the constraints and criteria of general programming problem with the membership function μ_{D_i} and μ_F respectively. w^* is the satisfying solution of model (8.18)–(8.20).

For the decision maker, since the optimal solution is not necessary at most time, satisfying solution may be enough to solve real-life problems. In the model MSD (8.1)–(8.3) and MMD (8.4)–(8.6), the crisp "distance measurements" (ξ_i and β_i) of observations in criteria and constraints are used to evaluate the classification model in application. To consider the flexibility of the choices for these measurements in obtaining a satisfying solution, we relax the crisp criteria and constraints to soft criteria and constraints. This means that we can allow the flexible boundary b for classification scalar to derive the result what we expect in reality. Based on this idea, we can build a FLP method with both soft criteria and constraints by the following steps. First, we define the membership functions for the MSD problem with soft criterion and constraints as follows:

$$\mu_{F_1}(w) = \begin{cases} 1, & \text{if } \sum_i \xi_i \leq y_{1L}, \\ \dfrac{\sum_i \xi_i - y_{1U}}{y_{1L} - y_{1U}}, & \text{if } y_{1L} < \sum_i \xi_i < y_{1U}, \\ 0, & \text{if } \sum_i \xi_i \geq y_{1U}, \end{cases} \tag{8.22}$$

$$\mu_{D_1}(w) = \begin{cases} 1, & \text{if } (w \cdot x_i) \leq b + \xi_i, \\ 1 - \dfrac{1}{d_1}[(w \cdot x_i) - (b + \xi_i)] & \text{if } b + \xi_i < (w \cdot x_i) < b + \xi_i + d_1, \\ 0, & \text{if } (w \cdot x_i) \geq b + \xi_i + d_1, \end{cases} \tag{8.23}$$

$$\mu_{D_2}(w) = \begin{cases} 1, & \text{if } (w \cdot x_i) \geq b - \xi_i, \\ 1 + \dfrac{1}{d_2}[(w \cdot x_i) - b + \xi_i] & \text{if } b - \xi_i - d_2 < (w \cdot x_i) < b - \xi_i, \\ 0, & \text{if } (w \cdot x_i) \leq b - \xi_i - d_2. \end{cases} \tag{8.24}$$

Then, the fuzzy MSD problem with soft criterion and constraints for (8.1)–(8.3) is constructed as follows:

$$\max \lambda \tag{8.25}$$

$$\text{s.t.} \quad \frac{\sum_i \xi_i - y_{1U}}{y_{1L} - y_{1U}} \geq \xi, \tag{8.26}$$

$$1 - \frac{(w \cdot x_i) - (b + \xi_i)}{d_1} \geq \lambda, \quad x_i \in B, \tag{8.27}$$

$$1 + \frac{(w \cdot x_i) - (b - \xi_i)}{d_2} \geq \lambda, \quad x_i \in G, \tag{8.28}$$

$$1 \geq \lambda > 0, \tag{8.29}$$

where x_i are given, w and b unrestricted, $\xi_i, d_1, d_2 > 0$ respectively.

$y_{1L} = \min \sum_i \xi_i$. This can be obtained from model (8.25)–(8.29) directly. Note that y_{1U} can be a fixed value between $\min \sum_i \xi_i$ and $\max \sum_i \xi_i$ subject to constraints of model (8.25)–(8.29).

For model (8.4)–(8.6), we similarly define the membership function as follows:

$$\mu_{F_2}(w) = \begin{cases} 1, & \text{if } \sum_i \beta_i \geq y_{2U}, \\ \frac{\sum_i \beta_i - y_{2L}}{y_{2U} - y_{2L}}, & \text{if } y_{2L} < \sum_i \beta_i < y_{2U}, \\ 0, & \text{if } \sum_i \beta_i \leq y_{2L}, \end{cases} \tag{8.30}$$

$$\mu_{D_3}(w) = \begin{cases} 1, & \text{if } (w \cdot x_i) \geq b - \beta_i, \\ 1 - \frac{1}{d_3}[(w \cdot x_i) - b + \beta_i)], & \text{if } b - \beta_i - d_3 < (w \cdot x_i) < b - \beta_i, \\ 0, & \text{if } (w \cdot x_i) \leq b - \xi_i - d_3, \end{cases} \tag{8.31}$$

$$\mu_{D_4}(w) = \begin{cases} 1, & \text{if } (w \cdot x_i) \leq b + \beta_i, \\ 1 - \frac{1}{d_4}[(w \cdot x_i) - b - \beta_i)], & \text{if } b + \beta_i < (w \cdot x_i) < b + \beta_i + d_4, \\ 0, & \text{if } (w \cdot x_i) \geq b + \beta_i + d_4. \end{cases} \tag{8.32}$$

Then, a fuzzy-linear programming for model (8.4)–(8.6) is built as below:

$$\max \lambda \tag{8.33}$$

$$\text{s.t.} \quad \frac{\sum_i \beta_i - y_{2L}}{y_{2U} - y_{2L}} \geq \lambda, \tag{8.34}$$

$$1 + \frac{(w \cdot x_i) - b + \beta_i}{d_3} \geq \lambda, \quad x_i \in B, \tag{8.35}$$

$$1 - \frac{(w \cdot x_i) - b - \beta_i}{d_4} \geq \lambda, \quad x_i \in G, \tag{8.36}$$

$$1 \geq \lambda > 0, \tag{8.37}$$

where x_i are given, w and b unrestricted, β_i, d_3, $d_4 > 0$ respectively.

Note that $y_{2U} = \max \sum_i \beta_i$, while y_{2L} is set a value between $\max \sum_i \beta_i$ and $\min \sum_i \beta_i$ subject to constraints of model (8.33)–(8.37).

To identifying a fuzzy model for model (8.7)–(8.9), we first relax it as follows:

$$\min \sum_i \xi_i \quad \text{and} \quad \min \sum_i \beta_i \tag{8.38}$$

$$\text{s.t.} \quad (w \cdot x_i) \geq b + \xi_i - \beta_i, \quad A_i \in B, \tag{8.39}$$

$$(w \cdot x_i) \leq b + \xi_i - \beta_i, \quad A_i \in B, \tag{8.40}$$

$$(w \cdot x_i) \geq b - \xi_i + \beta_i, \quad A_i \in G, \tag{8.41}$$

$$(w \cdot x_i) \leq b - \xi_i + \beta_i, \quad A_i \in G, \tag{8.42}$$

where x_1 are given, w and b are unrestricted, ξ_i and $\beta_i > 0$ respectively.

Then, we can combine (8.25)–(8.29) and (8.33)–(8.37) for fuzzy model of (8.38)–(8.42). Suppose that we set up $d_1 = d_2 = d_1^*$, $d_3 = d_4 = d_2^*$, and a fuzzy model will be:

$$\min \lambda \tag{8.43}$$

$$\text{s.t.} \quad \frac{\sum_i \xi_i - y_{1U}}{y_{1L} - y_{1U}} \geq \lambda, \tag{8.44}$$

$$\frac{\sum_i \beta_i - y_{2L}}{y_{2U} - y_{2L}} \geq \lambda, \tag{8.45}$$

$$1 + \frac{(w \cdot x_i) - (b + \xi_i - \beta_i)}{d_1^*} \geq \lambda, \quad x_i \in B, \tag{8.46}$$

$$1 - \frac{(w \cdot x_i) - (b + \xi_i - \beta_i)}{d_1^*} \geq \lambda, \quad x_i \in B, \tag{8.47}$$

$$1 + \frac{(w \cdot x_i) - (b - \xi_i + \beta_i)}{d_2^*} \geq \lambda, \quad x_i \in G, \tag{8.48}$$

$$1 - \frac{(w \cdot x_i) - (b - \xi_i + \beta_i)}{d_2^*} \geq \lambda, \quad x_i \in G, \tag{8.49}$$

$$1 \geq \lambda > 0, \tag{8.50}$$

where x_i are given, w and b unrestricted, ξ_i, $\beta_i > 0$ respectively. $d_i^* > 0$, $i = 1, 2$ are fixed in the computation. $y_{1L} = \min \sum_i \xi_i$ is obtained from model (8.1)–(8.3) directly, while y_{1U} is a fixed value between $\min \sum_i \xi_i$ and $\max \sum_i \xi_i$ with constraints of model (8.1)–(8.3). $y_{2U} = \max \sum_i \beta_i$, is obtained from the model (8.4)–(8.6) directly while y_{2L} is a fixed value between $\max \sum_i \beta_i$ and $\min \sum_i \beta_i$ with constraints of model (8.4)–(8.6).

There are differences between model (8.43)–(8.50) and model (8.7)–(8.9) or (8.38)–(8.42). First, instead of not the optimal solution, we obtain a satisfying solution based on the membership function from the fuzzy linear programming. Second, with the soft constraints to Model (8.7)–(8.9) or (8.38)–(8.42), the boundary b can be flexibly moved by the upper bound and the lower bound with the separated distance d_i, $i = 1, 2, 3, 4$ according the characteristics of the data.

8.3 FMCLP by Tolerances

In this section, we construct a right semi-trapezoidal fuzzy membership function for the fuzzy set y_{1L}, and let ξ^* ($\xi^* > 0$) be the initial values of y_{1L} which may be regarded as a decision maker's input parameter and its value should be smaller as possible as 0.01 or the similar one. Thus it is argued that if $\sum_i \xi_i < \xi^*$ then the objective of the ith observation or alternative is absolutely satisfied. At the same time, when we introduce the maximum tolerance factor p_1 ($p_1 > 0$) of the departing from, the object of the ith observation is also satisfied fuzzily if $\sum_i \xi_i \in (\xi^*, \xi^* + p_1)$ ($i = 1, 2, \ldots, l$). So, the monotonic and decreasing membership function may be expressed as:

$$
\mu_1 \left(\sum_i \xi_i \right) = \begin{cases} 1, & \sum_i \xi_i < \xi^*, \\ 1 - \frac{\sum_i \xi_i - \xi^*}{p_1}, & \xi^* \leq \sum_i \xi_i \leq \xi^* + p_1, \\ 0, & \sum_i \xi_i > \xi^* + p_1. \end{cases} \tag{8.51}
$$

Similarly, we build a left semi-trapezoidal fuzzy membership function for the fuzzy set y_{2U}, and let β^* ($\beta^* > 0$) be the initial values of y_{2U} which also is regarded as an input parameter and its value should be greater as possible as 300000 or the similar one. After we introduce another maximum tolerance factor p_2 ($p_2 > 0$) of the $\sum_i \beta_i$ departing from β^*, the monotonic and non-decreasing membership function may be given by:

$$
\mu_1 \left(\sum_i \beta_i \right) = \begin{cases} 1, & \sum_i \beta_i > \beta^*, \\ 1 - \frac{\sum_i \beta_i - \beta^*}{p_2}, & \beta^* - p_2 \leq \sum_i \xi_i \leq \beta^*, \\ 0, & \sum_i \xi_i < \xi^* - p_2. \end{cases} \tag{8.52}
$$

Now we use the above membership functions μ_1 and μ_2, for all $\lambda \in [0,1]$, the fuzzy decision set D of the FMCLP by tolerance may be defined as $D_\lambda = \{\mu_1(\sum_i \xi_i) \geq \lambda, \mu_1(\sum_i \beta_i) \geq \lambda, (w \cdot x_i) = b + \xi_i - \beta_i, \text{ for } x_i \in G, (w \cdot x_i) = b - \xi_i + \beta_i, \text{ for } x_i \in B, i = 1, 2, \ldots, l\}$. Then the problem is equivalent to the following model given by:

$$
\max \lambda \tag{8.53}
$$

$$
\text{s.t.} \quad w \in D_\lambda. \tag{8.54}
$$

Furthermore the (8.51)–(8.54) may be solved by the following fuzzy optimization problem:

$$\max \lambda \tag{8.55}$$

$$\text{s.t.} \quad \sum_i \xi_i \le \xi^* + (1 - \lambda) p_1, \tag{8.56}$$

$$\sum_i \beta_i \le \beta^* - (1 - \lambda) p_2, \tag{8.57}$$

$$(w \cdot x_i) = b + \xi_i - \beta_i, \quad \text{for } x_i \in G, \tag{8.58}$$

$$(w \cdot x_i) = b - \xi_i + \beta_i, \quad \text{for } x_i \in B, \tag{8.59}$$

$$\lambda \in [0, 1], \quad i = 1, 2, \dots, l, \tag{8.60}$$

where x_i are given, w and b are unrestricted, ξ_i and $\beta_i \ge 0$. Here it is noted that we have an optimal solution or a satisfied solution for each $\lambda \in [0, 1]$, thus in fact the solution with λ degree of membership is fuzzy.

8.4   Kernel-Based MCLP

Based on MCLP classifier, kernel-based multiple criteria linear programming (KM-CLP) is developed to solve nonlinear separable problem. As is commonly known, kernel function is a powerful tool to deal with nonlinear separable data set. By projecting the data into a high dimensional feature space, the data set will become more likely linear separable. Kernel-based MCLP method introduces kernel function into the original multiple criteria linear programming model to make it possible to solve nonlinear separable problem [241]. The idea to introduce kernel function into MCLP is originated from support vector machine (SVM).

In order to show how the kernel method works, we use the below problem as an example [54].

Suppose we are solving the classification problem in Fig. 8.3(a). The classification data set is $C = \{(x_i, y_i), \ i = 1, \dots, n\}$, where x_i is taken from the space $([x]_1, [x]_2)$, and we have $x_i = ([x_i]_1, [x_i]_2)^T$, $y_i \in \{-1, 1\}$. It is obvious that the best classification line is an ellipse in the space $([x]_1, [x]_2)$, shown in Fig. 8.3(b):

$$[w]_1 [x]_1^2 + [w]_2 [x]_2^2 + b = 0, \tag{8.61}$$

where $[w]_1$ and $[w]_2$ are coefficients in terms of the input data.

Now, the problem is how to get the two coefficients given the specific data set. On this matter, we notice the fact that if we replace the variable $x_i = ([x_i]_1, [x_i]_2)^T$ with $\overline{x_i} = ([\overline{x_i}]_1, [\overline{x_i}]_2)^T$ from a new feature space, where $[\overline{x}]_1 = [x]_1^2$, $[\overline{x}]_2 = [x]_2^2$ then the separation line will be:

$$[w]_1 [\overline{x}]_1 + [w]_2 [\overline{x}]_2 + b = 0. \tag{8.62}$$

With some linear classification methods, we can get $[w]_1$ and $[w]_2$.

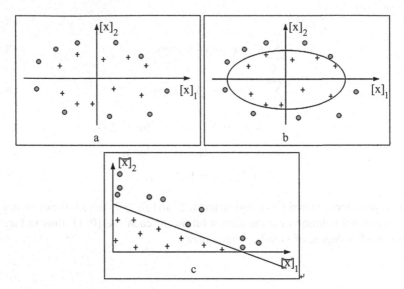

Fig. 8.3 A simple nonlinear classification example

The projection in this problem is:

$$\Phi: \quad \begin{aligned} [\overline{x}]_1 &= [x]_1^2, \\ [\overline{x}]_2 &= [x]_2^2, \end{aligned} \qquad (8.63)$$

where $\Phi : [x] \to [\overline{x}]$ is a nonlinear map from the input space to some feature space, usually with multi-dimensions. $[x] \in R^2$ is the independent variable with 2 attributes. $[\overline{x}] \in R^2$ is the new independent variable in the high-dimensional space with 2 attributes.

Similar to the above example, the basic way to build a nonlinear classification machine includes two steps: first a projection function transforms the data into a feature space, and then a linear classification method to classify them in the feature space. To implement the mapping process, a mapping function is necessary. But as is commonly known, the mapping function is always implicit. Thus, if the input data have many attributes, it is hard to perform such mapping operations.

If data have non-linear patterns, MCLP can be extended by imposing Kernel functions to estimate the non-linearity of the data structure [241]. A Kernel Based Multiple Criteria Linear Program (KBMCLP) was proposed as follows.

Suppose that the solution of (8.7)–(8.9) can be described in the form of $w = \sum_{i=1}^{n} \lambda_i y_i x_i$, which shows that w is a linear combination of $x_i \in X$. Assume that x_k is linearly dependent on the other vectors in feature space:

$$x_k = \sum_{i=1, i \neq k}^{l} \theta_i x_i, \qquad (8.64)$$

where the θ_i are scalar constants. So w can be written as

$$w = \sum_{i=1,i\neq k}^{l} \lambda_i y_i x_i + \lambda_k y_k \sum_{i=1,i\neq k}^{l} \theta_i x_i. \tag{8.65}$$

Now define $\lambda_k y_k \theta_i = \lambda_i y_i \overline{\theta_i}$ so that the above form can be written in the following form:

$$w = \sum_{i=1,i\neq k}^{l} \lambda_i (1+\overline{\theta_i}) y_i x_i = \sum_{i=1,i\neq k}^{l} \gamma_i y_i x_i, \tag{8.66}$$

where $\gamma_i = \lambda_i (1+\overline{\theta_i})$. Let X_m be a maximum linear independent subset from set X. In general, the dimension of the observation is much less than the number of the observations, i.e., $m \leq r \leq l$. The w is simplified as:

$$w = \sum_{i=1}^{m} \lambda_i y_i x_i. \tag{8.67}$$

A kernel function can be used on the maximum linear independent subset X_m for all observations. A KBMCLP, therefore, is formed as the following Model:

$$\min C \sum_{i=1}^{l} \xi_i - \sum_{i=1}^{l} \beta_i \tag{8.68}$$

$$\text{s.t.} \quad \sum_{l=1}^{m} \gamma_l y_l (x_l \cdot x_i) = b + y_i(\xi_i - \beta_i), \quad i = 1,\dots,l, \tag{8.69}$$

$$\xi_i \geq 0, \qquad \beta_i \geq 0, \qquad 0 < \gamma_i < C. \tag{8.70}$$

Note that in (8.68)–(8.70), the separating function has lost the meaning of a hyperplane since the MCLP formulation is used to handle non-linear data separation. Some variations of (8.68)–(8.70) can be further developed by using the MCQP structure with kernel functions.

8.5 Knowledge-Based MCLP

8.5.1 Linear Knowledge-Based MCLP

The classification principles of empirical methods, such as support vector machine, neural networks and decision tree etc., are learning directly from training samples which can draw out the classification rules or learn solely based on the training set. But when there are few training samples due to the obtaining difficulty or cost, these methods might be inapplicable. Actually, different from these empirical classification methods, another commonly used method in some area to classify the data is to use prior knowledge as the classification principle. Two well-known traditional methods are rule-based reasoning and expert system. In these methods, prior knowledge can take the form of logical rule which is well recognized by computer. However, these methods also suffer from the fact that pre-existing knowledge can not contain imperfections. Knowledge-incorporated multiple criteria linear programming classifier combines the above two classification principles to overcome the defaults of each approach [238]. Prior knowledge can be used to aid the training set to improve the classification ability; also training example can be used to refine prior knowledge.

Prior knowledge in some classifiers usually consists of a set of rules, such as, if A then $x \in G$ (or $x \in B$), where condition A is relevant to the attributes of the input data. One example of such form of knowledge can be seen in the breast cancer recurrence or nonrecurrence prediction. Usually, doctors can judge if the cancer recur or not in terms of some measured attributes of the patients. The prior knowledge used by doctors in the breast cancer dataset includes two rules which depend on two features of the total 32 attributes: tumor size (T) and lymph node status (L). The rules are:

- *If $L \geq 5$ and $T \geq 4$ Then RECUR and*
- *If $L = 0$ and $T \leq 1.9$ Then NONRECUR*

The conditions in the above rules can be written into such inequality as $Cx \leq c$, where C is a matrix driven from the condition, x represents each individual sample, c is a vector. For example, if each sample x is expressed by a vector $[x_1, \ldots, x_L, \ldots, x_T, \ldots, x_n]^T$, for the rule: *if $L \geq 5$ and $T \geq 4$ then RECUR*, it also means: *if $x_L \geq 5$ and $x_T \geq 4$, then $x \in RECUR$*, where x_L and x_T are the corresponding values of attributes L and T, n is the number of attributes. Then its corresponding inequality $Cx \leq c$ can be written as:

$$\begin{bmatrix} 0 & \ldots & -1 & \ldots & 0 & \ldots & 0 \\ 0 & \ldots & 0 & \ldots & -1 & \ldots & 0 \end{bmatrix} x \leq \begin{bmatrix} -5 \\ -4 \end{bmatrix}, \tag{8.71}$$

where x is the vector with 32 attributes including two features relevant with prior knowledge.

Similarly, the condition $L = 0$ and $T \leq 1.9$ can also be reformulated to be inequalities. With regard to the condition $L = 0$, in order to express it into the formulation of $Cx \leq c$, we must replace it with the condition $L \geq 0$ and $L \leq 0$. Then the condition $L = 0$ and $T \leq 1.9$ can be represented by two inequalities: $C^1 x \leq c^1$ and $C^2 x \leq c^2$, as follows:

$$\begin{bmatrix} 0 & \ldots & -1 & \ldots & 0 & \ldots & 0 \\ 0 & \ldots & 0 & \ldots & 1 & \ldots & 0 \end{bmatrix} x \leq \begin{bmatrix} 0 \\ 1.9 \end{bmatrix}, \tag{8.72}$$

$$\begin{bmatrix} 0 & \ldots & 1 & \ldots & 0 & \ldots & 0 \\ 0 & \ldots & 0 & \ldots & 1 & \ldots & 0 \end{bmatrix} x \leq \begin{bmatrix} 0 \\ 1.9 \end{bmatrix}. \tag{8.73}$$

Fig. 8.4 A classification example by prior knowledge

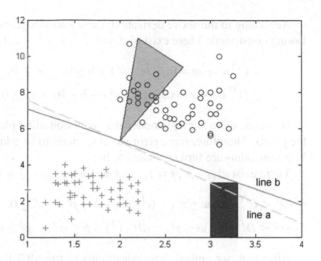

We notice the fact that the set $\{x \mid Cx \leq c\}$ can be viewed as polyhedral convex set. In Fig. 8.4, the triangle and rectangle are such sets.

The result RECUR or NONRECUR is equal to the expression $x \in B$ or $x \in G$. So according to the above rules, we have:

$$Cx \leq c \quad \Rightarrow \quad x \in G \text{ (or } x \in B). \tag{8.74}$$

In MCLP classifier, if the classes are linearly separable, then $x \in G$ is equal to $(x \cdot w) \geq b$, similarly, $x \in B$ is equal to $(x \cdot w) \leq b$. That is, the following implication must hold:

$$Cx \leq c \quad \Rightarrow \quad (x \cdot w) \geq b \text{ (or } (x \cdot w) \leq b). \tag{8.75}$$

For a given (w, b), the implication $Cx \leq c \Rightarrow x^{\mathrm{T}}w \geq b$ holds, this also means that $Cx \leq c$, $(x \cdot w) < b$ has no solution x. According to nonhomogeneous Farkas theorem, we can conclude that $C^{\mathrm{T}}u + w = 0$, $(c \cdot u) + b \leq 0$, $u \geq 0$, has a solution (u, w) [80].

The above statement is able to be added to constraints of MCLP. In this way, the prior knowledge in the form of some equalities and inequalities in constraints is embedded to the linear programming problem. Now, we are to explain how to embed such kind of knowledge into the original MCLP model.

Suppose there are a series of knowledge sets as follows:

- If $C^i x \leq c^i$, $i = 1, \ldots, k$ then $x \in G$,
- If $D^j x \leq d^j$, $j = 1, \ldots, l$ then $x \in B$.

The logical relationships between the above conditions $C^i x \leq c^i$, $i = 1, \ldots, k$ are "*and*". If there are "*or*" relationships between conditions, it will be a bit different. We will describe this kind of problem in the later part.

According to the above derivation, these knowledge can be converted to the following constraints: There exist u^i, $i = 1, \ldots, k$, v^j, $j = 1, \ldots, l$, such that:

$$C^{iT}u^i + w = 0, \quad (c^i \cdot u^i) + b \le 0, \quad u^i \ge 0, \quad i = 1, \ldots, k, \qquad (8.76)$$

$$D^{jT}v^i - w = 0, \quad (d^j \cdot v^j) - b \le 0, \quad v^j \ge 0, \quad j = 1, \ldots, l. \qquad (8.77)$$

However, there is no guarantee that such bounding planes precisely separate all the points. Therefore, some error variables need to be added to the above formulas. The constraints are further revised to be:

There exist $u^i, r^i, \rho^i, i = 1, \ldots, k$ and $v^j, s^j, \sigma^j, j = 1, \ldots, l$, such that:

$$-r^i \le C^{iT}u^i + w \le r^i, \quad (c^i \cdot u^i) + b \le \rho^i, \quad u^i \ge 0, \quad i = 1, \ldots, k, \qquad (8.78)$$

$$-s^j \le D^{jT}v^i - w \le s^j, \quad (d^j \cdot v^j) - b \le \sigma^j, \quad v^j \ge 0, \quad j = 1, \ldots, l. \qquad (8.79)$$

After that, we embed these constraints to the MCLP classifier, then obtain the linear knowledge-incorporated multiple criteria linear programming classifier:

$$\min \sum \alpha_i \quad \text{and} \quad \max \sum \beta_i \quad \text{and} \quad \min \sum (r_i + \rho_i) + \sum (s_j + \sigma_j) \qquad (8.80)$$

$$\text{s.t.} \quad (x_{11} \cdot w_1) + \cdots + (x_{1r} \cdot w_r) = b + \alpha_1 - \beta_1, \quad \text{for } A_1 \in B, \qquad (8.81)$$

$$\vdots$$

$$(x_{n1} \cdot w_1) + \cdots + (x_{nr} \cdot w_r) = b + \alpha_n - \beta_n, \quad \text{for } A_n \in G, \qquad (8.82)$$

$$-r^i \le C^{iT}u^i + w \le r^i, \quad i = 1, \ldots, k, \qquad (8.83)$$

$$(c^i \cdot u^i) + b \le \rho^i, \qquad (8.84)$$

$$-s^j \le D^{iT}v^j - w \le s^j, \quad j = 1, \ldots, l, \qquad (8.85)$$

$$(d^j \cdot v^i) - b \le \rho^j, \qquad (8.86)$$

$$\alpha_1, \ldots, \alpha_n \ge 0, \quad \beta_1, \ldots, \beta_n \ge 0, \quad (u^i, v^i, r^i, \rho^i, s^j, \sigma^j) \ge 0. \qquad (8.87)$$

In this model, all the inequality constraints are derived from the prior knowledge. The last objective $\min \sum (r^i + \rho^i) + \sum (s^j + \sigma^j)$ is about the slack error variables added to the original knowledge equality constraints, because there is no guarantee that the bounding plane of the convex sets will precisely separate the data. The last objective attempts to drive the error variables to zero. We want to get the best bounding plane (w, b) by solving this model to separate the two classes.

As we mentioned before, the logical relationships between the conditions $C^i x \le c^i$, $i = 1, \ldots, k$ are "and". If there are "or" relationships between conditions, it will be a different case. If there are "or" relationships, each condition in the "or" relationships should be dealt with and added as a constraint to the linear programming model individually. As a result of it, there will be several linear programming problems need to be solved individually. And the best solution will be generated from several optimal solutions of all the problems.

8.5.2 Nonlinear Knowledge and Kernel-Based MCLP

Consider model (7.18)–(7.22) and replace $(X_i \cdot X_j)$ by $K(X_i, X_j)$, then kernel-based multiple criteria linear programming (KMCLP) nonlinear classifier is formulated:

$$\min \ d_\xi^+ + d_\xi^- + d_\beta^+ + d_\beta^- \tag{8.88}$$

$$\text{s.t.} \ \ \xi^* + \sum_{i=1}^{n} \xi = d_\xi^- - d_\xi^+, \tag{8.89}$$

$$\beta^* - \sum_{i=1}^{n} \beta_i = d_\beta^- - d_\beta^+, \tag{8.90}$$

$$\lambda_1 y_1 K(X_1, X_1) + \cdots + \lambda_n y_n K(X_n, X_1) = b + \xi_1 - \beta_1, \quad \text{for } X_1 \in B, \tag{8.91}$$

$$\vdots$$

$$\lambda_1 y_1 K(X_1, X_n) + \cdots + \lambda_n y_n K(X_n, X_n) = b + \xi_n - \beta_n, \quad \text{for } X_n \in G, \tag{8.92}$$

$$\xi_1, \ldots, \xi_n \geq 0, \qquad \beta_1, \ldots, \beta_n \geq 0,$$
$$\lambda_1, \ldots, \lambda_n \geq 0, \qquad d_\xi^+, d_\xi^-, d_\beta^+, d_\beta^- \geq 0. \tag{8.93}$$

The above model can be used as a nonlinear classifier, where $K(X_i, X_j)$ can be any nonlinear kernel, for example RBF kernel The above model can be used as a nonlinear classifier, where $K(X_i, X_j)$ can be any nonlinear kernel, for example RBF kernel $k(x, x') = \exp(-q \|x - x'\|^2)$. ξ^* and β^* in the model need to be given in advance. With the optimal value of this model (λ, b, ξ, β), we can obtain the discrimination function to separate the two classes:

$$\lambda_1 y_1 K(X_1, z) + \cdots + \lambda_n y_n K(X_n, z) \leq b, \quad \text{then } z \in B, \tag{8.94}$$

$$\lambda_1 y_1 K(X_1, z) + \cdots + \lambda_n y_n K(X_n, z) \geq b, \quad \text{then } z \in G, \tag{8.95}$$

where z is the input data which is the evaluated target with r attributes.

Prior knowledge in some classifiers usually consist of a set of rules, such as, if A then $x \in G$ (or $x \in B$), where condition A is relevant to the attributes of the input data. For example,

- If $L \geq 5$ and $T \geq 4$ Then RECUR and
- If $L = 0$ and $T \leq 1.9$ Then NONRECUR,

where L and T are two of the total attributes of the training samples.

The conditions in the above rules can be written into such inequality as $Cx \leq c$, where C is a matrix driven from the condition, x represents each individual sample, c is a vector. In some works [80, 81, 241], such kind of knowledge was imposed to constraints of an optimization problem, thus forming the classification model with training samples and prior knowledge as well.

We notice the fact that the set $\{x \mid Cx \le c\}$ can be viewed as polyhedral convex set, which is a linear geometry in input space. But, if the shape of the region which consists of knowledge is nonlinear, for example, $\{x \mid \|x\|^2 \le c\}$, how to deal with such kind of knowledge?

Suppose the region is nonlinear convex set, we describe the region by $g(x) \le 0$. If the data is in this region, it must belong to class B. Then, such kind of nonlinear knowledge may take the form of:

$$g(x) \le 0 \quad \Rightarrow \quad x \in B, \qquad (8.96)$$

$$h(x) \le 0 \quad \Rightarrow \quad x \in G. \qquad (8.97)$$

Here $g(x) : R^r \to R^p$ $(x \in \Gamma)$ and $h(x) : R^r \to R^q$ $(x \in \Delta)$ are functions defined on a subset Γ and Δ of R^r which determine the regions in the input space. All the data satisfied $g(x) \le 0$ must belong to the class B and $h(x) \le 0$ must belong to the class G.

With KMCLP classifier, this knowledge equals to:

$$g(x) \le 0 \quad \Rightarrow \quad \lambda_1 y_1 K(X_1, x) + \cdots + \lambda_n y_n K(X_n, x) \le b \quad (x \in \Gamma), \quad (8.98)$$

$$h(x) \le 0 \quad \Rightarrow \quad \lambda_1 y_1 K(X_1, x) + \cdots + \lambda_n y_n K(X_n, x) \ge b \quad (x \in \Delta). \quad (8.99)$$

This implication can be written in the following equivalent logical form:

$$g(x) \le 0, \quad \lambda_1 y_1 K(X_1, x) + \cdots + \lambda_n y_n K(X_n, x) - b > 0 \text{ has no solution } x \in \Gamma,$$
$$\qquad (8.100)$$

$$h(x) \le 0, \quad \lambda_1 y_1 K(X_1, x) + \cdots + \lambda_n y_n K(X_n, x) - b < 0 \text{ has no solution } x \in \Delta.$$
$$\qquad (8.101)$$

The above expressions hold, then there exist $v \in R^p$, $r \in R^q$, $v, r \ge 0$ such that:

$$-\lambda_1 y_1 K(X_1, x) - \cdots - \lambda_n y_n K(X_n, x) + b + v^\mathsf{T} g(x) \ge 0 \quad (x \in \Gamma), \quad (8.102)$$

$$\lambda_1 y_1 K(X_1, x) + \cdots + \lambda_n y_n K(X_n, x) - b + r^\mathsf{T} h(x) \ge 0 \quad (x \in \Delta). \quad (8.103)$$

Add some slack variables on the above two inequalities, then they are converted to:

$$-\lambda_1 y_1 K(X_1, x) - \cdots - \lambda_n y_n K(X_n, x) + b + v^\mathsf{T} g(x) + s \ge 0 \quad (x \in \Gamma), \quad (8.104)$$

$$\lambda_1 y_1 K(X_1, x) + \cdots + \lambda_n y_n K(X_n, x) - b + r^\mathsf{T} h(x) + t \ge 0 \quad (x \in \Delta). \quad (8.105)$$

The above statement is able to be added to constraints of an optimization problem.

Suppose there are a series of knowledge sets as follows:

If $g(x) \leq 0$,   Then $x \in B$ $(g_i(x) : R^r \to R_i^p)$ $(x \in \Gamma_i, i = 1, \ldots, k)$,   (8.106)

If $h(x) \leq 0$,   Then $x \in G$ $(h_j(x) : R^r \to R_i^p)$ $(x \in \Delta_j, j = 1, \ldots, k)$.   (8.107)

We converted the knowledge to the following constraints:
There exist $v_i \in R_i^p$, $i = 1, \ldots, k$, $r_j \in R_j^q$, $j = 1, \ldots, l$, $v_i, r_j \geq 0$ such that:

$$-\lambda_1 y_1 K(X_1, x) - \cdots - \lambda_n y_n K(X_n, x) + b + v_i^T g_i(x) + s_i \geq 0 \quad (x \in \Gamma),$$
(8.108)

$$\lambda_1 y_1 K(X_1, x) + \cdots + \lambda_n y_n K(X_n, x) - b + r_j^T h_j(x) + t_j \geq 0 \quad (x \in \Delta).$$
(8.109)

These constraints can be easily imposed to KMCLP model (8.88)–(8.93) as the constraints acquired from prior knowledge. Nonlinear knowledge in KMCLP classifier:

$$\min (d_\xi^+ + d_\xi^- + d_\beta^+ + d_\beta^-) + C\left(\sum_{i=1}^{k} s_i + \sum_{j=1}^{l} t_j \right)$$
(8.110)

s.t. $\lambda_1 y_1 K(X_1, X_1) + \cdots + \lambda_n y_n K(X_n, X_1) = b + \xi_1 - \beta_1,$ for $X_1 \in B,$
(8.111)

$$\vdots$$

$$\lambda_1 y_1 K(X_1, X_n) + \cdots + \lambda_n y_n K(X_n, X_n) = b - \xi_n + \beta_n,$$

$$\text{for } X_n \in G,$$
(8.112)

$$\xi^* + \sum_{i=1}^{n} \xi = d_\xi^- - d_\xi^+,$$
(8.113)

$$\beta^* - \sum_{i=1}^{n} \beta_i = d_\beta^- - d_\beta^+,$$
(8.114)

$$-\lambda_1 y_1 K(X_1, x) - \cdots - \lambda_n y_n K(X_n, x) + b + v_i^T g_i(x) + s_i \geq 0,$$

$$i = 1, \ldots, k,$$
(8.115)

$$s_i \geq 0, \quad i = 1, \ldots, k,$$
(8.116)

$$\lambda_1 y_1 K(X_1, x) + \cdots + \lambda_n y_n K(X_n, x) - b + r_j^T h_j(x) + t_j \geq 0,$$

$$j = 1, \ldots, l,$$
(8.117)

$$t_j \geq 0, \quad j = 1, \ldots, l,$$
(8.118)

$$\xi_1, \ldots, \xi_n \geq 0, \qquad \beta_1, \ldots, \beta_n \geq 0, \qquad \lambda_1, \ldots, \lambda_n \geq 0,$$
(8.119)

$$(v_i, r_j) \geq 0,$$
(8.120)

$$d_\xi^+, d_\xi^-, d_\beta^+, d_\beta^- \geq 0.$$
(8.121)

In this model, all the inequality constraints are derived from the prior knowledge. The last objective $C(\sum_{i=1}^{k} s_i + \sum_{j=1}^{l} t_j)$ is about the slack error, which attempts to drive the error variables to zero. We notice the fact that if we set the value of parameter C to be zero, this means to take no account of knowledge. Then this model will be equal to the original KMCLP model. Theoretically, the larger the value of C, the greater impact on the classification result of the knowledge sets. Several parameters need to be set before optimization process. Apart from C we talked about above, the others are parameter of kernel function q (if we choose RBF kernel) and the ideal compromise solution ξ^* and β^*. We want to get the best bounding plane (λ, b) by solving this model to separate the two classes. And the discrimination function of the two classes is:

$$\lambda_1 y_1 K(X_1, z) + \cdots + \lambda_n y_n K(X_n, z) \leq b, \quad \text{then } z \in B, \qquad (8.122)$$

$$\lambda_1 y_1 K(X_1, z) + \cdots + \lambda_n y_n K(X_n, z) \geq b, \quad \text{then } z \in G, \qquad (8.123)$$

where z is the input data which is the evaluated target with r attributes. X_i represents each training sample. y_i is the class label of ith sample.

8.6 Rough Set-Based MCLP

8.6.1 Rough Set-Based Feature Selection Method

On account of the limitation which the MCLP model failed to make sure and remove the redundancy in variables or attributes set. That is to say the model is not good at giving judgment on attributes which are useful and important or unnecessary and unimportant relatively. However, rough set methods have an advantage in this aspect.

It is well known that rough set theory, which was developed by Z. Pawlak in 1980s, is a new mathematical analysis method used for dealing with fuzzy and uncertain information and discovering knowledge and rules hided in data or information [18, 161]. Besides, knowledge or attribute reduction is one of the kernel parts of rough set, and it can efficiently reduce the redundancy in knowledge base or attribute set [236].

For supervised learning a decision system or decision table may often be the form $\varpi = (U, C \cup D)$, where U is a nonempty finite set of objects called the universe, A is a nonempty finite set of attributes, D is the decision attribute. The elements of C are called conditional attributes or simple conditions.

And a binary relation which is reflexive (i.e. an object is in relation with itself xRx), symmetric (i.e. if xRy then yRx) and transitive (if xRy and yRz then xRz) is called an equivalence relation. The equivalence class of an element $x \in X$ consists of all objects $y \in X$ such that xRy.

Let $\varpi = (U, C \cup D)$ be an information system or decision table, then with any set $B \subseteq \varpi$ there is associated an equivalent relation $IND(B)$:

$$IND(B) = \{(x, x') \in B, \forall \omega \in B, \omega(x) = \omega(x')\}. \qquad (8.124)$$

Fig. 8.5 The flow of rough
set for attribute reduction

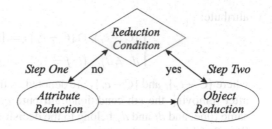

Here the value $IND(B)$ is called as B-indiscernibility relation. If the two-tuple $(x, x') \in IND(B)$, then the objects x and x' are indiscernible from each other by attributes from B. Then the equivalence classes of the B-indiscernibility relation are denoted $[x]_B$.

An equivalence relation induces a partitioning of the universe. These partitions can be used to build new subsets of the universe. These subsets that are most often of interest have the same value of the outcome attribute.

In addition, it is evident that the attribute ω ($\omega \in \varpi$) is reducible, for $\forall \omega \in \varpi$, if the equation (8.125) is tenable:

$$IND(\varpi) = IND(A - \{\omega\}). \tag{8.125}$$

Contrarily, the attribute ω is irreducible. Similarly, the attribute set ϖ is independent, otherwise, correlative and reducible. Further, if an attribute subset $B(B \subseteq \varpi)$ is independent and $IND(B) = IND(\varpi)$, the set B is a reduction of the set ϖ. What is more, the set B is the minimal reduction if it includes the least elements in all the reduction attribute subsets.

As far as classification is concerned, the positive region of the conditional attribute set C in the decision attribute set D may be defined as $Pos_C(D) = \bigcup_{[x]_D} P_-(X)$, where $P_-(X)$ is expressed as the upper approximation of the object. Therefore, the conditional attribute c ($c \in C$) in set is reducible about set if the equation (8.126) is tenable.

$$Pos_C(D) = Pos_{C-\{c\}}(D). \tag{8.126}$$

Generally speaking, for a decision table $\varpi = (U, C \cup D)$, the reduction of conditional attributes set C may be implemented by two steps 1 and 2 just like Fig. 8.5.

That is to say, given any object $x_i \in U$ ($i = 1, 2, \ldots, n$) in the universe U and the conditional attribute set $C = \{c_1, c_2, \ldots, c_m\}$ which is composed of them different attributes, so we have the processing flow for attribute reduction or feature selection as follows:

- **Step 1:** Attribute set reduction.
    For any attribute $c_r \in C$ ($r = 1, 2, \ldots, m$), if the reduction condition is untenable about the below expression (8.127), we will remove c_r as a redundant

attribute:

$$\begin{cases} \{C - c_r\}_i \cap \{C - c_r\}_j = \{C - c_r\}_i & \text{and} \\ d_i \neq d_j & (i, j = 1, 2, \ldots, n), \end{cases} \tag{8.127}$$

where $\{C - c_r\}_i$ and $\{C - c_r\}_j$ are denoted as the ith and jth objects respectively, after removing the rth conditional attribute c_r from the whole of set C. At the same time, and d_i and d_j belong to the decision attribute set D.

• **Step 2:** Object set reduction.

After the first step is finished, decision table may have l $(l \leq m)$ different attributes, thus we need still check whether the expression (8.127) is tenable or not again, if it is tenable, we will keep the attribute c_r similar to step 1, otherwise drop the redundant attribute and corresponding objects or data rows.

In a word, rough set is a powerful tool of data analysis with many merits as follows:

(a) No using the prior knowledge, traditional analysis methods (i.e. fuzzy set, probability and statistics method) are also used to process uncertain information, but it is necessary for them to provide additive information or prior knowledge. However, rough set only make use of information in data set.

(b) Expressing and processing uncertain information effectively, on the basis of equivalent relation and indiscernibility relation it can reduce redundant information and gain the minimal reduction of knowledge or attribute and discover simplifying knowledge and rules.

(c) Missing value, rough set can be avoided of the effects because of missing value in data set.

(d) High performance, it can rapidly and efficiently process the large number of data with many variables or attributes.

8.6.2 A Rough Set-Based MCLP Approach for Classification

Although rough set has many advantages just like the above mentioned, it is short of the fault tolerance and generalization in new data case. Besides, it only deals with discrete data. However, the MCLP model good at those aspects. In general, the MCLP model can gain the better compromise solution on condition that it got the better control of the trade-off of between minimizing the overlapping degree and maximizing the distance departed from boundary. That is to say it do not attempt to get optimal solution but to gain the better generalization by means of using regret measurement and to seek for non inferior solution. Nevertheless, the MCLP model only can deal with continuous and discrete data with numeric type. In addition, it is not immune to missing values, and the reduced conditional attribute subset can not be obtained through it. Especially, when the solved problem has a large number of features, the MCLP approach will occupy a lot of the system resources in time and space, even give the unsuccessful solutions. Therefore, it is required that we may construct a new method combining rough set and the MCLP model effectively for classification in data mining.

Although rough set has many advantages just like the above mentioned, it is short of the fault tolerance and generalization in new data case. Besides, it only deals with discrete data. However, the MCLP model good at those aspects.

In general, the MCLP model can gain the better compromise solution on condition that it got the better control of the trade-off of between minimizing the overlapping degree and maximizing the distance departed from boundary. That is to say it do not attempt to get optimal solution but to gain the better generalization by means of using regret measurement and to seek for non inferior solution. Nevertheless, the MCLP model only can deal with continuous and discrete data with numeric type. In addition, it is not immune to missing values, and the reduced conditional attribute subset can not be obtained through it. Especially, when the solved problem has a large number of features, the MCLP approach will occupy a lot of the system resources in time and space, even give the unsuccessful solutions.

Therefore, it is required that we may construct a new method combining rough set and the MCLP model effectively for classification in data mining.

According to the above analysis, we can find the difference and the existing mutual complementarities between them and need to combine rough set approach and the MCLP model in data mining so as to obtain even more satisfying methods to solve problems.

Generally, the MCLP model can not reduce the dimensions of input information space. Moreover, it result in too long training or learning time when the dimensions of input information space is too large, and to some extent the model will not get solutions of primal problems. However, rough set method can discover the hidden relation among data, remove redundant information and gain a better dimensionality reduction.

In practical applications, because rough set is sensible to the noise of data, the performance of model will become bad when we apply the results learned from data set without noise to data set containing noise. That is to say, rough set has the poor generalization. Nevertheless, the MCLP model provides with the good noise suppression and generalization.

Therefore, according to their characteristics of the mutual complementarities, the integration of rough set method and MCLP model will produce a new hybrid model or system. At the same time, rough set is used as a prefixion part which is responsible for data preprocessing and feature selection and the MCLP model will be regarded as the classification and prediction system where it uses the reduction information by rough set in the model or system. Consequently, the system structure of the integrated rough set method and MCLP model for classification and prediction may be presented as follows in Fig. 8.6.

Firstly, we derive the attributes or variables from the collected source data according to the classification requirements and quantify these attributes. Then a data set which is composed of quantified attributes or variables can be represented as a table, where each row represents a case, an event, an observation, or simply an object. Every column represents an attribute or a variable that can be measured for each

Fig. 8.6 The rough set-based
MCLP model for
classification

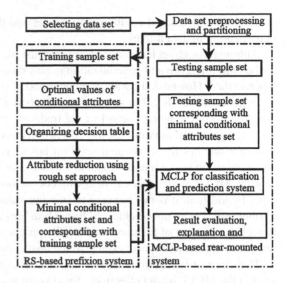

object, therefore this table is called a decision table (or a decision system, information system). In addition, attributes may be defined and categorized into conditional attributes and decision attributes.

After some preliminary exploratory data analysis, we may clean and correct the initial data set. And then data set is partitioned into two parts: training data set and testing data set. Here, for training data set, if it is necessary, we need to discretize all the continuous variables and merge all the unreasonable intervals of discrete variables by using discretization methods so that the optimal values of conditional attributes can be obtained. Then decision table is reorganized in accordance with the discretized variables.

Then, we may use rough set approach to find the minimal attribute reduction subset. Meanwhile, reduction in decision table includes both conditional attributes reduction and removing repetitious data rows. In succession, we need to create a new training set based on the minimal conditional attribute reduction subset and the corresponding data rows again, where the data set remains the important attributes which have great effects on performance of classification.

Finally, we use the training data set to learn and to train the MCLP model and obtain a classifier. Similarly we need to construct a new testing set based on the minimal conditional attribute subset and the corresponding data rows. And then, we use the data set to test the classifier learned from the above data set and get the results of prediction and give evaluations and explanations for these results. Alternatively, we may begin to train the MCLP model on the different training data sets respectively by using 10-fold cross-validation method. And the MCLP model is built in SAS system environment. Then the system rapidly gives the solutions of the models. Among these models, we may choose a stable one and apply it to testing data set and obtain the classification results and the corresponding evaluation indexes by computing. Of course, we may choose some good models to predict and get the average classification results on testing sets by using the ensemble method.

Fig. 8.7 (**a**) The primal
regression problem. (**b**) The
D^+ and D^- data sets

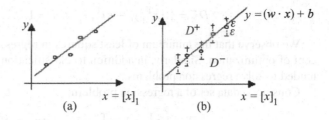

$$x = [x]_1$$

(a)

$$y = (w \cdot x) + b$$

$$x = [x]_1$$

(b)

8.7 Regression by MCLP

Prediction of continuous value is another important branch of data mining. It refers to using some variables in the data set to predict unknown or future values of other variables of interest. These values to be predicted are real values which are not limited to categorical values. Of all methods, regression is the most widely used method to solve such kind of problem. The key idea of regression is to discover the relationship between the dependent variable and one or more independent variables. In other words, the purpose of the regression problem is to determine a function $f(x)$ on the given data set with two kinds of attributes x (independent variables) and y (dependent variable), so that when given a new input x, we can infer the corresponding y with the regression function, where y is a real number. Figure 8.7(a) is a simple example to illustrate the regression problem.

To date, a lot of methods have been developed and widely used to solve the regression problem. For example, a commonly known statistical method least square regression has been successfully applied in many fields. It is first developed to fit the straight line by determining the coefficients of each independent variable, which can minimize the sum of squared error over all observations. In multiple regression and some nonlinear regression problems, least square can also work well. Different from traditional regression method, a new technique, support vector regression, is developed by Vapnik to solve multi-dimensional regression problem. Later, Bi and Bennett [19] initiated the idea that a regression problem can be transferred to a classification problem. According to this, ε-SVR model can be easily inferred [54]. Enlightened by this theory, the main idea of this research is to solve linear regression problem by multiple criteria linear programming classifier. And the key point of it is how to transform a regression problem into a classification problem.

Suppose the given sample set can be approximated by a certain regression function within precision ε, in order to transform a regression problem into a classification problem, we observe the fact that if we move the regression data upward and downward with width ε, the data moved upward, which form the D^+ data set, are naturally separated with the data moved downward, which form the D^- data set (see Fig. 8.7(b)). Hence, the linear classification problem of the D^+ and D^- data sets is constructed [1, 19, 147]. Using some classification method, we can find the discrimination hyperplane of the two classes. It is natural thought that a better hyperplane for separating D^+ and D^- is also a better ε-tube hyperplane of the original regression training set, so regarding it as the regression function is reasonable [54].

$$D^+ = \{((x_i^{\mathrm{T}}, y_i + \varepsilon)^{\mathrm{T}}, +1), \ i = 1, \ldots, n\}, \tag{8.128}$$

$$D^- = \{((x_i^T, y_i - \varepsilon)^T, -1), \ i = 1, \ldots, n\}. \tag{8.129}$$

We observe that the minimum of least squares in regression is based on the concept of optimization. Similarly, in addition to classification, MCLP can also be extended to solve regression problems.

Consider a data set of a regression problem:

$$T = \{(\mathbf{x}_1^T, z_1), (\mathbf{x}_2^T, z_2), \ldots, (\mathbf{x}_n^T, z_n)\}, \tag{8.130}$$

where $\mathbf{x}_i \in R^r$ are the input variables, and $z_i \in R$ is the output variable, which can be any real number. Define the G and B as "Good" (-1) and "Bad" $(+1)$ (a binary case), respectively. Then the corresponding S_{MCLP}^- and S_{MCLP}^+ data sets for MCLP regression model are constructed. With these datasets, the MCLP regression model is formalized as follows [237]:

$$\min \sum_{i=1}^{n}(\alpha_i - \alpha_i') - \max \sum_{i=1}^{n}(\beta_i - \beta_i') \tag{8.131}$$

$$\text{s.t.} \quad (x_{i1} \cdot w_1) + \cdots + (x_{ir} \cdot w_r) + (z_1 + \varepsilon)w_{r+1} = b - \xi_1 + \beta_1, \tag{8.132}$$

$$\vdots$$

$$(x_{k1} \cdot w_1) + \cdots + (x_{kr} \cdot w_r) + ((z_k + \varepsilon) \cdot w_{r+1}) = b - \xi_k + \beta_k,$$

$$\text{for all} \in B, \tag{8.133}$$

$$(x_{k+1,1} \cdot w_1) + \cdots + (x_{k+1,r} \cdot w_r) + ((z_{k+1} - \varepsilon) \cdot w_{r+1})$$

$$= b + \xi_{k+1}' - \beta_{k+1}', \tag{8.134}$$

$$\vdots$$

$$x_{n1}w_1 + \cdots + x_{nr}w_r + (z_n - \varepsilon)w_{r+1} = b + \xi_n' - \beta_n', \quad \text{for all} \in G, \tag{8.135}$$

$$\xi, \xi', \beta, \beta' \geq 0. \tag{8.136}$$

Aggregation of Bad samples:

$$S_{\mathrm{MCLP}}^+ = \{((\mathbf{x}_i^T, z_i + \varepsilon)^T, +1), \ i = 1, \ldots, n\}. \tag{8.137}$$

Aggregation of Good samples:

$$S_{\mathrm{MCLP}}^- = \{((\mathbf{x}_i^T, z_i - \varepsilon)^T, -1), \ i = 1, \ldots, n\}. \tag{8.138}$$

Experimental study of this MCLP regression model compared with other regression methods has shown its potential applicability.

Chapter 9
Multiple Criteria Quadratic Programming

9.1 A General Multiple Mathematical Programming

Consider a mathematical function $f(\xi)$ to be used to describe the relation of all overlapping ξ_i, while another mathematical function $g(\beta)$ represents the aggregation of all distances β_i. The classification accuracies depend on simultaneously minimize $f(\xi)$ and maximize $g(\beta)$. Thus, a general bi-criteria programming method for classification can be formulated as:

$$\min \ f(\xi) \tag{9.1a}$$

$$\max \ g(\beta) \tag{9.1b}$$

$$\text{s.t.} \quad (x_i \cdot w) - b - \xi_i + \beta_i = 0, \quad \forall y_i = -1; \tag{9.1c}$$

$$(x_i \cdot w) - b + \xi_i - \beta_i = 0, \quad \forall y_i = 1; \tag{9.1d}$$

$$\xi, \beta \geq 0, \tag{9.1e}$$

where x_i is given feature vector of the ith sample and y_i is the corresponding label, $i = 1, \ldots, l$. $\xi, \beta \in R^l$, $w \in R^n$ and $b \in R$ are variables.

We note that different forms of $f(\xi)$ and $g(\beta)$ will affect the classification criteria. $f(\xi)$ and $g(\beta)$ can be component-wise and non-decreasing functions. For example, in order to utilize the computational power of some existing nonlinear optimization software packages, a sub-model can be set up by using the l_p norm to represent $f(\xi)$ and l_q norm to represent $g(\beta)$ respectively. This means $f(\xi) = \|\xi\|_p^p$ and $g(\beta) = \|\beta\|_q^q$. Furthermore, to transform the bi-criteria problem of the general model into a single- criterion problem, we use weights $\sigma_\xi > 0$ and $\sigma_\beta > 0$ for $\|\xi\|_p^p$ and $\|\beta\|_q^q$, respectively. The values of σ_ξ and σ_β can be pre-defined in the process of identifying the optimal solution. Thus, the general model can be converted into a single criterion mathematical programming model as:

$$\min \ \sigma_\xi \|\xi\|_p^p - \sigma_\beta \|\beta\|_q^q \tag{9.2a}$$

$$\text{s.t.} \quad (x_i \cdot w) - b + \xi_i + \beta_i = 0, \quad \forall y_i = -1; \tag{9.2b}$$

Y. Shi et al., *Optimization Based Data Mining: Theory and Applications*,
Advanced Information and Knowledge Processing,
DOI 10.1007/978-0-85729-504-0_9, © Springer-Verlag London Limited 2011

$$(x_i \cdot w) - b - \xi_i - \beta_i = 0, \quad \forall y_i = 1; \tag{9.2c}$$

$$\xi, \beta \geq 0. \tag{9.2d}$$

Based on (9.2a)–(9.2d), mathematical programming models with any norm can be theoretically defined. This study is interested in formulating a linear or a quadratic programming model.

- **Case 1:** $p = q = 1$ (MCLP)

In this case, $\|\xi\|_1 = \sum_{i=1}^{l} \xi_i$ and $\|\beta\|_1 = \sum_{i=1}^{l} \beta_i$. Objective function in (9.2a)–(9.2d) can now be a linear objective function

$$\min \ \sigma_\xi \sum_{i=1}^{l} \xi_i - \sigma_\beta \sum_{i=1}^{l} \beta_i. \tag{9.3}$$

(9.2a)–(9.2d) turns to be a linear programming

$$\min \ \sigma_\xi \sum_{i=1}^{l} \xi_i - \sigma_\beta \sum_{i=1}^{l} \beta_i \tag{9.4a}$$

$$\text{s.t.} \quad (x_i \cdot w) - b - \xi_i + \beta_i = 0, \quad \forall y_i = -1; \tag{9.4b}$$

$$(x_i \cdot w) - b + \xi_i - \beta_i = 0, \quad \forall y_i = 1; \tag{9.4c}$$

$$\xi, \beta \geq 0. \tag{9.4d}$$

This model has been obtained by Shi *et al.* [183] and Freed and Glover [75]. There are many softwares which can solve linear programming problem very efficiently at present. We can use some of these softwares to solve the linear programming (9.4a)–(9.4d).

- **Case 2:** $p = 2, q = 1$ (MCVQP)

In this case, $\|\xi\|_2^2 = \sum_{i=1}^{l} \xi_i^2$ and $\|\beta\|_1 = \sum_{i=1}^{l} \beta_i$. Objective function in (9.2a)–(9.2d) can now be a convex quadratic objective

$$\min \ \sigma_\xi \sum_{i=1}^{l} \xi_i^2 - \sigma_\beta \sum_{i=1}^{l} \beta_i. \tag{9.5}$$

(9.2a)–(9.2d) turns to be a convex quadratic programming

$$\min \ \sigma_\xi \sum_{i=1}^{l} \xi_i^2 - \sigma_\beta \sum_{i=1}^{l} \beta_i \tag{9.6a}$$

$$\text{s.t.} \quad (x_i \cdot w) - b - \xi_i + \beta_i = 0, \quad \forall y_i = -1; \tag{9.6b}$$

$$(x_i \cdot w) - b + \xi_i - \beta_i = 0, \quad \forall y_i = 1; \tag{9.6c}$$

$$\xi, \beta \geq 0. \tag{9.6d}$$

It is well known that convex quadratic programming can be solved easily. We use the method developed by [166] to solve problem (9.6a)–(9.6d).

- **Case 3:** $p = 1, q = 2$ (MCCQP)

  In this case, $\|\xi\|_1 = \sum_{i=1}^{l} \xi_i$ and $\|\beta\|_2^2 = \sum_{i=1}^{l} \beta_i^2$. Objective function in (9.2a)–(9.2d) can now be a concave quadratic objective

$$\min \; \sigma_\xi \sum_{i=1}^{l} \xi_i - \sigma_\beta \sum_{i=1}^{l} \beta_i^2. \tag{9.7}$$

(9.2a)–(9.2d) turns to be a concave quadratic programming

$$\min \; \sigma_\xi \sum_{i=1}^{l} \xi_i - \sigma_\beta \sum_{i=1}^{l} \beta_i^2 \tag{9.8a}$$

$$\text{s.t.} \quad (x_i \cdot w) - b - \xi_i + \beta_i = 0, \quad \forall y_i = -1; \tag{9.8b}$$

$$(x_i \cdot w) - b + \xi_i - \beta_i = 0, \quad \forall y_i = 1; \tag{9.8c}$$

$$\xi, \beta \geq 0. \tag{9.8d}$$

Concave quadratic programming is an NP-hard problem. It is very difficult to find the global minimizer, especially for large problem. In order to solve (9.8a)–(9.8d) efficiently, we propose an algorithm, which converges to a local minimizer of (9.8a)–(9.8d).

In order to describe the algorithm in detail, we introduce some notation. Let $\omega = (w, \xi, \beta, b)$, $f(\omega) = \sigma_\xi \sum_{i=1}^{l} \xi_i - \sigma_\beta \sum_{i=1}^{l} \beta_i^2$, and

$$\Omega = \begin{cases} (w, \xi, \beta, b) : (x_i \cdot w) - b - \xi_i + \beta_i = 0, & \forall y_i = -1, \\ (x_i \cdot w) - b + \xi_i - \beta_i = 0, & \forall y_i = 1, \\ \xi_i \geq 0, \quad \beta_i \geq 0, \quad i = 1, \ldots, n \end{cases} \tag{9.9}$$

be the feasible region of (9.2a)–(9.2d).

Let $\chi_\Omega(\omega)$ be the index function of set Ω, i.e., $\chi_\Omega(\omega)$ is defined as follows

$$\chi_\Omega(\omega) = \begin{cases} 0, & \omega \in \Omega, \\ +\infty, & \omega \notin \Omega. \end{cases} \tag{9.10}$$

Then (9.8a)–(9.8d) is equivalent to the following problem

$$\min \; f(\omega) + \chi_\Omega(\omega). \tag{9.11}$$

Rewrite $f(\omega) + \chi_\Omega(\omega)$ as the following form

$$f(\omega) + \chi_\Omega(\omega) = g(\omega) - h(\omega) \tag{9.12}$$

where $g(\omega) = \frac{1}{2}\rho\|\omega\|^2 + \sigma_\xi \sum_{i=1}^{l} \xi_i + \chi_\Omega(\omega)$, $h(\omega) = \frac{1}{2}\rho\|\omega\|^2 + \sigma_\beta \sum_{i=1}^{l} \beta_i^2$ and $\rho > 0$ is a small positive number. Then $g(\omega)$ and $h(\omega)$ are convex functions. By applying the simplified DC algorithm in [5] to problem (9.11), we get the following algorithm

Algorithm 9.1 Given an initial point $\omega^0 \in R^{3l+1}$ and a parameter $\varepsilon > 0$, at each iteration $k \geq 1$, compute ω^{k+1} by solving the convex quadratic programming

$$\min_{\omega \in \Omega} \frac{1}{2}\rho\|\omega\|^2 + \sigma_\xi \sum_{i=1}^{l} \xi_i - (h'(\omega^k) \cdot \omega). \tag{9.13}$$

The stopping criterion is $\|\omega^{k+1} - \omega^k\| \leq \varepsilon$.

By standard arguments, we can prove the following theorem.

Theorem 9.2 *After finite number of iterations, Algorithm 9.1 terminates at a local minimizer of* (9.8a)–(9.8d).

• **Case 4:** $p = q = 2$ (MCQP)

In this case, $\|\xi\| = \sum_{i=1}^{l} \xi_i^2$ and $\|\beta\| = \sum_{i=1}^{l} \beta_i^2$ the objective function in (9.2a)–(9.2d) can now be an indefinite quadratic function

$$\min \sigma_\xi \sum_{i=1}^{l} \xi_i^2 - \sigma_\beta \sum_{i=1}^{l} \beta_i^2. \tag{9.14}$$

(9.2a)–(9.2d) turns to be an indefinite quadratic programming

$$\min \sigma_\xi \sum_{i=1}^{l} \xi_i^2 - \sigma_\beta \sum_{i=1}^{l} \beta_i^2 \tag{9.15a}$$

$$\text{s.t.}\quad (x_i \cdot w) - b - \xi_i + \beta_i = 0, \quad \forall y_i = -1; \tag{9.15b}$$

$$(x_i \cdot w) - b + \xi_i - \beta_i = 0, \quad \forall y_i = 1; \tag{9.15c}$$

$$\xi, \beta \geq 0. \tag{9.15d}$$

This is an indefinite quadratic programming, which is an NP-hard problem. So it is very difficult to find the global minimizer, especially for large problem. However, we note that this problem is a DC programming. Here we propose an algorithm for this model based on DC programming, which converges to a local minimizer of (9.15a)–(9.15d). Let $f_1(\omega) = \sigma_\xi \sum_{i=1}^{l} \xi_i^2 - \sigma_\beta \sum_{i=1}^{l} \beta_i^2$. Then (9.15a)–(9.15d) is equivalent to the following problem

$$\min f_1(\omega) + \chi_\Omega(\omega). \tag{9.16}$$

Rewrite $f_1(\omega) + \chi_\Omega(\omega)$ as the following form

$$f_1(\omega) + \chi_\Omega(\omega) = g_1(\omega) - h_1(\omega), \tag{9.17}$$

where $g(\omega) = \frac{1}{2}\rho\|\omega\|^2 + \sigma_\xi \sum_{i=1}^l \xi_i^2 + \chi_\Omega(\omega)$, $h(\omega) = \frac{1}{2}\rho\|\omega\|^2 + \sigma_\beta \sum_{i=1}^l \beta_i^2$ and $\rho > 0$ is a small positive number. Then $g_1(\omega)$ and $h_1(\omega)$ are convex functions. By apply the simplified DC algorithm in [5] to problem (9.16), we get the following algorithm

Theorem 9.3 *Given an initial point $\omega^0 \in R^{3n+1}$ and a parameter $\varepsilon > 0$, at each iteration $k \geq 1$, compute ω^{k+1} by solving the convex quadratic programming*

$$\min_{\omega \in \Omega} \frac{1}{2}\rho\|\omega\|^2 + \sigma_\xi \sum_{i=1}^n \xi_i^2 - (h_1'(\omega^k) \cdot \omega). \qquad (9.18)$$

The stopping criterion is $\|\omega^{k+1} - \omega^k\| \leq \varepsilon$.

From (Le Thi Hoai An and Pham Dinh Tao, 1997), we have the following theorem.

Theorem 9.4 *The sequence $\{\omega^k\}$ generated by Algorithm 9.1 converges to a local minimizer of* (9.15a)–(9.15d).

Remark 9.5 The purpose of this section is to propose a general optimization-based framework which unifies some existed optimization-based classification methods and to obtain some new models based on this general framework and to test the efficiency of these models by using some real problems. So we adopt some existed algorithms to solve these models. To propose some new and efficient methods for these models based on the structure of these models need to be studied further.

Remark 9.6 There are many deterministic and heuristic algorithms for global optimization, e.g., branch and bound, plane cutting, genetic algorithm, simulated annealing etc. Please see [160] and [112]. Here we adopt the simplified DC algorithm proposed by [5] to solve (9.8a)–(9.8d) and (9.15a)–(9.15d). The reasons are as follows: (i) The objective functions in (9.8a)–(9.8d) and (9.15a)–(9.15d) are DC functions. This algorithm exploits such structure; (ii) the algorithm is simple and implemented easily; (iii) as the authors indicated, the algorithm always converges to a global optimizer in numerical experiments.

9.2 Multi-criteria Convex Quadratic Programming Model

This section introduces a new MCQP model. Suppose we want to classify data in r-dimensional real space R^r into two distinct groups via a hyperplane defined by a kernel function and multiple criteria. Thus, we can establish the following linear inequalities for a linear separable dataset:

$$(x_i \cdot w) < b, \quad \forall y_i = -1; \qquad (9.19)$$

$$(x_i \cdot w) \geq b, \quad \forall y_i = 1. \tag{9.20}$$

To formulate the criteria and constraints for data separation, some variables need to be introduced. In the classification problem, $(x_i \cdot w)$ is the score for the ith data record. If all records are linear separable and an element x_i is correctly classified, then let $\frac{\beta_i}{\|w\|}$ be the distance from x_i to the hyperplane $(x \cdot w) = b$, and consider the linear system, $(x_i \cdot w) = b - \beta_i, \forall y_i = -1$ and $(x_i \cdot w) = b + \beta_i, \forall y_i = 1$. However, if we consider the case where the two groups are not linear separable because of mislabeled records, a "Soft Margin" and slack distance variable ξ_i need to be introduced. Previous equations now transforms to $(x_i \cdot w) = b + \xi_i - \beta_i, \forall y_i = -1$ and $(x_i \cdot w) = b - \xi_i + \beta_i, \forall y_i = 1$. To complete the definitions of β_i and ξ_i, let $\beta_i = 0$ for all misclassified elements and ξ_i equals to zero for all correctly classified elements. Incorporating the definitions of β_i and ξ_i, (9.19) and (9.20) can be reformulated as three different models (Medium, Strong and Weak).

- **Medium model**: $(x_i \cdot w) = b + \xi_i - \beta_i, \forall y_i = -1$ and $(x_i \cdot w) = b - \xi_i + \beta_i,$ $\forall y_i = 1$.
- **Strong model**: $(x_i \cdot w) = b - \delta + \xi_i - \beta_i, \forall y_i = -1$ and $(x_i \cdot w) = b + \delta - \xi_i + \beta_i,$ $y_i = 1, \delta > 0$ is a given scalar. $b - \delta$ and $b + \delta$ are two adjusted hyperplanes for strong model.
- **Weak model**: $(x_i \cdot w) = b + \delta + \xi_i - \beta_i, \forall y_i = 1$ and $(x_i \cdot w) = b - \delta - \xi_i + \beta_i,$ $y_i = 1, \delta > 0$ is a given scalar. $b + \delta$ and $b - \delta$ are two adjusted hyperplanes for weak model.

Denote the class which corresponding label $y_i = -1$ as G_1 and the class which corresponding label $y_i = -1$ as G_2. A loosing relationship of above models is given as:

Theorem 9.7

(i) *A feasible solution of Strong model is the feasible solution of Medium model and Weak model.*
(ii) *A feasible solution of Medium model is the feasible solution of Weak model.*
(iii) *If a data case x_i is classified in a given class G_j by Strong model, then it may be in G_j by using Medium model and Weak model.*
(iv) *If a data case x_i is classified in a given class G_j by model Medium model, then it may be in G_j by using Weak model.*

Proof Let $Feas_1$ = feasible area of Strong model, $Feas_2$ = feasible area of Medium model, and $Feas_3$ = feasible area of Weak model. $\forall w_s^* \in Feas_1$, $w_s^* \in Feas_2$ and $w_s^* \in Feas_3$. $\forall w_m^* \in Feas_2$, $w_m^* \in Feas_3$. Obviously, we will have $Feas_1 \subseteq Feas_2 \subseteq Feas_3$, so (i) and (ii) is true.

(iii) and (iv) are automatically true from the conclusion of (i) and (ii) for a certain value of $\delta > 0$. $\qquad\square$

Redefine w as w/δ, b as b/δ, ξ_i as ξ_i/δ, β_i as β/δ, and introduce

$$\delta' = \begin{cases} 1, & \text{Strong model,} \\ 0, & \text{Medium model,} \\ -1, & \text{Weak model.} \end{cases} \tag{9.21}$$

Define an $n \times n$ diagonal matrix Y which only contains "+1" or "−1" indicates the class membership. A "−1" in row i of matrix Y indicates the corresponding record $x_i \in G_1$ and a "+1" in row i of matrix Y indicates the corresponding record $x_i \in G_2$. The above three models can be rewritten as a single constraint:

$$Y(X^T w - eb) = \delta' + \xi - \beta, \tag{9.22}$$

where $e = (1, 1, \dots, 1)^T$, $\xi = (\xi_1, \dots, \xi_l)^T$ and $\beta = (\beta_1, \dots, \beta_l)^T$.

The proposed multi-criteria optimization problem contains three objective functions. The first mathematical function $f(\xi) = \|\xi\|_p^p = \sum_{i=1}^{l} |\xi_i|^p$ $(1 \le p \le \infty)$ describes the summation of total overlapping distance of misclassified records to the hyperplane $(x \cdot w) = b$. The second function $g(\beta) = \|\beta\|_q^q = \sum_{i=1}^{l} |\beta_i|^q$ $(1 \le q \le \infty)$ represents the aggregation of total distance of correctly separated records to the hyperplane $(x \cdot w) = b$. The distance between the two adjusted bounding hyperplanes is defined as $\frac{2}{\|w\|_s^s}$ in geometric view. In order to maximize this distance, a third function $h(w) = \frac{\|w\|_s^s}{2}$ should be minimized. The final accuracy of this classification problem depends on simultaneously minimize $f(\xi)$, minimize $h(w)$ and maximize $g(\beta)$. Thus, an extended Multi-criteria programming model for classification can be formulated as:

$$\min f(\xi), \quad \min h(w) \quad \text{and} \quad \max g(\beta) \tag{9.23a}$$

$$\text{s.t.} \quad Y(X^T w - eb) = \delta' + \xi - \beta, \tag{9.23b}$$

$$\xi, \beta \ge 0. \tag{9.23c}$$

where Y is a given $l \times l$ diagonal matrix, $e = (1, 1, \dots, 1)^T$, $\xi = (\xi_1, \dots, \xi_l)^T$, $\beta = (\beta_1, \dots, \beta_l)^T$, w and b are unrestricted.

Figure 9.1 shows the geometric view of our model. Squares indicate group "+" and dots represent group "−". In order to separate groups via the hyperplane $(x \cdot w) = b$, we set up two Max and one Min objectives. Max objective maximizes the sum of distance from the points to their boundary and the distance between the two adjusted boundaries. Min objective minimizes the total overlapping.

Furthermore, to transform the Multi-criteria classification model into a single-criterion problem, a weight vector $(\sigma_\xi, \sigma_\beta)$, weights $\sigma_\xi + \sigma_\beta = 1$, $\sigma_\xi > 0$ and $\sigma_\beta > 0$, is introduced for $f(\xi)$ and $g(\beta)$, respectively. The values of σ_ξ and σ_β can be arbitrary pre-defined in the process of identifying the optimal solution to indicate the relative importance of the three objectives. We assume $\sigma_\beta < \sigma_\xi$ considering minimizing misclassification rates has higher priority than maximizing distance of

Fig. 9.1 A two-class extended model

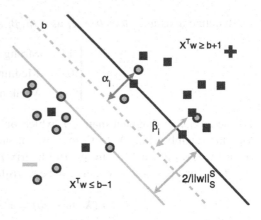

correctly separated records to the boundary in classification problems. The generalized model can be converted into a single-criterion mathematical programming model:

$$\min \frac{1}{2}\|w\|_s^s + \sigma_\xi \|\xi\|_p^p - \sigma_\beta \|\beta\|_q^q \tag{9.24a}$$

$$\text{s.t.}\quad Y(X^Tw - eb) = \delta'e - \xi + \beta, \tag{9.24b}$$

$$\xi, \beta \geq 0, \tag{9.24c}$$

where Y is a given $l \times l$ diagonal matrix, $e = (1, 1, \ldots, 1)^T$, $\xi = (\xi_1, \ldots, \xi_l)^T$, $\beta = (\beta_1, \ldots, \beta_l)^T$, w and b are unrestricted.

Please note that the introduction of β_i is one of the major differences between the proposed model and other existing Support Vectors approaches [206].

Without losing generality, let $s = 2$, $q = 1$ and $p = 2$ and make (9.24a)–(9.24c) a convex quadratic programming form, since it is much easier to find optimal solutions for convex quadratic programming form than any other forms of nonlinear programming. The constraints remain the same and the objective function becomes:

$$\min \frac{1}{2}\|w\|_2^2 + \sigma_\xi \sum_{i=1}^{l} \xi_i^2 - \sigma_\beta \sum_{i=1}^{l} \beta_i \tag{9.25a}$$

$$\text{s.t.}\quad Y(X^Tw - eb) = \delta'e - \xi + \beta, \tag{9.25b}$$

$$\xi, \beta \geq 0. \tag{9.25c}$$

Let $\eta_i = \xi_i - \beta_i$. According to our definition, $\eta_i = \xi_i$ for all misclassified records and $\eta_i = -\beta_i$ for all correctly separated records.

Add $\frac{\sigma_b}{2}b^2$ to the objective function of problem (9.25a)–(9.25c) and the weight σ_b is an arbitrary positive number and let $\sigma_b \ll \sigma_\beta$. Previous computation results shows that this change won't affect the optimal solution and adds strong convexity to the objection function. (9.25a)–(9.25c) becomes:

$$\min \frac{1}{2}\|w\|_2^2 + \frac{\sigma_\xi}{2}\sum_{i=1}^{l}\eta_i^2 + \sigma_\beta\sum_{i=1}^{l}\eta_i + \frac{\sigma_b}{2}b^2 \tag{9.26a}$$

$$\text{s.t.}\quad Y(X^T w - eb) = \delta'e - \eta, \tag{9.26b}$$

where Y is a given $n \times n$ diagonal matrix, $e = (1, 1, \ldots, 1)^T$, $\eta = (\eta_1, \ldots, \eta_l)^T$, η, w and b are unrestricted, $1 \le i \le l$.

The Lagrange function corresponding to (9.26a)–(9.26b) is

$$L(w, b, \eta, \theta) = \frac{1}{2}\|w\|_2^2 + \frac{\sigma_\xi}{2}\sum_{i=1}^{l}\eta_i^2 + \sigma_\beta\sum_{i=1}^{l}\eta_i + \frac{\sigma_b}{2}b^2$$
$$- \theta^T(Y(X^T w - eb) - e\delta' + \eta), \tag{9.27}$$

where $\theta = (\theta_1, \ldots, \theta_n)^T$, $\eta = (\eta_1, \ldots, \eta_n)^T$, $\theta_i, \eta_i \in \Re$.

According to Wolfe Dual Theorem,

$$\nabla_w L(w, b, \eta, \theta) = w - X^T Y\theta = 0, \tag{9.28a}$$

$$\nabla_b L(w, b, \eta, \theta) = \sigma_b b + e^T Y\theta = 0, \tag{9.28b}$$

$$\nabla_\eta L(w, b, \eta, \theta) = \sigma_\xi \eta + \sigma_\beta e - \theta = 0. \tag{9.28c}$$

Introduce the above three equations to the constraints of (9.26a)–(9.26b), we can get:

$$Y\left(XX^T Y\theta + \frac{1}{\sigma_b}e(e^T Y\theta)\right) + \frac{1}{\sigma_\xi}(\theta - \sigma_\beta e) = \delta'e \tag{9.29}$$

$$\Rightarrow \theta = \frac{(\delta' + \frac{\sigma_\beta}{\sigma_\xi})e}{\frac{I}{\sigma_\xi} + Y(XX^T + \frac{1}{\sigma_b}ee^T)Y}. \tag{9.30}$$

Proposition 9.8 *For some σ_ξ,*

$$\theta = \frac{(\delta' + \frac{\sigma_\beta}{\sigma_\xi})e}{\frac{I}{\sigma_\xi} + Y(XX^T + \frac{1}{\sigma_b}ee^T)Y}$$

exists.

Proof Let $H = Y[X - (\frac{1}{\sigma_b})^{\frac{1}{2}}e]$, we thus see that:

$$\theta = \frac{(\delta' + \frac{\sigma_\beta}{\sigma_\xi})e}{\frac{I}{\sigma_\xi} + HH^T} \tag{9.31}$$

$\forall H, \exists \sigma_\xi \geq \varepsilon > 0$, and $\forall i, j \in [1, \text{size of}(HH^T)]$, we may have $(\frac{I}{\sigma_\xi})_{i,i} > |(HH^T)_{i,i}|$ and the inversion of $(\frac{I}{\sigma_\xi} + HH^T)$ exists. So

$$\theta = \frac{(\delta' + \frac{\sigma_\beta}{\sigma_\xi})e}{\frac{I}{\sigma_\xi} + Y(XX^T + \frac{1}{\sigma_b}ee^T)Y}$$

exists.   □

The computation of optimal solution θ^* in (9.30) involves an inversion of a possibly massive matrix of order $(l+1) \times (l+1)$. When the size of dataset is moderate (e.g. $l \leq 100000$), the proposed solution, (9.30), is not difficult to compute. When both n ($n > 100000$) and l ($l > 100000$) are large, it is hard to get a solution. When the size of dataset is large (e.g. $l > 100000$) and the dimension is not high ($n \leq 100000$), we may use Sherman-Morrison formula $((A + UV^T)^{-1} = A^{-1} - \frac{A^{-1}UV^TA^{-1}}{1+V^TA^{-1}U})$ for matrix inversion and get:

$$\theta = \frac{(\delta' + \frac{\sigma_\beta}{\sigma_\xi})e}{\frac{I}{\sigma_\xi} + HH^T} = \sigma_\xi \left(\delta' + \frac{\sigma_\beta}{\sigma_\xi}\right)\left(I - H\frac{\sigma_\xi}{I + \sigma_\xi H^T H}H^T\right)e. \tag{9.32}$$

The expression in (9.32) only needs the inversion of an $(n+1) \times (n+1)$ matrix instead of an $(l+1) \times (l+1)$ matrix in (9.31), which is beneficial since normally $n \ll l$ in large-scale problem (for example, a typical problem may have $n = O(10^2)$ and $l = O(10^6)$).

Algorithm 9.9

Input: An $l \times n$ matrix X as the training dataset, an $l \times l$ diagonal matrix Y labels the class of each record.
Output: Classification accuracies for each group in the training dataset, score for every record, decision function

$$((x_i \cdot w^*) - b^*) \begin{cases} > 0 & \Rightarrow & x_i \in G_1, \\ \leq 0 & \Rightarrow & x_i \in G_2. \end{cases} \tag{9.33}$$

Step 1 Compute $\theta^* = (\theta_1, \ldots, \theta_l)^T$ by (9.32). $\sigma_\beta, \sigma_\xi, \sigma_b$ are chosen by cross validation.
Step 2 Compute $W^* = X^T Y \theta^*$, $b^* = -\frac{1}{\sigma_b}e^T Y \theta^*$.
Step 3 Classify a incoming x_i by using decision function

$$((x_i \cdot w^*) - b^*) \begin{cases} > 0 & \Rightarrow & x_i \in G_1, \\ \leq 0 & \Rightarrow & x_i \in G_2. \end{cases} \tag{9.34}$$

END

We may simplify (9.26a)–(9.26b) by setting $b = b_1$.

$$\min \ \frac{1}{2}\|w\|_2^2 + \frac{\sigma_\xi}{2}\sum_{i=1}^{l}\eta_i^2 + \sigma_\beta \sum_{i=1}^{l}\eta_i \tag{9.35a}$$

$$\text{s.t.} \quad Y(X^T w - eb_1) = \delta' e - \eta, \tag{9.35b}$$

where Y is a given $l \times l$ diagonal matrix, $e = (1, 1, \ldots, 1)^T$, $\eta = (\eta_1, \ldots, \eta_l)^T$, η, w are unrestricted, b_1 is a given scalar, $1 \le i \le l$. The Lagrange function corresponding to (9.35a)–(9.35b) is

$$L(w, \eta, \theta) = \frac{1}{2}\|w\|_2^2 + \frac{\sigma_\xi}{2}\sum_{i=1}^{l}\eta_i^2 + \sigma_\beta \sum_{i=1}^{l}\eta_i - \theta^T(Y(X^T w - eb_1) - e\delta' + \eta), \tag{9.36}$$

where $\theta = (\theta_1, \ldots, \theta_l)^T$, $\eta = (\eta_1, \ldots, \eta_l)^T$, $\theta_i, \eta_i \in R$.

According to Wolfe Dual Theorem,

$$\nabla_w L(w, b, \theta) = w - X^T Y\theta = 0, \tag{9.37}$$

$$\nabla_\eta L(w, b, \eta, \theta) = \sigma_\xi \eta + \sigma_\beta e - \theta = 0. \tag{9.38}$$

Introduce the above two equations to the constraints of (9.35a)–(9.35b), we can get:

$$Y((X \cdot X^T)Y\theta - eb_1) + \frac{1}{\sigma_\xi}(\theta - \sigma_\beta) = \delta' e \tag{9.39}$$

$$\Rightarrow \quad \theta = \frac{(Yb_1 + \delta' + \frac{\sigma_\beta}{\sigma_\xi})e}{\frac{I}{\sigma_\xi} + Y(X \cdot X^T)Y}. \tag{9.40}$$

Use Sherman-Morrison formula again for matrix inversion. Let $J = YW$, we get:

$$\theta = \frac{(Yb_1 + \delta' + \frac{\sigma_\beta}{\sigma_\alpha})e}{\frac{I}{\sigma_\xi} + J \cdot J^T} = \sigma_\xi \left(Yb_1 + \delta' + \frac{\sigma_\beta}{\sigma_\alpha}\right)\left(I - J\frac{\sigma_\xi}{I + \sigma_\beta J^T J}J^T\right)e. \tag{9.41}$$

9.3 Kernel Based MCQP

In order to find way to create non-linear classifiers, we may transform the original input space to high dimension feature space by a non-linear transformation. Thus, a non-linear hyperplane in original space may be linear in the high-dimensional feature space. For example, every inner product (x_i, x_j) in (9.26a)–(9.26b) can be replaced by a non-linear kernel function $K(x_i, x_j)$, which will extend the applicability of the proposed model to linear inseparable datasets. However, there are some difficulties to directly introduce kernel function to (9.26a)–(9.26b). Let $K(X, X^T) = \Phi(X)\Phi(X^T)$ be a kernel function. If we only replace XX^T with

$K(X, X^{\mathrm{T}})$, $w^* = \Phi(w^{\mathrm{T}})Y\theta^*$ need to be computed in order to get the decision function. However, the computation of $\Phi(X)$ or $\Phi(X^{\mathrm{T}})$ is almost impossible. In order to get ride of this problem, we replace w by $X^{\mathrm{T}}Y\theta$ and XX^{T} with $K(X, X^{\mathrm{T}})$ in model (9.26a)–(9.26b). Thus, the objective function in (9.42a)–(9.42b) is changed without influence of the optimal solution since σ_β, σ_ξ, σ_b are changeable.

$$\min \frac{1}{2}\|\theta\|_2^2 + \frac{\sigma_\xi}{2}\sum_{i=1}^{l}\eta_i^2 + \sigma_\beta\sum_{i=1}^{l}\eta_i + \frac{\sigma_b}{2}b^2 \tag{9.42a}$$

$$\text{s.t.} \quad Y(K(X, X^{\mathrm{T}})Y\theta - eb) = \delta'e - \eta. \tag{9.42b}$$

The Lagrange function corresponding to (9.42a)–(9.42b) is

$$L(\theta, b, \eta, \rho) = \frac{1}{2}\|\theta\|_2^2 + \frac{\sigma_\xi}{2}\sum_{i=1}^{l}\eta_i^2 + \sigma_\beta\sum_{i=1}^{l}\eta_i + \frac{\sigma_b}{2}b^2$$

$$- \rho^{\mathrm{T}}(Y(K(X, X^{\mathrm{T}})Y\theta - eb) - e\delta' + \eta), \tag{9.43}$$

where $\theta = (\theta_1, \ldots, \theta_l)^{\mathrm{T}}$, $\eta = (\eta_1, \ldots, \eta_l)^{\mathrm{T}}$, $\rho = (\rho_1, \ldots, \rho_l)^{\mathrm{T}}$, $\theta_i, \eta_i, \rho_i \in R$.

$$\nabla_\theta L(\theta, b, \eta, \rho) = \theta - YK(X, X^{\mathrm{T}})^{\mathrm{T}}Y\rho = 0, \tag{9.44a}$$

$$\nabla_b L(\theta, b, \eta, \rho) = \sigma_b b + e^{\mathrm{T}}Y\rho = 0, \tag{9.44b}$$

$$\nabla_\eta L(\theta, b, \eta, \rho) = \sigma_\xi \eta + \sigma_\beta e - \rho = 0. \tag{9.44c}$$

Introduce the above three equations to the constraints of (9.42a)–(9.42b), we can get:

$$Y(K(X, X^{\mathrm{T}})YYK(X, X^{\mathrm{T}})^{\mathrm{T}}Y\rho + \frac{1}{\sigma_b}e(e^{\mathrm{T}}Y\rho)) = \delta'e - \frac{1}{\sigma_\xi}(\rho - \sigma_\beta e) \tag{9.45}$$

$$\Rightarrow \quad \rho = \frac{(\delta' + \frac{\sigma_\beta}{\sigma_\xi})e}{\frac{I}{\sigma_\xi} + Y(K(X, X^{\mathrm{T}})K(X, X^{\mathrm{T}})^{\mathrm{T}} + \frac{1}{\sigma_b}ee^{\mathrm{T}})Y}. \tag{9.46}$$

Algorithm 9.10

Input: An $l \times n$ matrix X as the training dataset, an $l \times l$ diagonal matrix Y labels the class of each record, a kernel function $K(x_i, x_j)$.

Output: Classification accuracies for each group in the training dataset, score for every record, decision function.

Step 1 Compute $\rho^* = (\rho_1, \ldots, \rho_l)^{\mathrm{T}}$ by (9.46). σ_β, σ_ξ, σ_b are chosen by cross validation.

Step 2 Classify a incoming x_i by using decision function

$$\left(K(X_i, X^{\mathrm{T}})K(X, X^{\mathrm{T}})^{\mathrm{T}} + \frac{1}{\sigma_b}e^{\mathrm{T}}\right)Y\rho^* \begin{cases} > 0 & \Rightarrow \quad x_i \in G_1, \\ \leq 0 & \Rightarrow \quad x_i \in G_2. \end{cases} \tag{9.47}$$

END

However, the above model only gives classification results without any further explanation. Recently, lots of researchers focus on how to learn kernel functions from the data. Based on [134], we can prove the best kernel function always can be presented as the convex combination of several finite element basis kernel functions. In fact, references [12, 188] already provided some optimal kernel functions which are the convex combination of basis kernels. In practice, one the simple and effective choice is multiple kernel, which is a linear combination of several kernels. In detail,

$$K(x_i, x_j) = \sum_{d=1}^{r} \gamma_d K(x_{i,d}, x_{j,d}), \quad \gamma_d \geq 0, \tag{9.48}$$

where $K(\cdot, \cdot)$ is the pre-given basis kernel, $x_{i,d}$ represents the dth dimension of output vector x_i and γ_d is the feature coefficient of the dth dimension, $1 \leq d \leq r$. Using this expression, we can convert the feature selection problem into a parameter optimization problem. Kernel parameter $\gamma_d \neq 0$ means the corresponding feature classifier is chosen. Kernel parameter $\gamma_d = 0$ means the corresponding feature classifier does not contribute to the final classifier.

Similar to the analysis of (9.46), we use (9.48) to get the following multiple kernel solution:

$$\theta = \frac{[\delta' + \frac{\sigma_\beta}{\sigma_\xi}]e}{\frac{I}{\sigma_\xi} + Y[ee^T \sum_{d=1}^{r} \gamma_d K(x_{i,d}, x_{j,d}) + \frac{1}{\sigma_b} ee^T]Y}. \tag{9.49}$$

To make the model more explainable and provide more solid grounds for the decision makers to understand the classification results easier and make more reasonable decision, we need to choose feature for the model. We know L_1 norm possess the best sparsity. So the optimization problem for feature selection is constructed based on the idea of minimizing the L_1 norm of total error, which is used when derive the classic SVMs [193]. Therefore, feature selection is converted to a problem finding the best value of coefficient γ. After optimization, those features whose coefficients do not equal to zero are selected. In all, two stages are needed for training the final classifier. Firstly, feature coefficient γ is fixed, we solve problem (9.46) to get Lagrangian parameter θ_i. Then the Lagrangian parameter is fixed in the following LP to solve coefficient γ_d:

$$\min \ J(\gamma, \eta) = \sum_{i=1}^{l} \eta_i^2 + \lambda \|\gamma\|_1 \tag{9.50}$$

$$\text{s.t.} \ Y\left(e \sum_{j,d} \gamma_{j,d} \theta_j y_j K(w_{i,d}, w_{j,d}) + eb\right) \geq e - \eta, \tag{9.51}$$

where $\lambda > 0$ in (9.50) is a regularized parameter, whose role is to control the sparsity of feature subsets. Coefficient γ provides us lots of opportunity to find a satisfiable feature subsets in the whole feature space. More importantly, we can find the optimal feature subsets automatically by solving the above model and give an explainable results at the same time.

Use formula (9.42a)–(9.42b) to eliminate η in (9.50), we can get the following quadratic programming:

$$
\min \ J(\gamma, \eta) = \sum_{i=1}^{l} \left(1 + y_i b - y_i \sum_{d=1}^{r} \beta_d K_{i,d} \circ \theta \right)^2 + \lambda \sum_{d=1}^{r} \gamma_d
$$

$$
= \sum_{d=1}^{m} \sum_{\bar{d}=1}^{m} \gamma_d \gamma_{\bar{d}} \left(\sum_{i=1}^{l} y_i^2 (K_{i,d} \circ \theta)(K_{i,\bar{d}} \circ \theta) \right)^2
$$

$$
+ \sum_{d=1}^{r} \gamma_d \left(\lambda - 2 \sum_{i=1}^{n} (1 + y_i b) y_i K_{i,d} \circ \theta \right) + \sum_{i=1}^{n} (1 + y_i b)^2, \quad (9.52)
$$

where $K_{i,d} = [y_1 K(w_{i,d}, w_{1,d}), \ldots, y_j K(w_{i,d}, w_{j,d}), \ldots, y_n K(w_{i,d}, w_{n,d})]$.

We call the process of introducing multiple kernel to convert the feature selection problem into a kernel parameter learning problem as multiple kernel multi-objective classification. There are two coefficient need to be optimized: Lagrangian multiplier and feature parameter. This is the main difference between this method and MCCQP. From another point of view, the multiple kernel method for feature selection also can be viewed as a filter method. The classification results become more explainable by feature selection,which makes us find the important features exactly. At the same time, the complexity of the problem is reduced and the computational time is saved. We give the details of the algorithm in the following:

Algorithm 9.11

Step 1. Initialization. Give the initial values for model parameters σ_ξ, σ_β, σ_b, regularized parameter λ and kernel parameter σ^2. Set feature coefficient β_d as $\{\beta_d^{(0)} = 1 \mid d = 1, \ldots, r\}$ and $t = 1$.

Step 2. Compute Lagrangian coefficient $\theta^{(t)}$. Set the value of γ in (9.49) to $\gamma^{(t-1)}$, computer Lagrangian coefficient $\theta^{(t)}$.

Step 3. Compute feature coefficient $\gamma^{(t)}$. Used $\theta_j^{(t)}$ as the value of θ in model (9.50)–(9.51), solve the quadratic programming to get the feature coefficient $\gamma^{(t)}$.

Step 4. Termination test. Use $\theta_j^{(t)}$ and $\gamma^{(t)}$ computing the pre-given indicator. If the result does not converge, set $t := t + 1$ and go back to Step 2 redo the two stage optimization. If the result is not satisfied enough, set $t := t + 1$ and go back to the Step 1 to adjust the value of regularized parameter.

Step 5. Output the classification result. Compute b^*. For an input sample x, we get the classification results by the computation results of the following decision function:

$$
\text{sign}\left(\sum_{i=1}^{l} \sum_{d=1}^{r} \gamma_d^{(t)} \theta_i^{(t)} K(x_{i,d}, x_d) - b^* \right) \begin{cases} > 0, & x \in G_1, \\ < 0, & x \in G_2. \end{cases} \quad (9.53)
$$

Chapter 10
Non-additive MCLP

10.1 Non-additive Measures and Integrals

In order to model the interactions among attributes for classification, the non-additive measures are studied in this chapter. The non-additive measures provide very important information regarding the interactions among attributes and potentially useful for data mining [209, 210]. The concept of non-additive measures (also referred to as fuzzy measure theory) was initiated in the 1950s and have been well developed since 1970s [40, 55, 208]. Non-additive measures have been successfully used as a data aggregation tool for many applications such as information fusion, multiple regressions and classifications [92, 139, 208, 210]. The nonlinear integrals are the aggregation tools for the non-additive measures. The Choquet integral [40], a nonlinear integral, is utilized to aggregate the feature attributes with respect to the non-additive measure. Let finite set $X = \{x_1, \ldots, x_n\}$ denote the attributes in a multidimensional dataset. Several important non-additive measures are defined as the following definitions [93, 209]:

Definition 10.1 A *generalized non-additive measure* μ defined on X is a set function $\mu : \mathcal{P}(X) \to [0, \infty)$ satisfying

$$\mu(\emptyset) = 0. \tag{10.1}$$

μ is a *monotone non-additive measure* if it satisfies (10.1) and

$$\mu(E) \leq \mu(F) \quad \text{if } E \subseteq F \quad \text{(monotonicity)} \tag{10.2}$$

where $\mathcal{P}(X)$ denotes the power set of X and E, F are the elements in $\mathcal{P}(X)$.

Definition 10.2 A generalized non-additive measure is said to be *regular* if $\mu(X) = 1$.

Definition 10.3 A *signed non-additive measure* μ defined on X is a set function $\mu : \mathcal{P}(X) \to (-\infty, \infty)$ satisfying (10.1).

Y. Shi et al., *Optimization Based Data Mining: Theory and Applications*,
Advanced Information and Knowledge Processing,
DOI 10.1007/978-0-85729-504-0_10, © Springer-Verlag London Limited 2011

Nonlinear integrals are the aggregation tool for the non-additive measures μ_i in the set function μ. The studies of nonlinear integrals could be found in the literature [55, 209] from the integral of linearly weighted sum of the attributes (refers to as Lebesgue-like integral [209]) to nonlinear integrals. Considering the nonlinear relationships (particularly the interactions among attributes), the nonlinear integrals can be used as data aggregation tools. In the nonlinear integrals, the Choquet integral is more appropriate to be chosen for data mining applications because it provides very important information in the interaction among attributes in the database [210]. Thus, in this study, the Choquet integral with respect to the non-additive measure is chosen as the data aggregation tool.

Now let the values of $f = \{f(x_1), f(x_2) \ldots, f(x_n)\}$ denote the values of each attribute in the dataset; let μ be the non-additive measure. The general definition of the Choquet integral, with function $f : X \rightarrow (-\infty, \infty)$, based on signed non-additive measure μ, is defined as

$$(c) \int f \mathrm{d}\mu = \int_{-\infty}^{0} [\mu(F_\alpha) - \mu(X)] \mathrm{d}\alpha + \int_{0}^{\infty} \mu(F_\alpha) \mathrm{d}\alpha, \qquad (10.3)$$

where $F_\alpha = \{x \mid f(x) \geq \alpha\}$ is called α-cut set of f, for $\alpha \in (-\infty, \infty)$, and n is the number of attributes in the dataset.

10.2 Non-additive Classification Models

The MSD classification model could be extended with the Choquet integral with respect to the signed non-additive measure as [229]:

$$\min \sum_{j=1}^{m} \alpha_j \qquad (10.4)$$

$$\text{s.t.} \quad y_j \left((c) \int f \mathrm{d}\mu - b \right) \leq \alpha_j, \qquad (10.5)$$

$$\alpha_j \geq 0. \qquad (10.6)$$

The general algorithm used to calculate the Choquet integral is shown below.

Algorithm: Calculate the Choquet integral
Input: Dataset, non-additive measure μ.
Output: Choquet integral of each record in the dataset.

Step 1: Let $f = \{f(x_1), f(x_2), \ldots, f(x_n)\}$ denote the values of each attribute for a given record. Then, rearrange f into a non-decreasing order as: $f^* = \{f(x_1^*), f(x_2^*), \ldots, f(x_n^*)\}$, where $f(x_1^*) \leq f(x_2^*) \leq \cdots \leq f(x_n^*)$. Thus, the sequence of $(x_1^*, x_2^*, \ldots, x_n^*)$ is one of the possibilities from the permutation of $(x_1, x_2, \ldots, x_n)$ according to the ordering of the attribute values.

Step 2: Create the non-additive measure μ in terms of variables μ_i. Let $\mu = \{\mu_1, \mu_2, \ldots, \mu_{2^n}\}$, where $\mu_1 = \mu(\emptyset) = 0$.

Step 3: The value of the Choquet integral for the current data record is calculated by

$$(c) \int f \mathrm{d}\mu = \sum_{i=1}^{n} [f(x_i^*) - f(x_{i-1}^*)] \times \mu(\{x_i^*, x_{i+1}^*, \ldots, x_n^*\}), \qquad (10.7)$$

where $f(x_0^*) = 0$.

The above 3-step algorithm is easy to understand but hard to implement with a computer program because the non-additive measure μ is not properly indexed corresponding to the index of the attributes. To deal with this problem, the method proposed in [208] is used to calculate the Choquet integral. The method is presented as:

$$(c) \int f \mathrm{d}\mu = \sum_{j=1}^{2^n - 1} z_j \mu_j \qquad (10.8)$$

where

$$z_j = \begin{cases} \min_{i:\mathrm{frc}(\frac{j}{2^i}) \in [0.5, 1)} (f(x_i)) - \max_{i:\mathrm{frc}(\frac{j}{2^i}) \in [0, 0.5)} (f(x_i)), \\ \quad \text{if the above expression } > 0 \text{ or } j = 2^n - 1, \\ 0, \quad \text{otherwise.} \end{cases} \qquad (10.9)$$

$\mathrm{frc}(\frac{j}{2^i})$ is the fractional part of $\frac{j}{2^i}$ and the maximum operation on the empty set is zero. Let $j_n j_{n-1} \ldots j_1$ represent the binary form of j, the i in (10.9) is determined as follows:

$$\{i \mid \mathrm{frc}(\frac{j}{2^i}) \in [0.5, 1)\} = \{i \mid j_i = 1\} \quad \text{and}$$
$$\{i \mid \mathrm{frc}(\frac{j}{2^i}) \in [0, 0.5)\} = \{i \mid j_i = 0\}.$$

According to the definition in (10.3), the algorithm of calculating the Choquet integral requires pre-ordering of the attributes. However, the different scales of attributes affect the ordering because large scale attributes always appear to have larger values than the others. The data normalization techniques are used to reduce the scale difference among attributes such that they could compare to each other. The majority of the data normalization techniques treat the attributes equally, such as the typical min-max normalization and z-score normalization [101]. That is, the normalization process conducts the same strategy on each attribute. Here, in the Choquet integral, the weights and bias are introduced to the attributes to make them comparable in different scales of values. With this idea, the extended definition of the Choquet integral is given as:

$$(c) \int (\mathbf{a} + \mathbf{b}f) \mathrm{d}\mu \qquad (10.10)$$

where $\mathbf{a} = \{a_1, a_2, \ldots, a_n\}$, $\mathbf{b} = \{b_1, b_2, \ldots, b_n\}$ denote the corresponding bias and weights on the attributes.

This model could be solved by standard linear programming techniques such as simplex method. The optimization procedure of linear programming is to identify the non-additive measure μ. The values of $\mathbf{a}$ and $\mathbf{b}$ need to be determined before the calculation of the Choquet integral. In this research, the genetic algorithm is used for searching $\mathbf{a}$ and $\mathbf{b}$ according to the performance, such as classification accuracy. With the extended definition of the Choquet integral, Model (10.4)–(10.6) can be reformulated as the follows:

$$\min \sum_{j=1}^{m} \alpha_j \tag{10.11}$$

$$\text{s.t.} \quad y_j \left((c) \int (\mathbf{a} + \mathbf{b} f) \mathrm{d}\mu - b \right) \leq \alpha_j, \tag{10.12}$$

$$\alpha_j \geq 0. \tag{10.13}$$

A small artificial dataset is used to present the proposed classification model (10.11)–(10.13) and the geometric meaning of the extended Choquet integral as defined in (10.10) with respect to the signed non-additive measure in classification. Model (10.11)–(10.13) is applied on classifying the two dimensional artificial dataset which contains 200 linearly inseparable two-group data points (106 in G1 vs. 94 in G2). The results are also compared with linear MSD model and popular SVM (Support Vector Machine) classification models. MSD model separates the two different groups in the two dimensional space.

The MSD linear classification parameters obtained by linear programming are $a_1 = 0.985$, $a_2 = 0.937$ with setting of $b = 1$. This linear model cannot perfectly separate the data because the optimal value did not reach to zero, which is 2.60 in this case. The equation of line 1 in Fig. 10.1 is $0.985 f(x_1) + 0.937 f(x_2) = 1$. The classification results, in terms of sensitivity (G1 accuracy), specificity (G2 accuracy) and accuracy (as defined in [101]), are 83.0%, 76.6%, 80.0% respectively.

Model (10.11)–(10.13) is implemented in Java programming language environment by calling the optimization package for solving the non-additive measures and JGAP (package for Genetic Algorithm implementation [115]) for optimizing $\mathbf{a}$ and $\mathbf{b}$. The identified the non-additive measures μ are $\mu_1 = \mu(\{x_2\}) = 0$, $\mu_2 = \mu(\{x_1\}) = -5.13$, and $\mu_3 = \mu(\{x_1, x_2\}) = 12.82$. The weights of $\mathbf{a}$ and $\mathbf{b}$ are $a_1 = -0.41$, $a_2 = 0.64$, $b_1 = -0.35$, and $b_2 = 0.73$. The optimized boundary b is -16.38. Even through the optimal value of the linear programming is non-zero, says 70.8 in this case, the perfect separation is still achieved and the classification accuracy in terms of sensitivity, specificity, accuracy are all 100% shown as in Fig. 10.1. MSD linear classification model performs worst on this linearly inseparable dataset and it is similar to linear SVM. Polynomial SVM and RBF SVM are able to create more curved decision boundaries and perform better than linear models. The classification accuracies of MSD, decision tree, linear SVM, polynomial SVM and RBF SVM are 80%, 97.5%, 83%, 84.5% and 88% respectively. The

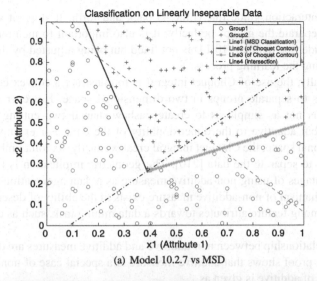

Classification on Linearly Inseparable Data

(a) Model 10.2.7 vs MSD

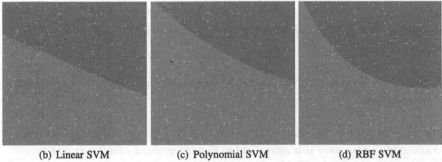

(b) Linear SVM   (c) Polynomial SVM   (d) RBF SVM

Fig. 10.1 Classification on linearly inseparable dataset

proposed Model (10.11)–(10.13) outperforms on this cashew-like two-dimensional dataset than other models. The geometrical meaning of using the Choquet integral with respect to signed non-additive measure as an aggregation tool for data modeling is to create one Choquet contour by two lines (Line 2 and Line 3 in Fig. 10.1) to separate the two different groups. Here, the intersection of Line 2 and Line 3 could be essentially presented by an invisible guide line (Line 4 in Fig. 10.1(a)), which is defined in following equation:

$$a_1 + b_1 f(x_1) = a_2 + b_2 f(x_2). \tag{10.14}$$

Line 2 and Line 3 must intersect on this guide line. Moreover, a_i and a_i are respectively used to determine the interception to Y axis and the slope of the guide line. The non-additive measure μ is used to determine the way those two lines are intercepted with the guide line where the slopes of those lines are controlled by μ. The learning process exactly looks like a butterfly flaps its wings (Line 2 and Line 3) crossing the interaction guide line (Line 4) back and forth. It stops at where the wings could separate the groups perfectly. It is also important to mention that the interaction guide line defined (10.14) is not fixed but being adjusted by the genetic algorithm during the learning process.

Geometrically, the use of Choquet integral is to construct a convex contour by two joint lines to separate groups in two dimensional space. However, the Choquet integral is not as simple as to create cashew-shaped two cutting lines for classification but adaptive to the dimensionality of the data. For example, in the three dimensional space, the Choquet integral creates exactly six joint umbrella-like shaped planes to separate the data [210]. The geometric implication is to emphasize the advantages of using non-additive measure as a data aggregation tool. The profound mechanism of non-additive measure exists in the ability of describing the interactions among feature attributes towards a data mining task, such as classification.

Now, the relationship between non-additive and additive measures are discussed. The following proof shows that additive measure is a special case of non-additive. The definition of additive is given as

Definition 10.4 μ is *additive* if $\forall E_1, E_2, E_1 \cap E_2 = \emptyset, E_1 \cup E_2 = E, \mu(E_1) + \mu(E_2) = \mu(E)$.

Then the following theorem is given to show additive classifier is a special case of non-additive.

Theorem 10.5 *Let μ be a non-additive measure. The Choquet integral with respect to non-additive measure μ is equivalent to linear weighted sum if μ is additive.*

Proof Let n be the number of attributes. When $n = 1$, it is obvious that the theorem is true. When $n = 2$, let the two attributes are x_1 and x_2, with $f(x_1)$ and $f(x_1)$ denoting their values. Suppose $f(x_1) < f(x_2)$, note $f(x_1^*) < f(x_2^*)$, then $\{x_1^*, x_2^*\} = \{x_1, x_2\}$. Since μ is additive, $\mu(x_1^*, x_2^*) = \mu(x_1^*) + \mu(x_2^*)$. The Choquet integral is

$$(c) \int f \mathrm{d}\mu = \sum_{i=1}^{n} [f(x_i^*) - f(x_{i-1}^*)] \mu(\{x_i^*, x_{i+1}^*, \ldots, x_n^*\})$$

$$= f(x_1^*) \mu(x_1^*, x_2^*) + [f(x_2^*) - f(x_1^*)] \mu(x_2^*)$$

$$= f(x_1^*)[\mu(x_1^*) + \mu(x_2^*)] + f(x_2^*) \mu(x_2^*) - f(x_1^*) \mu(x_2^*)$$

$$= f(x_1^*) \mu(x_1^*) + f(x_2^*) \mu(x_2^*).$$

Therefore, the Choquet integral is equivalent to linear weighted sum.

Now assume the theorem is true when $n = k - 1$, which means $(c)_{k-1} \int f \mathrm{d}\mu$ is linear weighted sum. Then, when $n = k$, then the following holds:

Fig. 10.2 Artificial dataset
classification

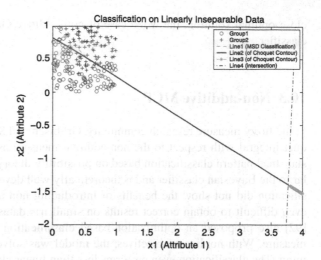

$$(c)_k \int f d\mu = \sum_{i=1}^{k} [f(x_i^*) - f(x_{i-1}^*)] \mu(\{x_i^*, x_{i+1}^*, \ldots, x_k^*\})$$

$$= f(x_1^*)\mu(x_1^*, \ldots, x_k^*) + [f(x_2^*) - f(x_1^*)]\mu(x_2^*, \ldots, x_k^*)$$

$$+ \cdots + [f(x_k^*) - f(x_{k-1}^*)]\mu(x_k^*)$$

$$= f(x_1^*)[\mu(x_1^*, \ldots, x_{k-1}^*) + \mu(x_k^*)]$$

$$+ [f(x_2^*) - f(x_1^*)][\mu(x_2^*, \ldots, x_{k-1}^*) + \mu(x_k^*)]$$

$$+ \cdots + [f(x_k^*) - f(x_{k-1}^*)]\mu(x_k^*)$$

$$= (c)_{k-1} \int f d\mu + f(x_1^*)\mu(x_k^*) + [f(x_2^*) - f(x_1^*)]\mu(x_k^*)$$

$$+ [f(x_3^*) - f(x_2^*)]\mu(x_k^*) + \cdots + f(x_k^*)\mu(x_k^*)$$

$$= (c)_{k-1} \int f d\mu + f(x_k^*)\mu(x_k^*).$$

Since $(c)_{k-1} \int f d\mu$ is linear weighted sum, therefore, $(c)_k \int f d\mu$ is also equivalent
to linear weighted sum.   □

Example When μ in Model (10.11)–(10.13) is set to be additive, the non-additive
classifier equals to linear classifier.

In the above two dimensional artificial dataset case, $\mu(\{x_1, x_2\}) = \mu(\{x_1\}) + \mu(\{x_2\})$. Thus, classifying by additive Model (10.11)–(10.13) is similar to the MSD
linear classifier. Figure 10.2 shows linear additive classifier is a special case of non-
additive.

Since additive measure is a special case of non-additive, non-additive classifier
outperforms additive classifier (i.e. MSD linear classifier). When both reached a

close optimal solution, the worst case of non-additive classifier equals to additive classifier.

10.3 Non-additive MCP

In the fuzzy measure research community, Grabisch and Sugeno [91] used the Choquet integral with respect to the non-additive measure as aggregation operator on statistical pattern classification based on possibility theory. The model is very similar to the Bayesian classifier and is theoretically well developed. However, the classification did not show the benefits of introducing non-additive measure and it is even difficult to obtain correct results on small iris dataset. Grabisch and Nicolas [92] also proposed an optimization-based classification model with non-additive measure. With nonlinear objectives, the model was solved by quadratic programming. The classification even performs less than linear classifier on iris dataset and competitive to fuzzy k-NN classifier on the other datasets. It pointed out that a better non-additive identification algorithm is needed.

In order to search solutions to improve the non-additive classifiers, firstly, it is important to address one issue in classic optimization-based models as described. In Model (10.11)–(10.13), the boundary b is not optimized but arbitrarily chosen. The practical solution is to either predetermine a constant value for b or implement a learning mechanism in the iterations, such as updating b with the average of the lowest and largest predicted scores [226, 229]. Xu et al. [226] also proposed the classification method by the Choquet integral projections. The famous Platt's Sequential Minimal Optimization (SMO) algorithm [165] for SVM classifier utilized the similar idea by choosing the average of the lower and upper multipliers in the dual problem corresponding to the boundary parameter b in the primal problem. Keerthi et al. [124] proposed the modified SMO with two boundary parameters to achieve a better and even faster solution.

The boundary b in MSD can be replaced with soft-margin $b \pm 1$ similar to SVM which constructs a separation belt in stead of a single cutting line. The soft margin by using $b \pm 1$ in optimization constrains coincidentally solved the degeneracy issue in mathematical programming. The value of b could be optimized by a simple linear programming technique. The improved model is a linear programming solvable problem with optimized b and signed non-additive measure. Now Model (10.11)–(10.13) is extended to Model (10.15)–(10.17) (soft-margin classifier with signed non-additive measure) as follows:

$$\min \sum_{j=1}^{m} \alpha_j \tag{10.15}$$

$$\text{s.t.} \quad y_j \left((c) \int (\mathbf{a} + \mathbf{b}f) \mathrm{d}\mu - b \right) \leq 1 + \alpha_j, \tag{10.16}$$

$$\alpha_j \geq 0. \tag{10.17}$$

Models (10.4)–(10.6), (10.11)–(10.13), (10.15)–(10.17) aim to minimize the empirical risk. However, the simple linear programming cannot easily produce a globe optimal solution than convex programming [126]. Thus, the model is extended to a convex quadratic programming form, as described as below:

$$\min \frac{1}{2}||\mu||^2 + C \sum_{i=1}^{m} \alpha_i \tag{10.18}$$

$$\text{s.t.} \quad y_j \left((c) \int (\mathbf{a} + \mathbf{b}f) d\mu - b \right) \leq 1 + \alpha_j, \tag{10.19}$$

$$\alpha_j \geq 0, \tag{10.20}$$

where C is the constant positive number used to balance the two objectives.

MCQP model in Chap. 9 with nonlinear objectives could also be extended with non-additive measures. The new model is described as (10.21)–(10.22):

$$\min \frac{1}{2}||\mu||^2 + W_\alpha \sum_{j=1}^{m} \eta_j^2 - W_\beta \sum_{j=1}^{m} \eta_j + b \tag{10.21}$$

$$\text{s.t.} \quad y_j \left((c) \int f d\mu - b \right) = 1 - \eta_j. \tag{10.22}$$

10.4 Reducing the Time Complexity

As mentioned earlier, the using of non-additive measure increases the computational cost because of the high time complexity caused by the power set operation. This might be another explanation on the difficulty of identifying non-additive measures in [91, 92]. In the literatures, there are two major solutions to reduce the number of non-additive measures. They are hierarchical Choquet integral [152] and the k-additive measure [94].

10.4.1 Hierarchical Choquet Integral

The hierarchical Choquet integral is a potentially useful approach to compromise the time complexity problem practically. Murofushi et al. [152] proposed the hierarchical decomposition theorems to reduce the number of coefficients in the Choquet integral with loss of some interactions. The essential idea is to properly group the attributes and calculate the Choquet integral within each group. In practice, the problem of searching the best grouping with limited loss of information has not been solved. Sugeno et al. [192] designed a genetic algorithm based subset selec-

Fig. 10.3 Hierarchical
Choquet integral with
grouping

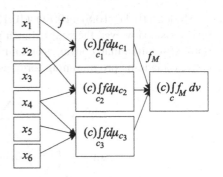

tion method to search good grouping for time series modeling with non-additive
measure but the cost of genetic algorithm is high and the complexity of solving
the problem has dramatically increased again. See an example of decomposition in
Fig. 10.3, in which the Hierarchical Choquet integral is grouping as $c = c_1 \cup c_2 \cup c_3$,
$c_1 = \{x_1, x_3\}$, $c_2 = \{x_2, x_4\}$, $c_3 = \{x_4, x_5, x_6\}$.

The best way to group those attributes probably is with the help of domain knowledge. However, the knowledge prior to the data mining does not always exist. Thus,
when no human experts are available, it is reasonable to statistically group the attributes into different groups according to their contributions towards the target attributes under a certain significance level.

10.4.2 Choquet Integral with Respect to k-Additive Measure

Besides the hierarchical Choquet integral, the k-additive measure may be used to
model the inter-relationships among any k attributes. Although there are many variations on the non-additive measures, such as Möbius transformation, Shapley interaction and the Banzhaf interaction [149]. No matter which one is chosen, any machine learning algorithm used to identify the non-additive measure μ will encounter
a computation cost of $O(2^n)$ because the power-set operation of $\mathbf{X}$ is involved. The
k-interactive measure is the compromised non-additive which only considers up to
k attributes interactions. It is also named k-order additive measure or k additive
measure in the early studies [94]. But the k-interactive measure is actually non-
additive measure. Thus, it is renamed as k-interactive measure according to its real
meaning. For example, when $k = 2$, there only exists interactions among any two
attributes. In this way, the time complexity of determining coefficients among attributes reduces from $O(2^n)$ to $O(2^k)$, where $k \ll n$. The study of using k-additive
measure in data mining may contribute to achieve even better classification in speed,
robustness, and scalability. For data mining applications, the signed k-interactive
measure is used because both positive and negative interactions are considered in
reality.

Definition 10.6 A *signed k-interactive measure* μ defined on X is a set function $\mu : \mathcal{P}(X) \to [0,\infty)$ satisfying

1. $\mu(\emptyset) = 0$,
2. $\mu(E) = 0$, if $|E| > k$,

where $|E|$ is the size of E, and k is the designated number of attributes having interactions.

Mikenina and Zimmermann [148] proposed the 2-additive classification with feature selections based on a pattern matching algorithm similar to [91]. The application is still limited to the data with small number of attributes. In this research, we apply k-additive measure to reduce the computation cost on the proposed classification models with optimized boundary.

Chapter 11
MC2LP

11.1 MC2LP Classification

11.1.1 Multiple Criteria Linear Programming

The compromise solution [107] in multiple criteria linear programming locates the best trade-offs between MMD and MSD for all possible choices.

$$\min \sum_i \xi_i \tag{11.1}$$

$$\max \sum_i \beta_i \tag{11.2}$$

$$\text{s.t.} \quad (x_i \cdot w) = b + \xi_i - \beta_i, \quad x_i \in M, \tag{11.3}$$

$$(x_i \cdot w) = b - \xi_i + \beta_i, \quad x_i \in N, \tag{11.4}$$

where x_i are given, w is unrestricted, $\xi_i \geq 0$, and $\beta_i \geq 0$, $i = 1, 2, \ldots, n$.

A boundary value b (cutoff) is often used to separate two groups, where b is unrestricted. Efforts to promote the accuracy rate have been largely restricted to the unrestricted characteristics of b (x given b is put into calculation to find coefficients w) according to the user's experience facing the real time data set. In such procedure, the goal of finding the optimal solution for classification question is replaced by the task of testing boundary b. If b is given, we can find a classifier using a optimal solution. The fixed cutoff value causes another problem that those cases that can achieve the ideal cutoff score would be zero. Formally, this means that the solutions obtained by linear programming are not invariant under linear transformations of the data. Alternative approach to solve this problem is to add a constant as ζ to all the values, but it will affect weight results and performance of its classification. Unfortunately, it cannot be implemented in [197]. Adding a gap between the two regions may overcome the above problem. However, if the score is falling into this gap, we must determine which class it should belong to [197].

Y. Shi et al., *Optimization Based Data Mining: Theory and Applications*,
Advanced Information and Knowledge Processing,
DOI 10.1007/978-0-85729-504-0_11, © Springer-Verlag London Limited 2011

To simplify the problem, we use linear combination of b^λ to replace of b. Then we can get the best classifier as $w^*(\lambda)$. Suppose we now have the upper boundary b_u and lower boundary b_l. Instead of finding the best boundary b randomly, we find the best linear combination for the best classifier. That is, in addition to considering the criteria space that contains the trade-offs of multiple criteria in (MSD), the structure of MC2 linear programming has a constraint-level space that shows all possible trade-offs of resource availability levels (i.e. the trade-off of upper boundary b_u and lower boundary b_l). We can test the interval value for both b_u and b_l by using the classic interpolation method such as Lagrange, Newton, Hermite, and Golden Section in real number $[-\infty, +\infty]$. It is not necessary to set negative and positive for b_l and b_u separately but it is better to set the initial value of b_l as the minimal value and the initial value of b_u as the maximum value. And then narrow down the interval $[b_l, b_u]$.

With the adjusting boundary, MSD and MMD can be changed from standard linear programming to linear programming with multiple constraints.

$$\min \ \sum_i \xi_i \tag{11.5}$$

$$\text{s.t.} \quad (x_i \cdot w) \leq \lambda_1 \cdot b_l + \lambda_2 \cdot b_u + \xi_i, \quad x_i \in B, \tag{11.6}$$

$$(x_i \cdot w) > \lambda_1 \cdot b_l + \lambda_2 \cdot b_u - \xi_i, \quad x_i \in G, \tag{11.7}$$

$$\lambda_1 + \lambda_2 = 1, \tag{11.8}$$

$$0 \leq \lambda_1, \lambda_2 \leq 1, \tag{11.9}$$

where x_i, b_u, b_l are given, w is unrestricted, $\xi_i \geq 0$.

$$\max \ \sum_i \beta_i \tag{11.10}$$

$$\text{s.t.} \quad (x_i \cdot w) \geq \lambda_1 \cdot b_l + \lambda_2 \cdot b_u - \beta_i, \quad x_i \in B, \tag{11.11}$$

$$(x_i \cdot w) < \lambda_1 \cdot b_l + \lambda_2 \cdot b_u + \beta_i, \quad x_i \in G, \tag{11.12}$$

$$\lambda_1 + \lambda_2 = 1, \tag{11.13}$$

$$0 \leq \lambda_1, \lambda_2 \leq 1, \tag{11.14}$$

where x_i, b_u, b_l are given, w is unrestricted, $\beta_i \geq 0$.

The above two programming is LP with multiple constraints. This formulation of the problem always gives a nontrivial solution.

A hybrid model that combines models of (MSD) and (MMD) model with multiple constraints level is given by:

$$\min \ \sum_i \xi_i \tag{11.15}$$

$$\max \ \sum_i \beta_i \tag{11.16}$$

Fig. 11.1 Overlapping case
in two-class separation

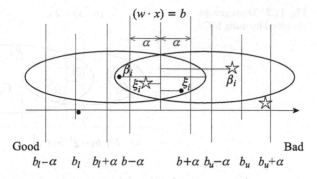

s.t. $\quad (x_i \cdot w) = \lambda_1 \cdot b_l + \lambda_2 \cdot b_u + \xi_i - \beta_i, \quad x_i \in M,$ $\qquad$ (11.17)

$\qquad (x_i \cdot w) = \lambda_1 \cdot b_l + \lambda_2 \cdot b_u - \xi_i + \beta_i, \quad x_i \in N,$ $\qquad$ (11.18)

$\qquad \lambda_1 + \lambda_2 = 1,$ $\qquad$ (11.19)

$\qquad 0 \leq \lambda_1, \lambda_2 \leq 1,$ $\qquad$ (11.20)

where x_i, b_u, b_l are given, w are unrestricted, $\xi_i \geq 0$ and $\beta_i \geq 0$, $i = 1, 2, \ldots, n$. Separating the MSD and MMD, (11.15)–(11.20) is reduced to LP problem with multiple constraints. Replacing the combination of b_l and b_u with the fixed b, (11.15)–(11.20) becomes MC problem.

A graphical representation of these models in terms of ξ is shown in Fig. 11.1.

For (11.15)–(11.20), theoretically, finding the ideal solution that simultaneously represents the maximal and the minimal is almost impossible. However, the theory of MC linear programming allows us to study the trade-offs of the criteria space. In the case, the criteria space is a two dimensional plane consisting of MSD and MMD. We use compromised solution of multiple criteria and multiple constraint linear programming to minimize the sum of ξ_i and maximize the sum of β_i simultaneously. Then the model can be rewritten as:

$$\max \ \gamma_1 \sum_i \xi_i + \gamma_2 \sum_i \beta_i \qquad (11.21)$$

s.t. $\quad (x_i \cdot w) = \lambda_1 \cdot b_l + \lambda_2 \cdot b_u - \xi_i - \beta_i, \quad x_i \in B,$ $\qquad$ (11.22)

$\qquad (x_i \cdot w) = \lambda_1 \cdot b_l + \lambda_2 \cdot b_u + \xi_i + \beta_i, \quad x_i \in G,$ $\qquad$ (11.23)

$\qquad \gamma_1 + \gamma_2 = 1,$ $\qquad$ (11.24)

$\qquad \lambda_1 + \lambda_2 = 1,$ $\qquad$ (11.25)

$\qquad 0 \leq \gamma_1, \gamma_2 \leq 1,$ $\qquad$ (11.26)

$\qquad 0 \leq \lambda_1, \lambda_2 \leq 1.$ $\qquad$ (11.27)

This formulation of the problem always gives a nontrivial solution and is invariant under linear transformation of the data. Both γ_1 and γ_2 are the weight parameters for MSD and MMD. λ_1, λ_2 are the weight parameters for b_l and b_u. They serve to

Fig. 11.2 Three groups
classified by using MC2

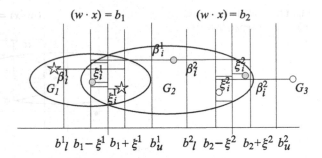

normalize the constraint-level and the criteria-level parameters. We note that the key
point of the two-class linear classification models is to use a linear combination of
the minimization of the sum of ξ_i or/and maximization of the sum of β_i to reduce
the two criteria problems into a single criterion. The advantage of this conversion is
to easily utilize all techniques of LP for separation, while the disadvantage is that it
may miss the scenario of trade-offs between these two separation criteria [107].

Then theoretically, (11.15)–(11.20), (11.21)–(11.27) can find the better classifier
than MSD or MMD. Using the software developed by Hao and Shi [103], we can
get the potential solution of MC2 for the massive data set. This algorithm has a
polynomial computational complexity of $O(F(m; n; L))$, where in (11.15)–(11.20)
A is an $m \times n$ matrix and $m, n \geq 2$, L is the number of binary bits required to story
all the data of the LP problem. L is also called the input length of the LP problem
and is known as a function of logarithm of m, n, c_j, a_{ij}, and b_i.

11.1.2 Different Versions of MC2

(11.15)–(11.20) can be extended easily to solve the following multi-class problem
in Fig. 11.2.

$$\min \sum_i \xi_i \tag{11.28}$$

$$\max \sum_i \beta_i \tag{11.29}$$

$$\text{s.t.} \quad (x_i \cdot w) = \lambda_1 \cdot b_l + \lambda_2 \cdot b_u + \xi_i - \beta_i, \quad x_i \in G_1, \tag{11.30}$$

$$\lambda_1^{k-1} \cdot b_l^{k-1} + \lambda_2^{k-1} \cdot b_u^{k-1} - \xi_i^{k-1} + \beta_i^{k-1} = (x_i \cdot w),$$

$$x_i \in G_k, \ k = 2, \ldots, s - 1, \tag{11.31}$$

$$(x_i \cdot w) = \lambda_1^k \cdot b_l^k + \lambda_2^k \cdot b_u^k - \xi_i^k + \beta_i^k, \quad x_i \in G_k, k = 2, \ldots, s - 1, \tag{11.32}$$

$$\lambda_1^{s-1} \cdot b_l^{s-1} + \lambda_2^{s-1} \cdot b_u^{s-1} - \xi_i^{s-1} + \beta_i^{s-1} = (x_i \cdot w), \quad x_i \in G_s, \tag{11.33}$$

$$\lambda_1^{k-1} \cdot b_l^{k-1} + \lambda_2^{k-1} \cdot b_u^{k-1} + \xi_i^{k-1} \leq \lambda_1^k \cdot b_l^k + \lambda_2^k \cdot b_u^k - \xi_i^k,$$

$$k = 2, \ldots, s - 1, \ i = 1, \ldots, n, \tag{11.34}$$

$$\lambda_1 + \lambda_2 = 1, \tag{11.35}$$

$$0 \le \lambda_1, \lambda_2 \le 1, \tag{11.36}$$

where x_i, b_u^k, b_l^k are given, w is unrestricted, $\xi_i^j \ge 0$ and $\beta_i^j \ge 0$.

Similar as the compromised solution approach using regret measurement, the best trade-off between MSD and MMD is identified for an optimal solution [107]. An MC2 model for multiple-class separation is presented as:

In compromised solution approach [179], the best trade-off between $-\sum_i \xi_i$ and $\sum_i \beta_i$ is identified as an optimal solution. To explain this, assume the "deal value" of $-\sum_i \xi_i$ be $\xi^* > 0$ and the "idea value" of $\sum_i \beta_i$ be $\beta^* > 0$. Then, if $-\sum_i \xi_i > \xi^*$, the regret measure is defined as $-d_a^+ = \sum_i \xi_i > \xi^*$; otherwise, it is 0. Thus, the relationship of these measures are

(i) $\sum_i \xi_i + \xi^* = d_a^- - d_a^+$,
(ii) $|\sum_i \xi_i + \xi^*| = d_a^- + d_a^+$, and
(iii) $d_a^-, d_a^+ \ge 0$.

Similarly, we derive [107, 180]

$$\beta^* - \sum_i \beta_i = d_\beta^- - d_\beta^+, \left| \beta^* + \sum_i \beta_i \right| = d_\beta^- + d_\beta^+ \quad \text{and} \quad d_\beta^- + d_\beta^+ \ge 0.$$

(11.28)–(11.36) can be rewritten as:

$$\min \sum_{y=1}^{s-1} (d_{aj}^- + d_{aj}^+ + d_{\beta j}^- + d_{\beta j}^+) \tag{11.37}$$

$$\text{s.t.} \quad \sum_i \xi_i^j + \xi_*^j = d_{aj}^- - d_{aj}^+, \quad j = 1, \dots, s-1, \tag{11.38}$$

$$\beta_*^j + \sum_i \beta_i^j = d_{\beta j}^- - d_{\beta j}^+, \quad j = 1, \dots, s-1, \tag{11.39}$$

$$(x_i \cdot w) = \lambda_1 \cdot b_l + \lambda_2 \cdot b_u + \xi_i - \beta_i, \quad x_i \in G_1, \tag{11.40}$$

$$\lambda_1^{k-1} \cdot b_l^{k-1} + \lambda_2^{k-1} \cdot b_u^{k-1} - \xi_i^{k-1} + \beta_i^{k-1} = (x_i \cdot w),$$
$$x_i \in G_k, k = 2, \dots, s-1, \tag{11.41}$$

$$(x_i \cdot w) = \lambda_1^k \cdot b_l^k + \lambda_2^k \cdot b_u^k - \xi_i^k + \beta_i^k, \quad x_i \in G_k, \, k = 2, \dots, s-1, \tag{11.42}$$

$$\lambda_1^{s-1} \cdot b_l^{s-1} + \lambda_2^{s-1} \cdot b_u^{s-1} - \xi_i^{s-1} + \beta_i^{s-1} = (x_i \cdot w), \quad x_i \in G_s, \tag{11.43}$$

$$\lambda_1^{k-1} \cdot b_l^{k-1} + \lambda_2^{k-1} \cdot b_u^{k-1} + \xi_i^{k-1} \le \lambda_1^k \cdot b_l^k + \lambda_2^k \cdot b_u^k - \xi_i^k,$$
$$k = 2, \dots, s-1, \, i = 1, \dots, n, \tag{11.44}$$

$$\lambda_1 + \lambda_2 = 1, \tag{11.45}$$

$$0 \le \lambda_1, \lambda_2 \le 1, \tag{11.46}$$

where x_i, b_u^k, b_l^k, ξ_*^j, ξ_i^j are given, w is unrestricted, ξ_i^j and β_i^j, d_{aj}^-, d_{aj}^+, $d_{\beta j}^-$, $d_{\beta j}^+ \geq$ 0.

In [179], a special MC2 problem is built as (11.47)–(11.49):

$$\max \ Z = \sum_i \gamma_i C^i x \tag{11.47}$$

$$\text{s.t.} \quad (x_i \cdot w) \leq \sum_i \lambda_i b_k, \tag{11.48}$$

$$\sum_k \lambda_k = 1. \tag{11.49}$$

Let u_i^0 be the upper bound and l_i^0 be the lower bound for the ith criterion $C^i X$ of (11.47)–(11.49) if x^* can be obtained from solving the following problem:

$$\max \ \xi \tag{11.50}$$

$$\text{s.t.} \quad \xi \leq \frac{C^i X - l_i^0}{u_i^0 - l_i^0}, \quad i = 1, 2, q, \tag{11.51}$$

$$(x_i \cdot w) \leq \sum_{i=1}^{p} \lambda_i b_k, \tag{11.52}$$

$$\sum_{k=1}^{p} \lambda_k = 1, \quad k = 1, 2, \ldots, p, \tag{11.53}$$

then $x^* \in X$ is a weak potential solution of (11.47)–(11.49) [100, 179]. According to this, in formulating a FLP problem, the objectives $(\min \sum_{i=1}^{n} \xi_i, \max \sum_{i=1}^{n} \beta_i)$ and constraints $(x_i \cdot w) = b + \xi_i - \beta_i$, $x_i \in G$; $(x_i \cdot w) = b - \xi_i + \beta_i$, $x_i \in B$ of (11.47)–(11.49) are redefined as fuzzy sets F and X with corresponding membership functions $\mu_F(x)$ and $\mu_X(x)$, respectively. In this case the fuzzy decision set D is defined as $D = F \cap X$, and the membership function is defined as $\mu_D(x_1) = \mu_F(x), \mu_X(x)$. In a maximization problem, x_1 is a "better" decision than x_2 if $\mu_D(x_1) \geq \mu_D(x_2)$. Thus, it can be considered appropriate to select x^* such as [107]:

$$\max \mu_D(x) = \max_x \min\{\mu_F(x), \mu_X(x)\} \tag{11.54}$$

$$= \min\{\mu_F(x^*), \mu_X(x^*)\}, \tag{11.55}$$

where $\max \mu_D(x)$ is the maximized solution.

Let y_{1L} be MSD and y_{2U} be MMD, the value of $\max \sum_{i=1}^{n} \xi_i$ is y_{1U} and the value of $\max \sum_{i=1}^{n} \xi_i$ is y_{2L}. Let $F_1 = w : y_{1L} \leq \sum_{i=1}^{n} \xi_i \leq y_{1U}$ and $F_2 = w : y_{2L} \leq \sum_{i=1}^{n} \beta_i \leq y_{2U}$. Their membership functions can be expressed respectively by [107]:

$$\mu_{F_1}(x) = \begin{cases} 1, & \text{if } \sum_{i=1}^{n} \xi_i \geq y_1 U, \\ \frac{\sum_{i=1}^{n} \xi_i - y_1 L}{y_1 U - y_1 L}, & \text{if } y_1 L \leq \sum_{i=1}^{n} \xi_i \leq y_1 U, \\ 0, & \text{if } \sum_{i=1}^{n} \xi_i \leq y_1 L. \end{cases} \tag{11.56}$$

Then the fuzzy set of the objective functions is $F = F_1 \cap F_2$ and its membership function is $\mu_F(x) = \min\{\mu_{F_1}(x), \mu_{F_2}(x)\}$. Using the crisp constraint set $w = \{x : (x_i \cdot w) = b + \xi_i - \beta_i, \ x_i \in G; (x_i \cdot w) = b - \xi_i + \beta_i, \ x_i \in B\}$, the fuzzy set of the decision problem is $D = F_1 \cap F_2 \cap w$, and its membership function is $\mu_D(x) = \mu_{F_1 \cap F_2 \cap w}(x)$ has shown that the "optimal solution" of $\max_x \mu_D(x) = \max_x \min\{\mu_{F_1}(x), \mu_{F_1}(x), \mu_w)(x)\}$ is an efficient solution of (11.15)–(11.20).

To explore this possibility, this paper proposes a heuristic classification to build scorecard by using the fuzzy linear programming for discovering the good and bad customers as follows:

$$\max \ \xi \tag{11.57}$$

$$\text{s.t.} \quad \xi \leq \frac{\sum \alpha_i - y_{1L}}{y_{1U} - y_{1L}}, \tag{11.58}$$

$$\xi \leq \frac{\sum \beta_i - y_{2L}}{y_{2U} - y_{2L}}, \tag{11.59}$$

$$(x_i \cdot w) = \lambda_1 \cdot b_l + \lambda_2 \cdot b_u + \xi_i - \beta_i, \quad x_i \in M, \tag{11.60}$$

$$(x_i \cdot w) = \lambda_1 \cdot b_l + \lambda_2 \cdot b_u - \xi_i + \beta_i, \quad x_i \in N, \tag{11.61}$$

where x_i, y_{1L}, y_{1U}, y_{2L}, y_{2U}, b_l, b_u are known, w is unrestricted, and $\alpha_i, \beta_i, \lambda_1, \lambda_2, \xi \geq 0, i = 1, 2, \ldots, n$.

11.1.3 Heuristic Classification Algorithm

To run the proposed algorithm below, we first create data warehouse for credit card analysis. Then we generate a set of relevant attributes from the data warehouse, transform the scales of the data warehouse into the same numerical measurement, determine the two classes of good and bad customers, classification threshold τ that is selected by user, training set and verifying set.

Algorithm: A credit scorecard by using MC2 in Fig. 11.3
Input: The training samples represented by discrete-valued attributes, the set of candidate attributes.
Output: Best b^* and parameters W^* for building a credit scorecard.

Fig. 11.3 A flowchart of
MC2 classification method

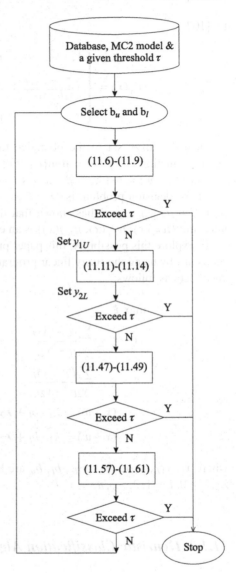

Method:

(1) Give a class boundary value b_u and b_l and use models (11.6)–(11.9), (11.11)–
    (11.14), and (11.47)–(11.49) to learn and compute the overall scores $x_i W$
    ($i = 1, 2, \ldots, n$) of the relevant attributes or dimensions over all observations
    repeatedly.

(2) If (11.6)–(11.9) exceeds the threshold τ, go to (6), else go to (3).

(3) If (11.11)–(11.14) exceeds the threshold τ, go to (6), else go to (4).

(4) If (11.47)–(11.49) exceeds the threshold τ, go to (6), else go to (5).

(5) If (11.57)–(11.61) exceeds the threshold τ, go to (6), else go to (1) to consider to give another cut off pair.
(6) Apply the final learned scores W^* to predict the unknown data in the verifying set.
(7) Find separation.

11.2 Minimal Error and Maximal Between-Class Variance Model

In this section, we will first explain Fisher's Linear Discriminant Analysis (LDA) from the view of multiple criteria programming. Then we formulate the Minimal Error and Maximal Between-class Variance (MEMBV) model by using the objective function of Fisher's LDA (maximizing the between-class variance) and the MC2LP model for relaxing the constraints.

To simplify the description, we introduce some notations first. Consider a two-group classification problem, group G_l has N_l instances which denoted by $W_1 = \{w_{1i}\}_{i=1}^{N_1}$, group G_2 has N_2 instances which denoted by $W_2 = \{w_{2j}\}_{j=1}^{N_2}$, classification models try to find a optional decision boundary b (which is determined by the projection direction ϖ), where W_1 and W_2 can be separated as far as possible.

Now we explain Fisher's LDA from the view of multiple objective programming. The mean vectors of G_1 and G_2 are $m_1 = \frac{1}{N_1} \sum_{i=1}^{N_1} w_{1i}$ and $m_2 = \frac{1}{N_2} \sum_{j=1}^{N_2} w_{2j}$ respectively, and the Euclidean distance between m_1 and m_2 can be denoted as $S_B = (m_1 - m_2)(m_1 - m_2)^T$. Meanwhile, the variances can be denoted as $S_1 = \sum_{i=1}^{N_1} (x_{1i} - m_1)(x_{1i} - m_1)^T$ for G_1 and $S_2 = \sum_{j=1}^{N_2} (x_{2j} - m_2)(x_{2j} - m_2)^T$ for G_2, and the whole variance of the training sample is the sum of S_1 and S_2, which can be denoted as $S_w = S_1 + S_2$. When projected on the direction vector w, the variances of G_1 and G_2 can be denoted as $\text{Var}_1 = \sum_{i=1}^{N_1} (wx_{1i} - wm_1)^2$ and $\text{Var}_2 = \sum_{i=1}^{N_2} (wx_{2i} - wm_2)^2$ respectively. Moreover, $\text{Var}_1 = w^T S_1 w$ and $\text{Var}_2 = w^T S_2 w$. So the "within-class variance" is the sum of Var_1 and Var_2, that is $w^T S_w w = w^T (S_1 + S_2)w$. And "between-class variance" can be denoted as $w^T S_B w = w^T (m_1 - m_2)(m_1 - m_2)^T w = w^T S_b w$. Finally, LDA tries to maximize the between-class variance $w^T S_b w$ and minimizing within-class variance $w^T S_w w$ as follows:

$$\max \ w^T S_B w, \tag{11.62}$$
$$\min \ w^T S_W w. \tag{11.63}$$

Combining (11.62) and (11.63), we get the formulation of the well-known Fisher's LDA model as follows:

$$\max \ J_F(w) = \frac{w^T S_B w}{w^T S_W w}. \tag{11.64}$$

Fisher's LDA is easy to understand and implement, and it has been widely used in classification and dimensionality reduction. However, since the objective function of LDA doesn't contain the training error, when the Fisher's criterion doesn't hold true, LDA can not achieve an optimal solution.

By using the objective function of Fisher's LDA (maximizing the "between-class variance"), we formulate the Minimal Error and Maximal Between-class Variance (MEMBV) model as follows:

$$\max \ w^T S_B w \tag{11.65}$$

$$\min \sum \xi_i \tag{11.66}$$

$$\text{s.t.} \quad (w \cdot x_i) - \xi_i \leq b, \quad x_i \in G_1, \tag{11.67}$$

$$(w \cdot x_i) + \xi_i \geq b, \quad x_i \in G_2, \tag{11.68}$$

$$\xi_i \geq 0, \tag{11.69}$$

where w is the projection direction, b is the classification boundary. When combining (11.65) and (11.66) into one single objective function by weight factor c, we get the MEMBV model as follows:

$$\min \sum \xi_i - c \cdot w^T S_B w \tag{11.70}$$

$$\text{s.t.} \quad (w \cdot x_i) - \xi_i \leq b, \quad x_i \in G_1, \tag{11.71}$$

$$(w \cdot x_i) + \xi_i \geq b, \quad x_i \in G_2, \tag{11.72}$$

$$\xi_i \geq 0. \tag{11.73}$$

As we discussed above, due to the complexity of the real world applications, the constraints in the MDMBV model may dynamically change, making the problem even more complex. To cope with this difficulty, by using the multiple criteria multiple constrain-levels theory, we extend the MEMBV model into the MC2LP model. More specifically, we extend the MEMBV model by relaxing the boundary b into a linear combination of the left limitation b_l and the right limitation b_r, which can be denoted as $\gamma_1 b_l + \gamma_2 b_r$, where $\gamma_1 + \gamma_2 = 1$.

$$\min \sum \xi_i - c \cdot w^T S_B w \tag{11.74}$$

$$\text{s.t.} \quad (w \cdot x_i) - \xi_i \leq \gamma_1 b_l + \gamma_2 b_r, \quad x_i \in G_1, \tag{11.75}$$

$$(w \cdot x_i) + \xi_i \geq \gamma_1 b_l + \gamma_2 b_r, \quad x_i \in G_2, \tag{11.76}$$

$$\xi_i \geq 0, \tag{11.77}$$

where both the matrix S_B and the vector ξ in the objective function are non-negative. When comparing the MC2LP model with the MEMBV model, we can also observe that, on one hand, MEMBV is a special case of MC2LP model, when γ_1 and γ_2 are stable, MC2LP model will degenerate into MEMBV model; on the other hand, since $-c \cdot w^T S_B w$ is a negative item int the objective function, both MEMBV and MC2LP are concave quadratic programming models, and we will discuss how to solve the MC2LP model in the next chapter.

Part III
Applications in Various Fields

Chapter 12
Firm Financial Analysis

12.1 Finance and Banking

Financial institutions and banks are among those industries that have relatively complete and accurate data and have first employed advanced analytic techniques on their data. Typical cases include stock investment, loan payment prediction, credit approval, bankruptcy prediction, and fraud detection.

Classification is one of most extensively used data mining methods in finance and banking. It can be considered as a two- or three-step process. In a two-step process, the first step constructs classification model based on historic data and the second step applies the classification model to unknown data to predict their class labels. A three-step process adds a model adjustment step between model construction and model usage to adapt the model better to data. Classification has wide applications, such as marketing promotion, credit card portfolio management, credit approval, network intrusion detection, and fraud detection.

Traditionally, discriminant analysis, linear and logistic regression, integer programming, decision trees, expert systems, neural networks, and dynamic models are commonly used techniques in finance and banking applications. Relatively few mathematical programming tools have been explored and applied to finance and banking analysis.

To test the applicability of preceding multiple-criteria mathematical models in finance and banking, we select one model, MCQP in Chap. 9 and apply it to three financial datasets. These datasets come from three countries and represent consumer credit card application data, credit approval data, and corporation bankruptcy data. The first dataset is a German credit risk dataset from UCI Machine Learning databases, which collects personal information (e.g., credit history, employment status, age, and housing). It contains 1000 records (700 Normal and 300 Bad) and 24 variables. The objective is to predict the credit risk of these records. The second set is an Australian credit approval dataset from See5 [169]. It has 690 records (383 Normal and 307 Bad) and 15 variables. This dataset concerns about credit card applicants' status: either good or bad. The third set is a Japanese firm bankruptcy set [132, 133], which collects bankrupt (bad) sample Japanese firms (37) and non-bankrupt (normal) sample Japanese firms (111) from 1989 to 1999 and each record

Y. Shi et al., *Optimization Based Data Mining: Theory and Applications*,
Advanced Information and Knowledge Processing,
DOI 10.1007/978-0-85729-504-0_12, © Springer-Verlag London Limited 2011

has 13 variables. Classification models build for this dataset can classify bankrupt and non-bankrupt firms. These three datasets represent different problems in finance and banking. For comparison purpose, the result of MCQP is compared with four well-know classification methods: SPSS linear discriminant analysis (LDA), Decision Tree based See5 [169], SVM light [118], and LibSVM [33].

Data mining and knowledge discovery consists of data selection, data cleaning, data transformation, data preprocessing, data mining or modeling, knowledge interpretation and presentation. Data selection, cleaning, transformation and preprocessing prepare data for modeling step. Knowledge interpretation and presentation summarizes results and presents in appropriate formats to end users. All steps are important; however, this chapter will focus on only data mining or modeling step.

The general classification process listed below summarizes the major steps in our classification experiments. Though this process concerns only about two-class problem, it can be extended to multiple class problems by changing the input and the decision function (Step 2). All the applications discussed in this chapter follow this general process.

12.2 General Classification Process

Input: The dataset $A = \{A_1, A_2, A_3, \ldots, A_n\}$, an $n \times n$ diagonal matrix Y, where

$$Y_{i,j} = \begin{cases} 1, & i \in \{\text{Bad}\}, \\ -1, & i \in \{\text{Normal}\}. \end{cases}$$

Output: Average classification accuracies of 10-fold cross-validation for Bad and Normal; decision scores for all records; decision function.
Step 1 Apply five classification methods: LDA, Decision Tree, SVM, MCQP (Chap. 9), to A using 10-fold cross-validation. The outputs are a set of decision functions, one for each classification method.
Step 2 Compute the classification accuracies using the decision functions.
END

Tables 12.1, 12.2 and 12.3 report the averages of 10-fold cross-validation results of the five classification methods for German set, Australian set, and Japanese set, respectively. Since different performance metrics measure different aspects of classifiers, we use five criteria: accuracy, KS score, type I and II errors, and correlation coefficient, to evaluate the model performance. Accuracy is one the most widely used classification performance metrics. It is the ratio of correct predicted records to the entire records or records in a particular class. Overall Accuracy = $(TN + TP)/(TP + FP + FN + TN)$, Normal Accuracy = $TN/(FP + TN)$, Bad Accuracy = $TP/(TP + FN)$, TN, TP, FN and FP represent true negative, true positive, false negative and false positive, respectively.

Type I error is defined as the percentage of predicted Normal records that are actually Bad records and Type II error is defined as the percentage of predicted Bad

Table 12.1   10-fold cross-validation results of German set

	Classification accuracy			Error rate		KS-score	Corr-coef
	Overall (%)	Normal (%)	Bad (%)	Type I (%)	Type II (%)		
Linear Discriminant Analysis							
Training	73.80	73.43	74.67	25.65	26.25	48.10	0.481
Test	72.20	72.57	71.33	28.32	27.77	43.90	0.439
See5							
Training	89.10	95.57	74.00	21.39	5.65	69.57	0.712
Test	72.20	84.00	44.67	39.71	26.37	28.67	0.312
SVM light							
Training	68.65	79.00	44.50	41.26	32.06	23.50	0.250
Test	66.50	77.00	42.00	42.96	35.38	19.00	0.203
LibSVM							
Training	93.25	100.0	77.50	18.37	0.00	77.50	0.795
Test	94.00	100.0	80.00	16.67	0.00	80.00	0.816
MCQP							
Training	73.86	74.91	71.42	27.62	26.00	46.33	0.464
Test	73.50	74.38	72.00	27.35	26.24	46.38	0.464

Table 12.2   10-fold cross-validation results of Australian set

	Classification accuracy			Error rate		KS-score	Corr-coef
	Overall (%)	Normal (%)	Bad (%)	Type I (%)	Type II (%)		
Linear Discriminant Analysis							
Training	86.09	80.94	92.51	8.47	17.08	73.45	0.739
Test	85.80	80.68	92.18	8.83	17.33	72.86	0.733
See5							
Training	90.29	91.64	88.60	11.06	8.62	80.24	0.803
Test	86.52	87.99	84.69	14.82	12.42	72.68	0.727
SVM light							
Training	55.22	23.76	94.46	18.90	44.66	18.22	0.258
Test	44.83	18.03	90.65	34.14	47.48	8.69	0.126
LibSVM							
Training	99.71	100.0	99.35	0.65	0.00	99.35	0.994
Test	44.83	86.89	27.10	45.62	32.61	13.99	0.174
MCQP							
Training	87.25	88.50	86.38	13.34	11.75	74.88	0.749
Test	86.38	87.00	85.52	14.27	13.20	72.52	0.725

Table 12.3 10-fold cross-validation results of Japanese set

	Classification accuracy			Error rate		KS-score	Corr-coef
	Overall (%)	Normal (%)	Bad (%)	Type I (%)	Type II (%)		
Linear Discriminant Analysis							
Training	73.65	72.07	78.38	23.08	26.27	50.45	0.506
Test	68.92	68.47	70.27	30.28	30.97	38.74	0.387
See5							
Training	93.24	96.40	83.78	14.40	4.12	80.18	0.808
Test	72.30	84.68	35.14	43.37	30.36	19.82	0.228
SVM light							
Training	100.0	100.0	100.0	0.00	0.00	100.0	1.000
Test	48.15	47.25	52.94	49.90	49.91	0.19	0.002
LibSVM							
Training	100.0	100.0	100.0	0.00	0.00	100.0	1.000
Test	50.46	49.45	55.88	47.15	47.49	5.33	0.053
MCQP							
Training	73.65	72.38	77.98	23.33	26.16	50.36	0.504
Test	72.30	72.30	72.47	27.58	27.65	44.77	0.448

records that are actually Normal records. In all three applications (credit risk, credit approval, and bankrupt prediction), Type I errors, which class a customer as good when they are bad, have more serious impact than Type II errors, which class a customer as bad when they are good.

$$\text{Type I error} = \frac{FN}{FN + TN}, \qquad \text{Type II error} = \frac{FP}{FP + TP}.$$

In addition, a popular measurement in credit risk analysis, KS score, is calculated. The KS (Kolmogorov–Smirnov) value measures the largest separation of cumulative distributions of Goods and Bads [46] and is defined as:

$$\text{KS value} = \max |Cumulative\ distribution\ of\ Bad - Cumulative\ distribution\ of\ Normal|.$$

Another measurement—Correlation coefficient, which falls into the range of $[-1, 1]$, is used to avoid the negative impacts of imbalanced classes. The correlation coefficient is -1 if the predictions are completely contrary to the real value, 1 if the predictions are 100% correct, and 0 if the predictions are randomly produced. The Correlation coefficient is calculated as follows:

$$\text{Correlation coefficient} = \frac{(TP \times TN) - (FP \times FN)}{\sqrt{(TP + FN)(TP + FP)(TN + FP)(TN + FN)}}.$$

Tables 12.1, 12.2 and 12.3 summarize the five metrics for training sets and test sets. Training results indicate how well the classification model fits the training set while test results reflect the real predicting power of the model. Therefore, test results determine the quality of classifiers. Among the five methods, LibSVM achieves the best performance on all five metrics for German set, but performs poorly on Australian set (27.1% accuracy for Bad; 45.62% Type I error; 32.61% Type II error; 13.99 KS; 0.174 correlation coefficient). LDA yields the best accuracies, error rates, KS scores, and correlation coefficient for Australian data and exhibits comparative results on other two sets. See5 has the best accuracies for Overall and Normal for Australian and Japanese sets. SVM light performs about average for German data, but poorly for Japanese and Australian sets. MCQP has excellent performance on five metrics for Japanese data and above average performance on other two sets.

The experimental study indicates that (1) MCQP can classify credit risk data and achieves comparable results with well-known classification techniques; (2) the performance of classification methods may vary when the datasets have different characteristics.

12.3 Firm Bankruptcy Prediction

Firm bankruptcy prediction is an interesting topic because many stakeholders such as bankers, investors, auditors, management, employees, and the general public are interested in assessing bankruptcy risk. The MCLP model (Model 2) is used as a data mining tool for bankruptcy prediction. Using the well-known 1990s Japanese financial data (37 Bankrupt and 111 Non-Bankrupt), Kwak, Shi and Cheh [132, 133] showed that MCLP can be used as a standard to judge the well-known prediction methods. According to the empirical results in Table 12.4, Ohlson's [157] predictor variables (Panel B) perform better than Altman's [4] predictor variables (Panel A), based on the overall prediction rates of 77.70% and 72.97%, respectively. To determine whether a combined set of variables provides better bankruptcy prediction results, MCLP is applied by using Ohlson's [157] and Altman's [4] variables together. The results of Panel C in Table 12.4 are inferior to those using just Ohlson's [157] variables (overall (Type I) prediction rate of 74.32% (51.35%) in Panel C versus 77.70% (70.27%) in Panel B). These results support the superiority of using Ohlson's nine variables to predict bankruptcy using Japanese financial data.

Kwak et al. [132, 133] also tested the use of the MCLP model for bankruptcy prediction using U.S. data from a sample period of 1992 to 1998 (133 Bankrupt and 1021 Non-Bankrupt). The results showed that MCLP performed better than either the multiple discriminant analysis (MDA) approach of Altman [4] or the Logit regression of Ohlson [157] as in Table 12.5. If we consider that the costs of Type I errors outweigh the costs of Type II errors in a bankruptcy situation, our focus of prediction error rate should be on Type I errors. The prediction rate of Type I errors is increasing and the prediction rate of Type II errors is decreasing in Altman's model compared with the model with more numbers of control firms. This is an interesting

Table 12.4 Predictability of MCLP using various financial predictor variables on bankruptcy for Japanese firms

Panel A: Altman's [4] five predictor variables

All years	N	Number correct	Percent correct	Percent error
Type-I	37	6	16.22	83.78
Type-II	111	102	91.89	8.11
Overall prediction rate			72.97%	27.03%

Panel B: Ohlson's [157] nine predictor variables

All years	N	Number correct	Percent correct	Percent error
Type-I	37	26	70.27	29.73
Type-II	111	89	80.18	19.82
Overall prediction rate			77.70%	22.30%

Panel C: Combination of Altman's [4] and Ohlson's [157] five predictor variables

All years	N	Number correct	Percent correct	Percent error
Type-I	37	19	51.35	48.65
Type-II	111	91	81.98	18.02
Overall prediction rate			74.32%	25.26%

Table 12.5 Predictability of MCLP on bankruptcy methods for USA firms

Year		Altman predictor variables			Ohlson predictor variables		
		N	Number correct	Percent correct (%)	N	Number correct	Percent correct (%)
1992	Type-I	21	15	71	21	11	52
	Type-II	21	17	81	70	66	94
1993	Type-I	15	13	87	15	13	87
	Type-II	15	13	87	74	72	97
1994	Type-I	12	8	67	12	9	75
	Type-II	12	10	83	63	62	98
1995	Type-I	8	8	100	8	6	75
	Type-II	8	8	100	58	57	98
1996	Type-I	15	14	93	15	8	53
	Type-II	15	14	93	96	93	97
1997	Type-I	12	10	83	12	7	58
	Type-II	12	11	92	70	69	99
1998	Type-I	9	7	78	9	4	44
	Type-II	9	8	89	89	87	98
All data	Type-I	126	109	87	91	81	89
	Type-II	91	62	68	521	492	94

result which suggests that Altman's original model could be upward-biased, and that he should use more control firms to represent the real-world situation of bankruptcy.

From the results of the overall predication rate in terms of Type-I errors, Table 12.5 shows that MCLP on Ohlson's variables is similar to Altman's variables (87% vs. 89%) using more control firms. The percentages are higher than these of two studies (see [4, 157]). This sustains our findings in this study. From the above results, the MCLP approach performs better than both Altman's method and Ohlson's method.

Chapter 13
Personal Credit Management

13.1 Credit Card Accounts Classification

The most commonly used methods in predicting credit card defaulters are credit scoring models. Based on their applications in credit management, the scoring models can be classified into two categories. The first category concerns about *application scores* which can help to decide whether or not to issue credit to a new credit applicant. The second category concerns *behavior scores* which can help to forecast future behavior of an existing account for risk analysis. Specifically, behavior scores are used to determine "raising or lowering the credit limit; how the account should be treated with regard to promotional or marketing decisions; and when action should be taken on a delinquent account" [172]. Since they focus on different aspects of credit card management, these two scores require different methods to implement.

Behavior scoring models utilize various techniques to identify attributes that can effectively separate credit cardholders' behaviors. These techniques include linear discriminant analysis (LDA), decision trees, expert systems, neural networks (NN), and dynamic models [172]. Among them, linear discriminant analysis has been regarded as the most commonly used technique in credit card bankrupt prediction.

Since Fisher [73] developed a discriminant function, linear discriminant analysis had been applied to various managerial applications. It can be used for two purposes: to classify observations into groups or to describe major differences among the groups. For the purpose of classification, LDA builds a discriminant function of group membership based on observed attributes of data objects. The function is generated from a training dataset for which group membership (or label) is known. If there are more than two groups, a set of discriminant functions can be generated. This generated function is then applied to a test dataset for which group membership is unknown, but has the same attributes as the training dataset. The fundamental of LDA can be found in [6, 122, 150].

Although LDA has well established theory and been widely accepted and used in credit card accounts prediction, it has certain drawbacks. Eisenbeis [64] pointed out eight problems in applying discriminant analysis for credit scoring. These problems range from group definition to LDA's underlying assumptions. Kolesar and

Showers [125] also suggested that LDA "produces solutions that are optimal for a particular decision problem when the variables have a special multivariate normal distribution". Violations of these assumptions, such as equal covariance matrices and multivariate normal distributions, happen frequently in real-life applications and raise questions about the validity of using LDA in credit card accounts classification. Therefore, alternative classification methods that are not restricted by statistical assumptions behoove to be explored.

The real-life credit card dataset used in this chapter is provided by First Data Corporation (FDC), the largest credit card service industry in the world. The raw data came originally from a major US bank which is one of FDC's clients. It contains 3,589 records and 102 variables (38 original variables and 64 derived variables) describing cardholders' behaviors. The data were collected from June 1995 to December 1995 (seven months) and the cardholders were from twenty-eight states in USA. This dataset has been used as a classic working dataset in FDC for various data analyses to support the bank's business intelligence. Each record has a class label to indicate its credit status: either Good or Bad. Bad indicates a bankrupt credit card account and Good indicates a good status account. The 38 original variables can be divided into four categories: balance, purchase, payment, cash advance, in addition to related variables. The category variables represent raw data of previous six or seven consecutive months. The related variables include interest charges, date of last payment, times of cash advance, and account open date. The detailed description of these variables is given in Table 13.1.

The 64 derived variables are created from the original 38 variables to reinforce the comprehension of card-holder's behaviors, such as times over-limit in the last two years, calculated interest rate, cash as percentage of balance, purchase as percentage to balance, payment as percentage to balance, and purchase as percentage to payment. Table 13.2 gives a brief description of these attributes (variables).

Table 13.3 illustrates a sample record from Good credit card accounts group with 64 derived attributes and their corresponding values.

Table 13.4 shows a sample record from Bad credit card accounts group with 64 derived attributes and their corresponding values.

For the purpose of credit card classification, the 64 derived variables were chosen to compute the model since they provide more precise information about credit card accounts' behaviors. The dataset is randomly divided into one training dataset (200 records) and one verifying dataset (3,389 records). The training dataset has class label for each record and is used to calculate the optimal solution. The verifying dataset, on the other hand, has no class labels and is used to validate the predicting accuracy of MCQP. The predicting accuracy of a classification method is not judged by the accuracy of training dataset, but the accuracy of verifying dataset. The goal of classification is to apply solutions obtained from the training phase to predict future unknown data objects.

The objective of this research is to produce a "black list" of the credit cardholders. This means we seek a classifier that can identify as many Bad records as possible. This strategy is a basic one in credit card business intelligence. Theoretically speaking, we shall first construct a number of classifiers and then choose one more Bad

Table 13.1 Original attributes of credit card dataset

Variables	Description		Variables	Description	
1	Balance	Jun 95	20	Cash advance	Jul 95
2		Jul 95	21		Aug 95
3		Aug 95	22		Sep 95
4		Sep 95	23		Oct 95
5		Oct 95	24		Nov 95
6		Nov 95	25		Dec 95
7		Dec 95	26	Interest charge: Mechanize	Dec 95
8	Purchase	Jul 95	27	Interest charge: Cash	Dec 95
9		Aug 95	28	Number of purchase	Dec 95
10		Sep 95	29	Number of cash advance	Dec 95
11		Oct 95	30	Cash balance	Nov 95
12		Nov 95	31	Cash balance	Dec 95
13		Dec 95	32	Number of over limit in last 2 years	
14	Payment	Jul 95	33	Credit line	
15		Aug 95	34	Account open date	
16		Sep 95	35	Highest balance in last 2 years	
17		Oct 95	36	Date of last payment	
18		Nov 95	37	Activity index	Nov 95
19		Dec 95	38	Activity index	Dec 95

records. The research procedure in this paper has four steps. The first step is *data cleaning*.Within this step, missing data cells were excluded and extreme values that were identified as outliers were removed from the dataset. The second step is *data transformation*. The dataset was transformed according to the format requirements of LINGO 8.0, which is a software tool for solving nonlinear models used by this research [137]. The third step is *model formulation and classification*. A two-group MCQP model is formulated, which will be elaborated in Sect. 13.3, and applied to the training dataset to obtain optimal solutions. The solutions are then applied to the verifying dataset within which class labels were removed to compute scores for each record. Based on these scores, each record is predicted as either Bad or Good account. By comparing the predicted labels with original labels of records, the classification accuracy of MCQP model can be determined. If the classification accuracy is acceptable by data analysts, this solution will be used to predict future credit card records. Otherwise, data analysts need to modify the boundary and attributes values to get another set of optimal solutions. The fourth step is *results presentation*. The acceptable classification results were summarized in tables and figures using Excel and presented to end users.

Table 13.2 Derived attributes of credit card dataset

Variables	Description
1	Balance of Nov 95
2	Max balance of Jul 95 to Dec 95
3	Balance difference between Jul 95 to Sep 95 and Oct 95 to Dec 95
4	Balance of Dec 95 as percent of max balance of Jul 95 to Dec 95
5	Average payment of Oct 95 to Dec 95
6	Average payment of Jul 95 to Dec 95
7	Payment of Dec 95 ($\leq$\$20)
8	Payment of Dec 95 (>\$20)
9	Payment of Nov 95 ($\leq$\$20)
10	Payment of Nov 95 (>\$20)
11	Payment of Oct 95 to Dec 95 minus Payment of Jul 95 to Sep 95 (Min)
12	Payment of Oct 95 to Dec 95 minus Payment of Jul 95 to Sep 95 (Max)
13	Purchase of Dec 95
14	Purchase of Nov 95
15	Purchase of Dec 95 as percent of max purchase of Jul 95 to Dec 95
16	Revolve balance between Jul 95 to Sep 95 and Oct 95 to Dec 95
17	Max minus Min revolve balance between Jul 95 to Dec 95
18	Cash advance of Jul 95 to Dec 95
19	Cash as percent of balance of Jul 95 to Dec 95 (Max)
20	Cash as percent of balance of Jul 95 to Dec 95 (Min)
21	Cash as percent of balance of Jul 95 to Dec 95 (Indicator)
22	Cash advance of Dec 95
23	Cash as percent of balance of Jul 95 to Dec 95
24	Cash as percent of balance of Oct 95 to Dec 95
25	Cash as percent of payment of Jul 95 to Dec 95 (Min)
26	Cash as percent of payment of Jul 95 to Dec 95 (Log)
27	Revolve balance to payment ratio Dec 95
28	Revolve balance to payment ratio Nov 95
29	Revolve balance to payment ratio Oct 95 to Dec 95
30	Revolve balance to payment ratio Jul 95 to Dec 95
31	Revolve balance to payment ratio Jul 95 to Sep 95 minus Oct 95 to Dec 95 (Max)
32	Revolve balance to payment ratio Jul 95 to Sep 95 minus Oct 95 to Dec 95 (Min)
33	Revolve balance to payment ratio Jul 95 to Dec 95, Max minus Min (>35)
34	Revolve balance to payment ratio Jul 95 to Dec 95, Max minus Min ($\leq$35)
35	Purchase as percent of balance Dec 95
36	Purchase as percent of balance Oct 95 to Dec 95
37	Purchase as percent of balance Jul 95 to Dec 95
38	Purchase as percent of balance Jul 95 to Sep 95 minus Oct 95 to Dec 95
39	Purchase as percent of balance, Max minus Min, Jul 95 to Dec 95

Table 13.2 (Continued)

Variables	Description
40	Purchase as percent of payment, Jul 95 to Dec 95
41	Purchase as percent of payment, Nov 95
42	Purchase as percent of payment, Jul 95 to Sep 95 minus Oct 95 to Dec 95 (Max)
43	Purchase as percent of payment, Jul 95 to Sep 95 minus Oct 95 to Dec 95 (Min)
44	Purchase as percent of payment, Dec 95 as percent of Jul 95 to Dec 95
45	Purchase as percent of payment, Max minus Min, Jul 95 to Dec 95
46	Interest charge Dec 95
47	Interest charge Dec 95 as percent of credit line
48	Calculated interest rate ($\leq 5\%$)
49	Calculated interest rate ($>5\%$)
50	Number of months since last payment
51	Number of months since last payment squared
52	Number of purchases, Dec 95
53	Number of cash advances, Dec 95
54	Credit line
55	Open to buy, Dec 95
56	Over limit indicator of Dec 95
57	Open to buy, Nov 95
58	Utilization, Dec 95
59	Number of times delinquency in last two years
60	Residence state category
61	Transactor indicator
62	Average payment of revolving accounts
63	Last balance to payment ratio
64	Average OBT revolving accounts

13.2 Two-Class Analysis

13.2.1 Six Different Methods

Using the real-life credit card dataset in the above section, we first conduct the MCQP classification. Then, we compare the performance of MCQP with multiple criteria linear programming (MCLP) (Chap. 7), linear discriminant analysis (LDA), decision tree (DT), support vector machine (SVM), and neural network (NN) methods in terms of predictive accuracy.

The MCQP classification consists of training process and verifying process. Given the training dataset, the classifier or optimal solution of MCQP depends on the choice of weights σ_ξ and σ_β. The verifying dataset will be calculated according to the resulting classifier. To identify the better classifier, we assume that $\sigma_\xi + \sigma_\beta = 1$

Table 13.3 A sample record of Good group with 64 derived attributes

Attributes:	1	2	3	4	5	6	7	8
Value:	9.18091	0.53348	7.3	1	5.32301	5.32301	5.32301	0
Attributes:	9	10	11	12	13	14	15	16
Value:	5.32301	0	0	0	6.00859	5.08153	3.70601	7.051758
Attributes:	17	18	19	20	21	22	23	24
Value:	1	1	0	6.9277	0	1	0	1.7027
Attributes:	25	26	27	28	29	30	31	32
Value:	0.81301	4.14654	46.3659	45.9328	46.2161	45.8386	0	0.7548
Attributes:	33	34	35	36	37	38	39	40
Value:	0	3.1354	1.39887	0.65921	0.99082	−1.521	10.29	1.13407
Attributes:	41	42	43	44	45	46	47	48
Value:	1	−0.7253	0	1	5	7.19811	1.16252	1.37683
Attributes:	49	50	51	52	53	54	55	56
Value:	0	−1	1	3.68888	0	8	7.28235	0
Attributes:	57	58	59	60	61	62	63	64
Value:	7.48997	2.61352	0	1	0	1.77071	7.00021	4.1325

and $0 \leq \sigma_\xi$, $\sigma_\beta \leq 0$. Let $\sigma_\xi = 1, 0.75, 0.625, 0.5, 0.375, 0.25, 0$ and $\sigma_\beta = 0, 0.25,$ $0.375, 0.5, 0.625, 0.75, 1$, then we obtained a number of classifiers as in Table 13.5.

In Table 13.5, we define "Type I error" to be the percentage of predicted Good records which are actually Bad records and "Type II error" to be the percentage of predicted Bad records which are actually Good records. Then, these two errors can be used to measure the accuracy rates of different classifiers. For example, in Table 13.5, the value 3.03% of Type I error for training dataset when $\sigma_\xi = 0.375$ and $\sigma_\beta = 0.625$ indicates that 3.03% of predicted Good records are actually Bad records. Similarly, the value 3.96% of Type II error for training dataset when $\sigma_\xi = 0.375$ and $\sigma_\beta = 0.625$ indicates that 3.96% of predicted Bad records are actually Good records. As we discussed in Sect. 13.2.3, misclassified Bad accounts contribute to huge lost in credit card businesses and thus creditors are more concern about Type I error than Type II error. Also, since the accuracy of verifying dataset is the real indicator of prediction quality of classification techniques, classification techniques with lower verifying Type I errors are judged as superior than those with higher verifying Type I errors. Of course, we are by no means saying that Type II error is not important. Actually, misclassification of Good customers as Bad ones (Type II error) may cause increased customer dissatisfaction and appropriate corrective actions should be considered.

In this dataset, we see that except the extreme cases $\sigma_\xi = 1, 0$ and $\sigma_\beta = 0, 1$, all other values of σ_ξ and σ_β offer the same classification results. This shows that

Table 13.4 A sample record of Bad group with 64 derived attributes

Attributes:	1	2	3	4	5	6	7	8
Value:	8.48182	1.36142	0	2.13355	4.9151	4.63149	4.61512	0
Attributes:	9	10	11	12	13	14	15	16
Value:	4.65396	0	0	67.333	0	0	0	6.803711
Attributes:	17	18	19	20	21	22	23	24
Value:	1	0	0	3.23942	0	1	0	0
Attributes:	25	26	27	28	29	30	31	32
Value:	0	0	46.7845	45.2997	34.8588	46.7744	−15.1412	0
Attributes:	33	34	35	36	37	38	39	40
Value:	0	26.4752	0	0	0	0	0	0
Attributes:	41	42	43	44	45	46	47	48
Value:	1	0	0	0	0	6.63935	1.27433	1.58426
Attributes:	49	50	51	52	53	54	55	56
Value:	0	0	0	0	0	6	7.08867	0
Attributes:	57	58	59	60	61	62	63	64
Value:	7.06798	3.04318	0	1	0	2.13219	6.83884	3.4208

the MCQP classification model is stable in predicting credit cardholders' behaviors. In the training process, 96 data records out of 100 were correctly classified for the "Good" group and 97 data records out of 100 were correctly classified for the "Bad" group. The classification rates are 96% for the "Good" group and 97% for the "Bad" group. In the verifying process, 2,890 data records out of 3,090 were correctly classified for the "Good" group and 290 data records out of 299 were correctly classified for the "Bad" group. The catch rates are 93.53% for the "Good" group and 96.99% for the "Bad" group.

Now, we conducted an experiment by applying four classification techniques: multiple criteria linear programming (MCLP), linear discriminant analysis (LDA), decision tree (DT), support vector machine (SVM), and neural network (NN) methods to the same credit card dataset and contrasting the results with MCQP's.

Although the multiple criteria linear programming (MCLP) is a linear version of MCQP, the classifier is determined by a compromise solution, not an optimal solution [127, 162, 182, 183]. The MCLP has been implemented by C++ [128]. In linear discriminant analysis (LDA), the commercial statistical software SPSS 11.0.1 Windows Version was chosen. SPSS is one of the most popular and easy-to-use statistical software packages for data analysis. Decision Tree has been widely regarded as an effective tool in classification [169]. We adopted the commercial decision tree software C5.0 (the newly updated version of C4.5) to test the classification accuracy of the two groups. The software of support vector machine (SVM) for this dataset was Chinese Modeling Support Vector Machines produced by the Training Center

Table 13.5 MCQP classification with different weights

	Bad		Good	
	Correctly identified	Type II error	Correctly identified	Type I error
$\sigma_\xi = 1, \sigma_\beta = 0$				
Training	67	4.29%	97	25.38%
Verifying	283	71.24%	2389	0.67%
$\sigma_\xi = 0.75, \sigma_\beta = 0.25$				
Training	97	3.96%	96	3.03%
Verifying	290	40.82%	2890	0.31%
$\sigma_\xi = 0.625, \sigma_\beta = 0.375$				
Training	97	3.96%	96	3.03%
Verifying	290	40.82%	2890	0.31%
$\sigma_\xi = 0.5, \sigma_\beta = 0.5$				
Training	97	3.96%	96	3.03%
Verifying	290	40.82%	2890	0.31%
$\sigma_\xi = 0.375, \sigma_\beta = 0.625$				
Training	97	3.96%	96	3.03%
Verifying	290	40.82%	2890	0.31%
$\sigma_\xi = 0.25, \sigma_\beta = 0.75$				
Training	97	3.96%	96	3.03%
Verifying	290	40.82%	2890	0.31%
$\sigma_\xi = 1, \sigma_\beta = 0$				
Training	No optimal solution			
Verifying				

of China Meteorological Administration [37]. Finally, we conducted a comparison of the MQLP method to neural networks (NN). In this comparison, we used codes through a typical back-propagation algorithm for the two-group classification [228]. All of the comparisons are given in Table 13.6. In this table, the result of MCQP is based on the stable case discussed in the previous subsection.

With the objective of producing a "black list" of the credit cardholders, we see that for the verifying sets of Table 13.6, MCQP provides the highest number of correctly predicted Bad records (290 records out of 299) as well as the lowest Type I error (0.31%) among all six techniques. Therefore, MCQP is the winner in this experiment. In addition, from Table 13.6, we can summarize the following observations:

Table 13.6 Comparisons of MCQP with others

	Bad		Good	
	Correctly identified	Type II error	Correctly identified	Type I error
Multi-Criteria Quadratic Programming				
Training	97	3.96%	96	3.03%
Verifying	290	40.82%	2890	0.31%
Multi-Criteria Linear Programming				
Training	84	16.00%	84	16.00%
Verifying	219	74.80%	2440	3.17%
Linear Discriminant Analysis				
Training	98	4.85%	95	2.06%
Verifying	288	45.35%	2851	0.38%
Decision Tree				
Training	97	3.00%	97	3.00%
Verifying	274	53.08%	2780	0.89%
Support Vector Machine				
Training	100	0.00%	100	0.00%
Verifying	184	61.51%	2796	3.95%
Neural Network				
Training	75	6.25%	95	20.83%
Verifying	245	19.41%	3031	1.75%

(1) Besides MCQP, the order of better producing a black list is LDA, DT, NN, MCLP and SVM.
(2) NN exhibits the best verifying Type II error (19.41%) among the six techniques used in this experiment, followed by MCQP and LDA.
(3) All six techniques present a much lower Type I errors than Type II errors.
(4) Although SVM displays a perfect training Type I and II errors (0%), its verifying Type I and II errors are much higher (3.95% and 61.51%).

13.2.2 Implication of Business Intelligence and Decision Making

From the aspect of business intelligence and decision making, data mining consists of four stages: (1) Selecting, (2) Transforming, (3) Mining, and (4) Interpreting [182]. If Sects. 13.2.3–13.4 of this chapter are related to the first three stages, then data interpretation is very critical for credit card portfolio management. A poor interpretation analysis may lead to missing useful information, while a good

Table 13.7 Cumulative
distributions: Bad vs. Good

Range	cum_Bad	cum_Good	KS 1 vs. KS 2
−0.80782	0.00%	0.00%	0.00
−0.60695	0.00%	0.06%	0.06
−0.40608	0.00%	0.32%	0.32
−0.20522	0.00%	0.97%	0.97
−0.00435	0.00%	2.33%	2.33
0.196523	0.00%	5.40%	5.40
0.397392	0.00%	9.58%	9.58
0.598262	0.00%	17.99%	17.99
0.799131	0.67%	33.88%	33.21
1	3.01%	59.32%	56.31
1.09346	**22.41%**	**93.53**	**71.12**
1.18692	45.82%	95.37%	49.55
1.280381	65.55%	96.96%	31.41
1.373841	82.94%	98.32%	15.37
1.467301	91.97%	99.06%	7.09
1.560761	97.32%	99.48%	2.16
1.654222	99.00%	99.64%	0.65
1.747682	99.67%	99.84%	0.17
1.841142	99.67%	99.87%	0.20
1.934602	100.00%	100.00%	0.00

analysis can provide a comprehensive picture for effective decision making. Even though data mining results from MCQP can be flexibly interpreted and utilized for credit cardholders' retention and promotion, we now apply a popular method, called Kolmogorov-Smirnov (KS) value in US credit card industry to interpret the MCQP classification [46]. For the problem of the two-group classification in this paper, the KS value is given as:

$$\text{KS value} = \max |\text{Cumulative distribution of Good}$$
$$- \text{Cumulative distribution of Bad}|.$$

Based on the predicted Bad records and Good records using MCQP from the verifying dataset in Table 13.6, we construct their cumulative distributions respectively as in Table 13.7. Note that the "range" is derived from the MCQP score calculated from classifier (optimal solution) **W** in the training dataset. The graphical presentation of KS value is given in Fig. 13.1.

The KS measurement has empirically demonstrated the high correlation to predict Bad records in terms of Good records. Suppose that the verifying dataset of 3,389 records has unknown label and Table 13.7 or Fig. 13.1 are the predicted spending behaviors of these credit cardholders in the near future, then the knowledge provided in Table 13.7 or Fig. 13.1 can be used for various business decision making.

Fig. 13.1 KS value of 71.12 on verifying records

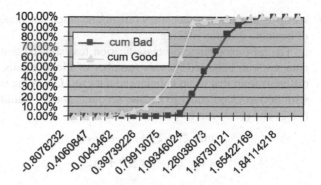

As an example, when we identify all credit cardholders whose score under or equal to 1.09346 (see Table 13.7), we will find that 67 of these people (22.41% × 299) intend to be Bad customers and 2,898 of them (93.53% × 3,099) could be Good customers. Since these 2,965 cardholders are 87.49% of the total 3,389 records, we can certainly make necessary managerial decision based on this business intelligence. For instance, we might implement the credit policy to reduce the credit limit of 67 predicted Bad records and increase the credit limit to 2,965 predicted Good records or send them promotion coupons.

13.2.3 FMCLP Analysis

In this subsection, the FDC dataset with 65 derived attributes and 1000 records is first used to train the FMCLP (Sect. 8.1) classifier. Then, the training solution is employed to predict the spending behaviors of another 5000 customers from different states. Finally, the classification results are compared with that of the MCLP method [127, 128], neural network method [98, 228], and decision tree method [52, 169].

There are two kinds of accuracy rates involved in this section. The first one is the *absolute accuracy rate* for Bad (or Good) which is the number of actual Bad (or Good) identified divided by the total number of Bad (or Good). The second is called *catch rate*, which is defined as the actual number of caught Bad and Good divided by the total number of Bad and Good. Let r_g be the absolute accuracy rate for Good and r_b be the absolute accuracy rate for Bad. Then the catch rate c_r can be written as:

$$c_r = \frac{r_g \times \text{the total number of Good} + r_b \times \text{the total number of Bad}}{\text{the total number of Good \& Bad}}.$$

The difference of two rates is that the absolute accuracy rate measures the separation power of the model for an individual class while the catch rate represents the overall degree of correctness when the model is used. A threshold τ in this section is set up against absolute accuracy rate or/and catch rate depending on the requirement of business decision making.

The past experience on classification test showed that the training results of a data set with balanced records (number of Good equals number of Bad) may be different from that of an unbalanced data set. Given the unbalanced 1000 accounts with 860 as Good and 140 as Bad accounts for the training process, model (13.1), (13.2) and (13.3) can be built as follows:

$$\min \xi_1 + \xi_2 + \cdots + \xi_{1000}$$

$$\text{s.t.} \quad (x_1 \cdot w) \le b + \xi_1,$$

$$\vdots$$

$$(x_{140} \cdot w) \le b + \xi_{140},$$

$$(x_{141} \cdot w) \ge b - \xi_{141},$$

$$\vdots \tag{13.1}$$

$$(x_{1000} \cdot w) \ge b - \xi_{1000},$$

$$\alpha_i \ge 0,$$

$$w = (w_1, \ldots, w_{65}) \text{ is unrestricted,}$$

$$x_i = (x_{i1}, \ldots, x_{i65}) \text{ is given.}$$

Let $\quad \min \xi_1 + \xi_2 + \cdots + \xi_{1000} = y_{1L}$ and

$$\max \xi_1 + \xi_2 + \cdots + \xi_{1000} = y_{1U} = 1000.$$

$$\min \beta_1 + \beta_2 + \cdots + \beta_{1000}$$

$$\text{s.t.} \quad (x_1 \cdot w) \ge b - \beta_1,$$

$$\vdots$$

$$(x_{140} \cdot w) \ge b - \beta_{140},$$

$$(x_{141} \cdot w) \le b + \beta_{141},$$

$$\vdots \tag{13.2}$$

$$(x_{1000} \cdot w) \le b + \beta_{1000},$$

$$\beta_i \ge 0,$$

$$w = (w_1, \ldots, w_{65}) \text{ is unrestricted,}$$

$$x_i = (x_{i1}, \ldots, x_{i65}) \text{ is given.}$$

Let $\quad \max \beta_1 + \beta_2 + \cdots + \beta_{1000} = y_{2U}$ and

$$\min \beta_1 + \beta_2 + \cdots + \beta_{1000} = y_{2L} = 10.$$

Then,

$$\max \ \eta$$

$$\text{s.t.} \quad \eta \le \frac{\xi_1 + \xi_2 + \cdots + \xi_{1000} - 1000}{y_{1U} - 1000},$$

$$\eta \le \frac{\beta_1 + \beta_2 + \cdots + \beta_{1000} - 10}{y_{2U} - 10},$$

$$(x_1 \cdot w) = b + \xi_1 - \beta_1,$$

$$\vdots$$

$$(x_{140} \cdot w) = b + \xi_{140} - \beta_{140}, \qquad\qquad (13.3)$$

$$(x_{141} \cdot w) = b - \xi_{141} + \beta_{141},$$

$$\vdots$$

$$(x_{1000} \cdot w) = b - \xi_{1000} + \beta_{1000},$$

$$\eta, \xi_i, \beta_i \ge 0,$$

$$w = (w_1, \ldots, w_{65}) \text{ is unrestricted,}$$

$$x_i = (x_{i1}, \ldots, x_{i65}) \text{ is given.}$$

A well-known commercial software package, LINDO has been used to perform the training and predicting processes [137]. Table 13.8 shows learning results of the above FMCLP method for different values of the boundary b. If the threshold τ of finding the absolute accuracy rate of bankruptcy accounts (Bad) is predetermined as 0.85, then the situations when $b = -3.00$ and -4.50 are satisfied as better classifiers according to FMCLP Algorithm. Both classifiers (solutions) are resulted from model (13.3) (see Table 13.8). The catch rates of these two classifiers is not as high as the classifier of model (13.3) when $b = -1.10$ or -0.50. If we use classifier of model (13.3) when $b = -3.00$, then the prediction for the bankruptcy accounts (Bad) among 5000 records by using the absolute accuracy rate is 88% (Table 13.9).

A balanced data set was formed by taking 140 Good accounts from 860 of the 1000 accounts used before and combining with the 140 accounts. The records of 280 were trained and some of results are summarized in Table 13.10, where we see that the best catch rate .907 is at $b = -1.1$ of model (13.3). Although the best absolute accuracy rate for Bad accounts is .907143 at $b = -1.1$ of model (13.3), the predicting result on Bad accounts for 5000 records is 87% at $b = -0.50$ and 1.10 of model (13.3) in Table 13.11. If the threshold τ for all of the absolute accuracy rate of Bad accounts, the absolute accuracy rate of Good accounts and catch rate is set up as 0.9 above, then model (13.3) with $b = -1.10$ is the only one to produce the satisfying classifier by FMCLP Algorithm. However, the predicting result of this case is not the best one (see Table 13.11).

Generally, in both balanced and unbalanced data sets, model (13.1) is better than models (13.2) and (13.3) in identifying Good accounts, while model (13.2) identifies

Table 13.8 Learning results of unbalanced 1000 records

Different b value	Absolute accuracy rate (13.1)		Absolute accuracy rate (13.2)		Absolute accuracy rate (13.3)		Catch rate by (13.1)	Catch rate by (13.2)	Catch rate by (13.3)
	Good	Bad	Good	Bad	Good	Bad			
−0.50	1	0	.932558	.121429	.981395	.021429	.86	.819	.847
−1.10	1	0	.870930	.207143	.862791	.307143	.86	.778	.785
−2.00	1	0	1	0	.575581	.142857	.86	.86	.515
−3.00	1	0	.732558	.657143	.324419	.900000	.86	.722	.405
−4.50	1	0	.598837	.528571	.641860	.878571	.86	.589	.675

Table 13.9 Predicting results of 5000 records via unbalanced training

Different b value	Absolute accuracy rate (13.1)		Absolute accuracy rate (13.2)		Absolute accuracy rate (13.3)	
	Good	Bad	Good	Bad	Good	Bad
−0.50	83.1%	29%	95%	12%	98%	2.2%
−1.10	83.1%	29%	88%	19%	83%	29%
−2.00	83.1%	29%	83%	29%	51%	28%
−3.00	83.1%	29%	77%	56%	32%	88%
−4.50	83.1%	29%	62%	53%	69%	78%

more Bad accounts than both model (13.1) and (13.3). Model (13.3) has a higher catch rate in balanced data set than unbalanced data set compared with models (13.1) and (13.2). Therefore, if the data set is balanced, it is meaningful to implement FMCLP Algorithm proposed for credit card bankruptcy analysis. This conclusion, however, may not be true for all kinds of data sets because of the different data structure and data feature.

Three known classification techniques, decision tree, neural network and multiple criteria linear programming (MCLP) have been used to run the 280 balanced data set and test (or predict) the 5000 credit card-holder records in a major US bank. These results are compared with the FMCLP approach discussed above (see Table 13.12 and Table 13.13). The software of decision tree is the commercial version called C5.0 (the newly updated version of C4.5) [14] while software for both neural network and MCLP were developed at the Data Mining Lab, University of Nebraska at Omaha [128, 228]. Note that in both Table 13.12 and Table 13.13, the column T_g and T_b respectively represent the number of Good and Bad accounts identified by a method, while the rows of Good and Bad represent the actual numbers of the accounts.

In Table 13.12, the final training result on decision tree was produced by C5.0. The configuration used for training the neural network result includes a back prorogation algorithm, one hidden layer with 16 hidden nodes, random initial weight, sigmoid function, and 8000 training periods. The boundary value of b in both MCLP and FMCLP methods were −1.10. As we see, the best training comparison on Good (non-bankruptcy) accounts is the decision tree with 138 out of 140 (98.57%) while the best for Bad (bankruptcy) accounts is the MCLP method with 133 out of 140 (95%). However, the FMCLP method has equally identified 127 out of 140 (90.71%) for Good and Bad. The neural network method underperformed others in the case.

Table 13.13 shows the predicting (or testing) results on 5000 records by using the classifiers based on the results of 280 balanced data sets. The MCLP method outperforms others in terms of predicting Good accounts with 3160 out of 4185 (75.51%), but the FMCLP method proposed in this paper is the best for predicting Bad accounts with 702 out of 815 (86.14%). If the business strategy of making black list on Bad accounts is chosen, then the FMCLP method should be used to conduct the data mining project. Therefore, the proposed FMCLP method demonstrated its

Table 13.10 Learning results of unbalanced 1000 records

Different b value	Absolute accuracy rate (13.1)		Absolute accuracy rate (13.2)		Absolute accuracy rate (13.3)		Catch rate by (13.1)	Catch rate by (13.2)	Catch rate by (13.3)
	Good	Bad	Good	Bad	Good	Bad			
−2.00	.771429	.571429	.571429	.821429	.9	.9	.67	.425	.904
−1.10	.992857	.65	.65	.821429	.907143	.907143	.82	.425	.907
−0.50	.992857	.657143	.657143	.821429	.9	.9	.83	.425	.900
0.50	1	.657143	.657143	.807143	.892857	.892857	.83	.429	.896
1.10	.992857	.65	.65	.821429	.9	.9	.82	.425	.904

Table 13.11 Predicting results of 5000 records via unbalanced training

Different b value	Absolute accuracy rate (13.1)		Absolute accuracy rate (13.2)		Absolute accuracy rate (13.3)	
	Good	Bad	Good	Bad	Good	Bad
−2.00	75.34%	43.57%	6.81%	82.45%	59.50%	86.87%
−1.10	66.98%	66.14%	6.81%	82.45%	59.68%	86.13%
−0.50	65.89%	65.89%	6.81%	82.95%	59.74%	87%
0.50	71.54%	65.03%	6.81%	82.45%	59.43%	86.87%
1.10	67%	66.14%	6.81%	82.45%	59.43%	87%

Table 13.12 Learning comparisons on balanced 280 records

	T_g	T_b	Total
Decision tree			
Good	138	2	140
Bad	13	127	140
Total	151	129	280
Neural network			
Good	116	24	140
Bad	14	126	140
Total	130	150	280
MCLP			
Good	134	6	140
Bad	7	133	140
Total	141	139	280
FMCLP			
Good	127	13	140
Bad	13	127	140
Total	140	140	280

advantages over the MCLP method and has a certain significance to be an alternative tool to the other well-known data mining techniques in classification.

13.3 Three-Class Analysis

13.3.1 Three-Class Formulation

The reason to separate credit card-holder behavior into multi-class is to meet the needs of advanced credit card portfolio management. Comparing with two-class separation, multi-class method enlarges the difference between "Good" and "Bad"

Table 13.13 Learning comparisons on balanced 280 records

	T_g	T_b	Total
Decision tree			
Good	2180	2005	4185
Bad	141	674	815
Total	2321	2679	5000
Neural network			
Good	2814	1371	4185
Bad	176	639	815
Total	2990	2010	5000
MCLP			
Good	3160	1025	4185
Bad	484	331	815
Total	3644	1356	5000
FMCLP			
Good	2498	1687	4185
Bad	113	702	815
Total	2611	2389	5000

cardholders behavior. This enlargement increases not only the accuracy of separation, but also efficiency of credit card portfolio management. For example, by considering the number of months where the account has been over-limit during the previous two years, we can define "Good" as the cardholders have less than 3 times over-limit; "Normal" as 4–12 times over-limit, and "Bad" as 13 or more times over-limit. By using the prediction distribution for each behavior class and inner relationship between these classes, credit card issuers can establish their credit limit policies for various cardholders.

Based on the two-class MCLP model, a three-class MCLP model can be developed. Three-class separation can use two boundaries, b_1 and b_2, to separate class 1 (G1: Bad), class 2 (G2: Normal), and class 3 (G3: Good). Credit cardholders behaviors are represented as: **1** stands for class 1 (Bad), **2** stands for class 2 (normal), and **3** stands for class 3 (Good).

Given a set of r variables about the cardholders $\xi = (\xi_1, \ldots, \xi_r)$, let $\xi_i = (\xi_{1i}, \ldots, \xi_{ir})$ be the sample of data for the variables, where $i = 1, \ldots, n$ and n is the sample size. We want to find the coefficients for an appropriate subset of the variables, denoted by $w = (w_1, \ldots, w_r)$, and boundary b_1 to separate G1 from G2 and G3, boundary b_2 to separate G3 from G2 and G1. This can be done as following:

$$(x_i \cdot w) \leq b_1, \quad x_i \in G1;$$
$$b_1 \leq (x_i \cdot w) \leq b_2, \quad x_i \in G2; \quad \text{and}$$
$$(x_i \cdot w) \geq b_2, \quad x_i \in G3.$$

Similarly to two-class model, we apply two measurements for better separation of Good, Normal, and Bad.

Let ξ_i^1 be the overlapping degree and ξ^1 be $\max \xi_i^1$ with respect of x_i within G1 and G2, and ξ_i^2 be the overlapping degree and ξ^2 be $\max \xi_i^2$ with respect of x_i within G2 and G3. Let β_i^1 be the distance from x_i within G1 and G2 to its adjusted boundaries $((x_i \cdot w) = b_1 - \xi^1$, or $(x_i \cdot w) = b_1 + \xi^1)$, and β_i^2 be the distance from x_i within G2 and G3 to its adjusted boundaries $((x_i \cdot w) = b_1 - \xi^2$, and $(x_i \cdot w) = b_1 + \xi^2)$. Our goal is to reach the maximization of β_i^1 and β_i^2, and the minimization of ξ_i^1 and ξ_i^2 simultaneously. After putting ξ_i^1, ξ_i^2, β_i^1 and β_i^2 into the above constraints, we have:

$$\min \sum_i (\xi_i^1 + \xi_i^2) \quad \text{and} \quad \max \sum_i (\beta_i^1 + \beta_i^2)$$

$$\text{s.t.} \quad (x_i \cdot w) = b_1 - \xi_i^1 + \beta_i^1, \quad x_i \in G1,$$

$$(x_i \cdot w) = b_1 + \xi_i^1 - \beta_i^1 + b_2 - \xi_i^2 + \beta_i^2, \quad x_i \in G2,$$

$$(x_i \cdot w) = b_2 - \xi_i^2 + \beta_i^2, \quad x_i \in G3,$$

$$b_1 + \xi_i^1 \le b_2 - \xi_i^2,$$

where x_i are given, w, b_1, and b_2 are unrestricted, and ξ_i^1, ξ_i^2, β_i^1, and $\beta_i^2 \ge 0$.

Note that the constraint $b_1 + \xi_i^1 \le b_2 - \xi_i^2$ guarantees the existence of three classes.

Refer to Chap. 7, a three-class MCLP model is reformulated as:

$$\min d_{\xi_1}^- + d_{\xi_1}^+ + d_{\beta_1}^- + d_{\beta_1}^+ + d_{\xi_2}^- + d_{\xi_2}^+ + d_{\beta_2}^- + d_{\beta_2}^+$$

$$\text{s.t.} \quad \xi_*^1 + \sum_i \xi_i^1 = d_{\xi_1}^- - d_{\xi_1}^+,$$

$$\beta_*^1 - \sum_i \beta_i^1 = d_{\beta_1}^- - d_{\beta_1}^+,$$

$$\xi_*^2 + \sum_i \xi_i^2 = d_{\xi_2}^- - d_{\xi_2}^+,$$

$$\beta_*^2 - \sum_i \beta_i^2 = d_{\beta_2}^- - d_{\beta_2}^+,$$

$$b_1 + \xi_i^1 \le b_2 - \xi_i^2,$$

class 1 (Good): $\quad (x_i \cdot w) = b_1 - \xi_i^1 + \beta_i^1, \quad x_i \in G1,$

class 2 (Normal): $\quad (x_i \cdot w) = b_1 + \xi_i^1 - \beta_i^1 + b_2 - \xi_i^2 + \beta_i^2, \quad x_i \in G2,$

class 3 (Bad): $\quad (x_i \cdot w) = b_2 - \xi_i^2 + \beta_i^2, \quad x_i \in G3,$

where x_i are given, $b_1 \le b_2$, w, b_1, and b_2 are unrestricted, and ξ_i^1, ξ_i^2, β_i^1, and $\beta_i^2 \ge 0$.

Table 13.14 Previous experience with 12 customers

Applicants	Credit customer	Responses	
		Quest 1	Quest 2
Class 1 (Poor risk)	1	1	3
	2	2	5
	3	3	4
	4	4	6
Class 2 (Fair risk)	5	5	7
	6	6	9
	7	7	8
	8	7	7
	9	9	9
Class 3 (Good risk)	10	6	2
	11	6	4
	12	8	3

13.3.2 Small Sample Testing

In order to test the feasibility of the proposed three-class model, we consider the task of assigning credit applications to risk classifications adopted from the example of Freed and Glover [75]. An applicant is to be classified as a poor, fair, or good credit risk based on his/her responses to two questions appearing on a standard credit application. Table 13.14 shows previous experience with 12 customers.

Because this example is rather small and simple, we applied well-known LINDO (Linear, Interactive and Discrete Optimizer) computer software to conduct a series of tests on our three-class model. In order to use LINDO program, the three-class model is adjusted to satisfy its input style.

The first test is constructed as following:

$$\min\ d^-_{\xi_1} + d^+_{\xi_1} + d^-_{\beta_1} + d^+_{\beta_1} + d^-_{\xi_2} + d^+_{\xi_2} + d^-_{\beta_2} + d^+_{\beta_2}$$

$$\text{s.t.}\quad d^-_{\xi_1} - d^+_{\xi_1} - \xi^1_1 - \xi^1_2 - \xi^1_3 - \xi^1_4 - \xi^1_5 - \xi^1_6 - \xi^1_7 - \xi^1_8 - \xi^1_9 - \xi^1_{10} - \xi^1_{11}$$

$$-\xi^1_{12} = 0.1,$$

$$d^-_{\beta_1} - d^+_{\beta_1} + \beta^1_1 + \beta^1_2 + \beta^1_3 + \beta^1_4 + \beta^1_5 + \beta^1_6 + \beta^1_7 + \beta^1_8 + \beta^1_9 + \beta^1_{10} + \beta^1_{11}$$

$$+\beta^1_{12} = 10,$$

$$d^-_{\xi_2} - d^+_{\xi_2} - \xi^2_1 - \xi^2_2 - \xi^2_3 - \xi^2_4 - \xi^2_5 - \xi^2_6 - \xi^2_7 - \xi^2_8 - \xi^2_9 - \xi^2_{10} - \xi^2_{11}$$

$$-\xi^2_{12} = 0.2,$$

$$d^-_{\beta_2} - d^+_{\beta_2} + \beta^2_1 + \beta^2_2 + \beta^2_3 + \beta^2_4 + \beta^2_5 + \beta^2_6 + \beta^2_7 + \beta^2_8 + \beta^2_9 + \beta^2_{10} + \beta^2_{11}$$

$$+\beta^2_{12} = 9,$$

$$\xi_1^1 + \xi_1^2 < 1,$$

$$\xi_2^1 + \xi_2^2 < 1,$$

$$\xi_3^1 + \xi_3^2 < 1$$

$$\xi_4^1 + \xi_4^2 < 1,$$

$$\xi_5^1 + \xi_5^2 < 1,$$

$$\xi_6^1 + \xi_6^2 < 1,$$

$$\xi_7^1 + \xi_7^2 < 1,$$

$$\xi_8^1 + \xi_8^2 < 1,$$

$$\xi_9^1 + \xi_9^2 < 1,$$

$$\xi_{10}^1 + \xi_{10}^2 < 1,$$

$$\xi_{11}^1 + \xi_{11}^2 < 1,$$

$$\xi_{12}^1 + \xi_{12}^2 < 1,$$

$$w_1 + 3w_2 + \xi_1^1 - \beta_1^1 = 1.5,$$

$$2w_1 + 5w_2 + \xi_2^1 - \beta_2^1 = 1.5,$$

$$3w_1 + 4w_2 + \xi_3^1 - \beta_3^1 = 1.5,$$

$$4w_1 + 6w_2 + \xi_4^1 - \beta_4^1 = 1.5,$$

$$5w_1 + 7w_2 - \xi_5^1 + \beta_5^1 + \xi_5^2 - \beta_5^2 = 4,$$

$$6w_1 + 9w_2 - \xi_6^1 + \beta_6^1 + \xi_6^2 - \beta_6^2 = 4,$$

$$7w_1 + 8w_2 - \xi_7^1 + \beta_7^1 + \xi_7^2 - \beta_7^2 = 4,$$

$$7w_1 + 7w_2 - \xi_8^1 + \beta_8^1 + \xi_8^2 - \beta_8^2 = 4,$$

$$9w_1 + 9w_2 - \xi_9^1 + \beta_9^1 + \xi_9^2 - \beta_9^2 = 4,$$

$$6w_1 + 2w_2 + \xi_{10}^2 - \beta_{10}^2 = 2.5,$$

$$6w_1 + 4w_2 + \xi_{11}^2 - \beta_{11}^2 = 2.5,$$

$$8w_1 + 3w_2 + \xi_{12}^2 - \beta_{12}^2 = 2.5,$$

where the values of boundary b_1 and b_2, ξ_*^1, β_*^1, ξ_*^2, β_*^2 are set to 1.5, 2.5, 0.1, 10, 0.2, and 9 respectively, and the other variables ≥ 0.

The results (optimal coefficients values of w) $\binom{w_1^*}{w_2^*}$ are $\binom{0.28125}{0.40625}$.

Applying these coefficients into the model, we get the final separation of twelve credit applicants. Class 2 (Fair risk) separates from other two classes very well, while applicants of class 1 (Poor risk) and class 3 (Good risk) are nearly in the same range.

Since this separation is not satisfactory, we decide to try another way in which the values of boundary b_1 and b_2 are restricted as: $b_1 > 2.5$, $b_2 > 5$. The second model becomes:

$$\min \ d_{\xi_1}^- + d_{\xi_1}^+ + d_{\beta_1}^- + d_{\beta_1}^+ + d_{\xi_2}^- + d_{\xi_2}^+ + d_{\beta_2}^- + d_{\beta_2}^+$$

s.t.
$$d_{\xi_1}^- - d_{\xi_1}^+ - \xi_1^1 - \xi_2^1 - \xi_3^1 - \xi_4^1 - \xi_5^1 - \xi_6^1 - \xi_7^1 - \xi_8^1 - \xi_9^1 - \xi_{10}^1 - \xi_{11}^1$$
$$- \xi_{12}^1 = 0.1,$$

$$d_{\beta_1}^- - d_{\beta_1}^+ + \beta_1^1 + \beta_2^1 + \beta_3^1 + \beta_4^1 + \beta_5^1 + \beta_6^1 + \beta_7^1 + \beta_8^1 + \beta_9^1 + \beta_{10}^1 + \beta_{11}^1$$
$$+ \beta_{12}^1 = 10,$$

$$d_{\xi_2}^- - d_{\xi_2}^+ - \xi_1^2 - \xi_2^2 - \xi_3^2 - \xi_4^2 - \xi_5^2 - \xi_6^2 - \xi_7^2 - \xi_8^2 - \xi_9^2 - \xi_{10}^2 - \xi_{11}^2$$
$$- \xi_{12}^2 = 0.2,$$

$$d_{\beta_2}^- - d_{\beta_2}^+ + \beta_1^2 + \beta_2^2 + \beta_3^2 + \beta_4^2 + \beta_5^2 + \beta_6^2 + \beta_7^2 + \beta_8^2 + \beta_9^2 + \beta_{10}^2 + \beta_{11}^2$$
$$+ \beta_{12}^2 = 9,$$

$$\xi_1^1 + \xi_1^2 + b_1 - b_2 < 0,$$
$$\xi_2^1 + \xi_2^2 + b_1 - b_2 < 0,$$
$$\xi_3^1 + \xi_3^2 + b_1 - b_2 < 0,$$
$$\xi_4^1 + \xi_4^2 + b_1 - b_2 < 0,$$
$$\xi_5^1 + \xi_5^2 + b_1 - b_2 < 0,$$
$$\xi_6^1 + \xi_6^2 + b_1 - b_2 < 0,$$
$$\xi_7^1 + \xi_7^2 + b_1 - b_2 < 0,$$
$$\xi_8^1 + \xi_8^2 + b_1 - b_2 < 0$$
$$\xi_9^1 + \xi_9^2 + b_1 - b_2 < 0,$$
$$\xi_{10}^1 + \xi_{10}^2 + b_1 - b_2 < 0,$$
$$\xi_{11}^1 + \xi_{11}^2 + b_1 - b_2 < 0,$$
$$\xi_{12}^1 + \xi_{12}^2 + b_1 - b_2 < 0,$$
$$w_1 + 3w_2 + \xi_1^1 - \beta_1^1 - b_1 = 0,$$
$$2w_1 + 5w_2 + \xi_2^1 - \beta_2^1 - b_1 = 0,$$
$$3w_1 + 4w_2 + \xi_3^1 - \beta_3^1 - b_1 = 0,$$
$$4w_1 + 6w_2 + \xi_4^1 - \beta_4^1 - b_1 = 0,$$
$$5w_1 + 7w_2 - \xi_5^1 + \beta_5^1 + \xi_5^2 - \beta_5^2 - b_1 - b_2 = 0,$$
$$6w_1 + 9w_2 - \xi_6^1 + \beta_6^1 + \xi_6^2 - \beta_6^2 - b_1 - b_2 = 0,$$
$$7w_1 + 8w_2 - \xi_7^1 + \beta_7^1 + \xi_7^2 - \beta_7^2 - b_1 - b_2 = 0,$$
$$7w_1 + 7w_2 - \xi_8^1 + \beta_8^1 + \xi_8^2 - \beta_8^2 - b_1 - b_2 = 0,$$
$$9w_1 + 9w_2 - \xi_9^1 + \beta_9^1 + \xi_9^2 - \beta_9^2 - b_1 - b_2 = 0,$$
$$6w_1 + 2w_2 + \xi_{10}^2 - \beta_{10}^2 - b_2 = 0,$$
$$6w_1 + 4w_2 + \xi_{11}^2 - \beta_{11}^2 - b_2 = 0,$$
$$8w_1 + 3w_2 + \xi_{12}^2 - \beta_{12}^2 - b_2 = 0,$$

$$b_1 > 2.5, \qquad b_2 > 5,$$

where the values of $\xi_*^1, \beta_*^1, \xi_*^2, \beta_*^2$ are set to 0.1, 10, 0.2, and 9 respectively, and the other variables ≥ 0.

At this time, the coefficients values of w, $\binom{w_1^*}{w_2^*} = \binom{1.095238}{0.404762}$.

We can see that this time class 1 (Poor risk) applicants separated fairly good from other two classes, however, the overlapping degree between class 2 (Fair risk) and class 3 (Good risk) increases. The following four tests are very similar to the second one, except that restricted conditions of boundary one and two are changed. In test three, restricted conditions are set as $b_1 < 7$, $b_2 > 15$; in test four, $b_1 < 5$ and $b_2 > 15$; in test five, $b_1 < 5$ and $b_2 > 20$; in test six, $b_1 < 3$ and $b_2 > 15$. The results of these tests show that overlapping between classes, especially class 2 and class 3, cannot be decreased or eliminated if the values of boundaries are restricted as nonnegative. Since LINDO program doesn't accept negative slacks, we cannot directly set boundary one and two less than zero. Thus, from the seventh test, we introduced $b_1 = u_1 - u_2$ to represent boundary one, $b_2 = v_1 - v_2$ to represent boundary two, $x_1 - y_1$ to represent x_1, and $x_2 - y_2$ to represent x_2. Although $u_1, u_2, v_1, v_2, x_1, y_1, x_2$, and y_2 are all nonnegative variables, by allowing their values to be flexible, $u_1 - u_2, v_1 - v_2, x_1 - y_1$, and $x_2 - y_2$ can be either negative or positive. For example, if we let $u_1 < 5$ and $u_2 < 25$, then $u_1 - u_2 < -20$. After testing another four tests, we observed that when the difference between u_1 and u_2 increases, or when the difference between v_1 and v_2 decreases, the instances of overlap decrease. The test ten is analyzed here as a representative for these four tests. The model of test ten is given as:

$$\min d_{\xi_1}^- + d_{\xi_1}^+ + d_{\beta_1}^- + d_{\beta_1}^+ + d_{\xi_2}^- + d_{\xi_2}^+ + d_{\beta_2}^- + d_{\beta_2}^+,$$

$$\text{s.t.} \quad d_{\xi_1}^- - d_{\xi_1}^+ - \xi_1^1 - \xi_2^1 - \xi_3^1 - \xi_4^1 - \xi_5^1 - \xi_6^1 - \xi_7^1 - \xi_8^1 - \xi_9^1 - \xi_{10}^1 - \xi_{11}^1$$
$$- \xi_{12}^1 = 0.1,$$

$$d_{\beta_1}^- - d_{\beta_1}^+ + \beta_1^1 + \beta_2^1 + \beta_3^1 + \beta_4^1 + \beta_5^1 + \beta_6^1 + \beta_7^1 + \beta_8^1 + \beta_9^1 + \beta_{10}^1 + \beta_{11}^1$$
$$+ \beta_{12}^1 = 10,$$

$$d_{\xi_2}^- - d_{\xi_2}^+ - \xi_1^2 - \xi_2^2 - \xi_3^2 - \xi_4^2 - \xi_5^2 - \xi_6^2 - \xi_7^2 - \xi_8^2 - \xi_9^2 - \xi_{10}^2 - \xi_{11}^2$$
$$- \xi_{12}^2 = 0.2,$$

$$d_{\beta_2}^- - d_{\beta_2}^+ + \beta_1^2 + \beta_2^2 + \beta_3^2 + \beta_4^2 + \beta_5^2 + \beta_6^2 + \beta_7^2 + \beta_8^2 + \beta_9^2 + \beta_{10}^2 + \beta_{11}^2$$
$$+ \beta_{12}^2 = 9,$$

$$\xi_1^1 + \xi_1^2 + u_1 - u_2 - v_1 + v_2 < 0,$$
$$\xi_2^1 + \xi_2^2 + u_1 - u_2 - v_1 + v_2 < 0,$$
$$\xi_3^1 + \xi_3^2 + u_1 - u_2 - v_1 + v_2 < 0,$$
$$\xi_4^1 + \xi_4^2 + u_1 - u_2 - v_1 + v_2 < 0,$$
$$\xi_5^1 + \xi_5^2 + u_1 - u_2 - v_1 + v_2 < 0,$$
$$\xi_6^1 + \xi_6^2 + u_1 - u_2 - v_1 + v_2 < 0,$$

$$\xi_7^1 + \xi_7^2 + u_1 - u_2 - v_1 + v_2 < 0,$$

$$\xi_8^1 + \xi_8^2 + u_1 - u_2 - v_1 + v_2 < 0,$$

$$\xi_9^1 + \xi_9^2 + u_1 - u_2 - v_1 + v_2 < 0,$$

$$\xi_{10}^1 + \xi_{10}^2 + u_1 - u_2 - v_1 + v_2 < 0,$$

$$\xi_{11}^1 + \xi_{11}^2 + u_1 - u_2 - v_1 + v_2 < 0,$$

$$\xi_{12}^1 + \xi_{12}^2 + u_1 - u_2 - v_1 + v_2 < 0,$$

$$w_1 + 3w_2 + \xi_1^1 - \beta_1^1 - u_1 - u_2 = 0,$$

$$2w_1 + 5w_2 + \xi_2^1 - \beta_2^1 - u_1 - u_2 = 0,$$

$$3w_1 + 4w_2 + \xi_3^1 - \beta_3^1 - u_1 - u_2 = 0,$$

$$4w_1 + 6w_2 + \xi_4^1 - \beta_4^1 - u_1 - u_2 = 0,$$

$$5w_1 + 7w_2 - \xi_5^1 + \beta_5^1 + \xi_5^2 - \beta_5^2 - u_1 + u_2 - v_1 + v_2 = 0,$$

$$6w_1 + 9w_2 - \xi_6^1 + \beta_6^1 + \xi_6^2 - \beta_6^2 - u_1 + u_2 - v_1 + v_2 = 0,$$

$$7w_1 + 8w_2 - \xi_7^1 + \beta_7^1 + \xi_7^2 - \beta_7^2 - u_1 + u_2 - v_1 + v_2 = 0,$$

$$7w_1 + 7w_2 - \xi_8^1 + \beta_8^1 + \xi_8^2 - \beta_8^2 - u_1 + u_2 - v_1 + v_2 = 0,$$

$$9w_1 + 9w_2 - \xi_9^1 + \beta_9^1 + \xi_9^2 - \beta_9^2 - u_1 + u_2 - v_1 + v_2 = 0,$$

$$6w_1 + 2w_2 + \xi_{10}^2 - \beta_{10}^2 - v_1 - v_2 = 0,$$

$$6w_1 + 4w_2 + \xi_{11}^2 - \beta_{11}^2 - v_1 - v_2 = 0,$$

$$8w_1 + 3w_2 + \xi_{12}^2 - \beta_{12}^2 - v_1 - v_2 = 0,$$

$$u_1 < 5,$$

$$u_2 < 25,$$

$$v_1 > 20,$$

$$v_2 > 30,$$

where the values of $\xi_*^1, \beta_*^1, \xi_*^2, \beta_*^2$ are set to 0.1, 10, 0.2, and 9 respectively, and the other variables ≥ 0.

Through these tests, we observe that the extent of class 1 (Poor risk) is $(-5\text{–}16)$, the extent of class 2 (Fair risk) is $(23\text{–}63)$, and the extent of class 3 (Good risk) is $(54\text{–}86)$. Therefore, class 1 has no overlap with both class 2 and class 3, while class 2 has only one applicant overlapped with class 3 (applicant number 9). This is basically consistent with the result of Freed and Glover. We further observe that:

(1) The proposed three-group model provides a feasible alternative for classification problems;
(2) Misclassification can be decreased by adjusting the values of boundary one and boundary two.

In addition to the experimental study of the small sample, the following real-life data testing demonstrates the potential applicability of the proposed three-group model.

13.3.3 Real-Life Data Analysis

This subsection summarizes the empirical study of the three-class model that we have conducted by using a large real-life data from the credit data of FDC. We chose the SAS LP (linear programming) package to perform all of computations. There is a training data set (a 1000-sample) and a verifying data set (5000-sample) in total. Both were randomly selected from 25,000 credit card records. The proposed three-class model with 64 credit variables, $\xi_*^1 = 0.1$, $\beta_*^1 = 10$, $\xi_*^2 = 0.2$, $\beta_*^2 = 9$ has been tested on the training data set and then used in the statistical cross-validation study on the verifying data set. The results are two Excel charts: one is for training data set and another for the verifying data set. The Excel charts show cumulative distribution of three classes of credit cardholders' behaviors (class 1: Bad, class 2: Normal, and class 3: Good). These two Excel charts and further elaboration of the SAS programs are in the next few paragraphs.

Based on the three-class model and FDC's Proprietary Score system, the SAS programs are developed and consist of four steps. Each step has one or more SAS programs to carry out the operations. The first step is to use ReadCHD.sas to convert raw data into SAS data sets. Before raw data can be processed by the SAS procedures, they have to be transformed into SAS data sets format. The outputs are two kinds of SAS data sets: training and verifying data sets. The training data set is used to test the model, while the verifying data set is used to check how stable the model is. Each data set has 38 original variables and 64 derived variables (see Tables 13.1 and 13.2) to describe cardholders' behaviors. The 38 original variables can be divided into four categories and some other related variables. The four categories are balance, purchase, payment, and cash advance. Other related variables include credit line (represent as CHDCRLN in the program), account open date (represent as CHDOPNDT in the program), date of last payment (represent as CHDTLZNT in the program), highest balance (represent as CHHGBLLF in the program), times over limit last 2 years (represent as CHDHLNZ in the program), etc.

After converting raw data into the training and verifying data sets, the program classDef.sas in the second step is used to separate credit card accounts within the training and verifying data sets into three classes: Bad, Normal or Good. The separation is based on the variable-CHDHLNZ, which records the number of months the account has been over limit during the previous two years. Three classes are defined as: class 1 (Bad, 13+ times over limit), class 2 (Normal, 4–12 times over limit), and class 3 (Good, 0–3 times over limit). The result of this step is the class variable, which represents the class status of each record within the training and verifying data sets. For example, the class value of records equals to 1 if the credit card-holder has more than 13 times over limit during the previous two years.

Then, the third step is the main stage that implements the three-class model. G3Model.sas is the program performing this task. The program sets up the constraints matrix using the training data set established from the second step. The constraints matrix is constructed according to the proposed three-class model. The PROC LP (procedural linear programming) is then called to calculate this MCLP model. The LP output file provides the coefficients solution for the three-class model.

The fourth step converts the solutions (coefficient data sets) into Excel charts for comparison. The program Score.sas calls the program Compareclasses.sas to make the charts. Coefficients in LP output file are pulled out and transpose into a single record by using SAS procedure Transpose. Scores in both training and verifying data sets are calculated from data and coefficients. The calculated scores for the training and verifying data sets are sent to the Compareclasses.sas program to get a final Excel chart to show the results of separation.

The cumulative distributions show that class 1 (Bad) has the least percentage in the whole sample, while class 3 (Good) contains the majority of credit cardholders. class 2 (Normal) is located between classes 1 and 3. The separation of three classes is satisfactory.

13.4 Four-Class Analysis

13.4.1 Four-Class Formulation

In the four-class separation, we use term: "charge-off" to predict the cardholders' behaviors. According to this idea, four classes are defined as: Bankrupt charge-off accounts, Non-bankrupt charge-off accounts, Delinquent accounts, and Current accounts. Bankrupt charge-off accounts are accounts that have been written off by credit card issuers because of cardholders' bankrupt claims. Non-bankrupt charge-off accounts are accounts that have been written off by credit card issuers due to reasons other than bankrupt claims. The charge-off policy may vary among authorized institutions. Normally, an account will be written off when the receivable has been overdue for more than 180 days or when the ultimate repayment of the receivable is unlikely (e.g., the card-holder cannot be located) (http://www.info.gov.hk/hkma/eng/press/2001/20010227e6.htm). Delinquent accounts are accounts that have not paid the minimum balances for more than 90 days. Current accounts are accounts that have paid the minimum balances or have no balances.

This separation is more precise than two-class and three-class models in credit card portfolio management. For instance, bankrupt charge-off and non-bankrupt charge-off accounts are probably both classified as "Bad" accounts in two or three-group separations. This model, however, will call for different handling against these accounts.

A four-class model has three boundaries, b_1, b_2, and b_3, to separate four classes. Each class is represented by a symbol as follows:

- G1 stands for (Bankrupt charge-off account),
- G2 stands for (Non-bankrupt charge-off account),
- G3 stands for (Delinquent account), and
- G4 stands for (Current account).

Given a set of r variables about the cardholders $x = (x_1, \ldots, x_i)$, let $x_i = (x_{i1}, \ldots, x_{ir})$ be the development sample of data for the variables, where $i =$

$1, \ldots, n$ and n is the sample size. We try to determine the coefficients of the variables, denoted by $w = (w_1, \ldots, w_r)$, boundary b_1 to separate G1 from G2, G3, and G4, boundary b_3 to separate G4 from G1, G2, and G3, boundary b_2 to separate G2 and G3. This separation can be represented by:

$$(x_i \cdot w) \le b_1, \quad x_i \in G1,$$
$$b_1 \le (x_i \cdot w) \le b_2, \quad x_i \in G2,$$
$$b_2 \le (x_i \cdot w) \le b_3, \quad x_i \in G3,$$
$$b_3 \le (x_i \cdot w), \quad x_i \in G4.$$

Similar to two-class and three-class models, we apply two measurements for better separations. Let ξ_i^1 be the overlapping degree with respect of x_i within G1 and G2, ξ_i^2 be the overlapping degree with respect of x_i within G2 and G3, ξ_i^3 be the overlapping degree with respect of x_i within G3 and G4. Let β_i^1 be the distance from x_i within G1 and G2 to its adjusted boundaries $((x_i \cdot w) = b_1 - \xi_i^1$, and $(x_i \cdot w) = b_1 + \xi_*^1)$, β_i^2 be the distance from x_i within G1 and G2 to its adjusted boundaries $((x_i \cdot w) = b_2 - \xi_*^2$, and $(x_i \cdot w) = b_2 + \xi_*^2)$, β_i^3 be the distance from x_i within G1 and G2 to its adjusted boundaries $((x_i \cdot w) = b_3 - \xi_i^3$, and $(x_i \cdot w) = b_3 + \xi_*^3)$. We want to reach the maximization of β_i^1, β_i^2, and β_i^3 and the minimization of ξ_i^1, ξ_i^2 and ξ_i^3 simultaneously. After putting ξ_i^1, ξ_i^2, ξ_i^3, β_i^1, β_i^2, and β_i^3 into the above four-class separation, we have:

$$\min \sum_i (\xi_i^1 + \xi_i^2 + \xi_i^3) \quad \text{and} \quad \max \sum_i (\beta_i^1 + \beta_i^2 + \beta_i^3)$$

s.t.

G1: $(x_i \cdot w) = b_1 + \xi_i^1 - \beta_i^1, \quad x_i \in G1$ (Bankrupt charge-off),

G2: $(x_i \cdot w) = b_1 + \xi_i^1 - \beta_i^1, \quad (x_i \cdot w) = b_2 + \xi_i^2 - \beta_i^2, \quad x_i \in G2$

(Non-bankrupt charge-off),

G3: $(x_i \cdot w) = b_2 + \xi_i^2 - \beta_i^2, \quad (x_i \cdot w) = b_3 + \xi_i^3 - \beta_i^3, \quad x_i \in G3$

(Delinquent),

G4: $(x_i \cdot w) = b_3 + \xi_i^3 - \beta_i^3, \quad x_i \in G4$ (Current),

$$b_1 + \xi_i^1 \le b_2 - \xi_i^2,$$
$$b_2 + \xi_i^2 \le b_3 - \xi_i^3,$$

where x_i, are given, w, b_1, b_2, b_3 are unrestricted, and ξ_i^1, ξ_i^2, ξ_i^3, β_i^1, β_i^2, and $\beta_i^3 \ge 0$.

The constraints $b_1 + \xi_i^1 \le b_2 - \xi_i^2$ and $b_2 + \xi_i^2 \le b_3 - \xi_i^3$ guarantee the existence of four groups by enforcing b_1 lower than b_2, and b_2 lower than b_3. Then we apply the compromise solution approach [183, 184] to reform the model. We assume the ideal value of $-\sum_i \xi_i^1$ be $\xi_*^1 > 0$, $-\sum_i \xi_i^2$ be $\xi_*^2 > 0$, $-\sum_i \xi_i^3$ be $\xi_*^3 > 0$, and the ideal value of $\sum_i \beta_i^1$ be $\beta_*^1 > 0$, $\sum_i \beta_i^2$ be $\beta_*^2 > 0$, $\sum_i \beta_i^3$ be $\beta_*^3 > 0$.

Then, the above four-group model is transformed as:

$$\min\ d_{\xi_1}^- + d_{\xi_1}^+ + d_{\beta_1}^- + d_{\beta_1}^+ + d_{\xi_2}^- + d_{\xi_2}^+ + d_{\beta_2}^- + d_{\beta_2}^+ + d_{\xi_3}^- + d_{\xi_3}^+ + d_{\beta_3}^- + d_{\beta_1}^+$$

$$\text{s.t.}\quad \xi_*^1 + \sum_i \xi_i^1 = d_{\xi_1}^- - d_{\xi_1}^+, \qquad \beta_*^1 - \sum_i \beta_i^1 = d_{\beta_1}^- - d_{\beta_1}^+,$$

$$\xi_*^2 + \sum_i \xi_i^2 = d_{\xi_2}^- - d_{\xi_2}^+, \qquad \beta_*^2 - \sum_i \beta_i^2 = d_{\beta_2}^- - d_{\beta_2}^+,$$

$$\xi_*^3 + \sum_i \xi_i^3 = d_{\xi_3}^- - d_{\xi_3}^+, \qquad \beta_*^3 - \sum_i \beta_i^3 = d_{\beta_3}^- - d_{\beta_3}^+,$$

$$b_1 + \xi_i^1 \le b_2 - \xi_i^2, \qquad b_2 + \xi_i^2 \le b_3 - \xi_i^3,$$

G1: $(x_i \cdot w) = b_1 + \xi_i^1 - \beta_i^1,\quad x_i \in G1$ (Bankrupt charge-off),

G2: $(x_i \cdot w) = b_1 + \xi_i^1 - \beta_i^1,\quad (x_i \cdot w) = b_2 + \xi_i^2 - \beta_i^2,\quad x_i \in G2$

(Non-bankrupt charge-off),

G3: $(x_i \cdot w) = b_2 + \xi_i^2 - \beta_i^2,\quad (x_i \cdot w) = b_3 + \xi_i^3 - \beta_i^3,\quad x_i \in G3$

(Delinquent),

G4: $(x_i \cdot w) = b_3 + \xi_i^3 - \beta_i^3,\quad x_i \in G4$ (Current),

where x_i, are given, $b_1 \le b_2 \le b_3$, w, b_1, b_2, b_3 are unrestricted, and ξ_i^1, ξ_i^2, ξ_i^3, β_i^1, β_i^2, and $\beta_i^3 \ge 0$.

13.4.2 Empirical Study and Managerial Significance of Four-Class Models

The FDC credit data is again used to perform the four-class model. A training set of 160 card account samples from 25,000 credit card records is used to test the control parameters of the model for the best class separation. A verifying data set with 5,000 accounts is then applied. Four groups are defined as: Bankrupt charge-off accounts (the number of over-limits ≥ 13), Non-bankrupt charge-off accounts ($7 \le$ the number of over-limits ≤ 12), Delinquent accounts ($2 \le$ the number of over-limits ≤ 6), and Current accounts ($0 \le$ the number of over-limits ≤ 2).

After several learning trials for different sets of boundary values, we found the values: $b_1 = 0.05$, $b_2 = 0.8$, $b_3 = 1.95$ (without changing ξ_*^1, ξ_*^2, ξ_*^3, β_*^1, β_*^2, and β_*^3) brought the best separation, in which "cum G1" refers to cumulative percentage of G1 (Bankrupt charge-off accounts); "cum G2" refers to cumulative percentage of G2 (Non-bankrupt charge-off accounts); "cum G3" refers to cumulative percentage of G3 (Delinquent accounts); and "cum G4" refers to cumulative percentage of G4 (Current accounts).

This training set has total 160 samples. G1 has been correctly identified for 90% (36/40), G2 90% (36/40), G3 85% (34/40) and G4 97.5% (39/40). In addition to these absolute classifications criteria, the KS Score is calculated by KS value $=$ max |Cum. distribution of Good – Cum. distribution of Bad|.

The KS values are 50 for G1 vs. G2, 62.5 for G2 vs. G3 and 77.5 for G3 vs. G4.

We observe some relationships between boundary values and separation:

(1) Optimal solution and better separation can be achieved by changing the value of b_j, ξ_*^j and β_*^j.
(2) Once a feasible area is found, the classification result will be similar for the solution in that area.
(3) The definition of group and data attributes will influence the classification result.

When applying the resulting classifier to predict the verifying data set, we can predict the verifying set by the classifier as G1 for 43.4% (23/53), G2 for 51% (84/165), G3 for 28% (156/557) and G4 for 68% (2872/4225). The predicted KS values are 36.3 for G1 vs. G2, 21.6 for G2 vs. G3 and 50.7 for G3 vs. G4. These results indicate that the predicted separation between G3 and G4 is better than others.

In multi-group classification, a better result will be achieved in the separation between certain two groups. According to the definition of group and data attributes, the best KS score usually appears in the separation between the first group and second group or the separation between the last group and other groups. In the model, the last group is defined as individuals with perfect credit performance and other groups are defined as individuals who have some stains in their credit histories. As a result, the separations indicated that the distance between the last group (Good) and other groups is larger and a better KS score for that separation. It means that in practice it is easier to catch good ones or bad ones, but it is much more difficult to discriminant different levels of credit performance in between.

Chapter 14
Health Insurance Fraud Detection

14.1 Problem Identification

Health care frauds cost private health insurance companies and public health insurance programs at least $51 billion in calendar year 2003 [196]. Health insurance fraud detection is an important and challenging task. Traditional heuristic-rule based fraud detection techniques cannot identify complex fraud schemes. Such a situation demands more sophisticated analytical methods and techniques that are capable of detecting fraud activities from large databases. Traditionally, insurance companies use human inspections and heuristic rules to detect fraud. As the number of electronic insurance claims increases each year, it is difficult to detect insurance fraud in a timely manner by manual methods alone. In addition, new types of fraud emerge constantly and SQL operations based on heuristic rules cannot identify those new emerging fraud schemes. Such a situation demands more sophisticated analytical methods and techniques that are capable of detecting fraud activities from large databases. This chapter describes the application of three predictive models: MCLP, decision tree, and Naive Bayes classifier, to identify suspicious claims to assist manual inspections. The predictive models can label high-risk claims and help investigators to focus on suspicious records and accelerate the claim-handling process. The software tools used in this study are base SAS, SAS Enterprise Miner (EM) and C++. The tree node of SAS EM was used to compute decision tree solution. A C++ program was developed to construct and solve the MCLP model, Naive Bayes classifier was implemented in base SAS.

14.2 A Real-Life Data Mining Study

The data are from a US insurance company and provide information about policy types, claim/audit, policy producer, and client. The dataset has over 18,000 records with 103 variables. The variables include a mixture of numeric, categorical, and date

Y. Shi et al., *Optimization Based Data Mining: Theory and Applications*, 233
Advanced Information and Knowledge Processing,
DOI 10.1007/978-0-85729-504-0_14, © Springer-Verlag London Limited 2011

Table 14.1 Decision tree:
chi-square test

Actual	Predicted		Total	Accuracy
	Normal	Abnormal		
Normal	18510	12	18522	99.94%
Abnormal	316	37	353	10.48%
Total	18826	49	18875	

Table 14.2 Decision tree:
entropy reduction

Actual	Predicted		Total	Accuracy
	Normal	Abnormal		
Normal	18458	64	18522	99.47%
Abnormal	102	251	353	71.1%
Total	18560	315	18875	

Table 14.3 Decision tree:
Gini reduction

Actual	Predicted		Total	Accuracy
	Normal	Abnormal		
Normal	18496	26	18522	99.86%
Abnormal	213	140	353	39.66%
Total	18709	166	18875	

types. Each record has a target variable indicates the class label: either Normal or Abnormal. Abnormal indicates potential fraud claims.

Unlike the finance and credit cardholders' behavior datasets, this dataset is not cleaned and preprocessed. Thus this application conducted data preparation, data transformation, and variable selection before the General Classification Process (see Sect. 12.1). Data preparation removes irrelevant variables and missing values; correlation analysis is used to select attributes that have strong relationship with the target attribute; and transformation prepares attributes in appropriate forms for the predictive models. SAS EM tree node accommodates both numeric and character data types. MCLP requires all input variables to be numeric. Categorical variables were represented by N binary derived variables; N is the number of values of the original variable. For Naive Bayes classifier, all numeric data were transformed to character by equal-depth binning.

The Tree node in SAS EM provides three criteria for tree splitting: chi-square test, entropy reduction, and Gini reduction. Each criterion uses different techniques to split tree and establish a different decision tree model. Since each technique has its own bias, it may potentially get stronger solution if the results from three models are integrated. The method of integrating results from multiple models is called *ensemble*. The ensemble node creates a new model by averaging the posterior probabilities from chi-square, entropy, and Gini models. Tables 14.1, 14.2, 14.3, and 14.4 summarize the output of test dataset for four decision tree models: chi-square test,

Table 14.4 Ensemble: chi-square, entropy reduction, and Gini reduction

Actual	Predicted		Total	Accuracy
	Normal	Abnormal		
Normal	18519	3	18522	99.98%
Abnormal	227	126	353	35.69%
Total	18746	129	18875	

Table 14.5 MCLP classification result

Actual	Predicted		Total	Accuracy
	Normal	Abnormal		
Normal	7786	10736	18522	57.96%
Abnormal	204	149	353	42.21%
Total	7990	10885	18875	

Table 14.6 Naive Bayes classification result

Actual	Predicted		Total	Accuracy
	Normal	Abnormal		
Normal	17949	573	18522	96.91%
Abnormal	2	351	353	99.43%
Total	17951	924	18875	

entropy reduction, Gini reduction, and ensemble. For each model, the confusion matrix is reported. Tables 14.5 and 14.6 summarize classification results of MCLP and Naive Bayes classifier.

The results of NB, decision tree and MCLP indicate that probability-based methods, such as NB, outperform decision tree and MCLP for this dataset. The major reason is that many variables that are strongly associated with the target variable in this dataset are categorical. In order to improve the classification results of decision tree and MCLP, advanced transformation schemes should be used.

Chapter 15
Network Intrusion Detection

15.1 Problem and Two Datasets

Network intrusion refers to inappropriate, incorrect, or anomalous activities aimed at compromise computer networks. The early and reliable detection of network attacks is a pressing issue of today's network security. Classification methods are one the major tools in network intrusion detection. A successful network intrusion detection system needs to have high classification accuracies and low false alarm rates. Figure 15.1 describes a simplified network intrusion detection system with data mining nodes. In this system, network audit data are collected by sensors and stored in databases. Some data preprocessing steps, such as variable selection, transformation, data cleaning, and aggregation, can be implemented as in-database procedures. The preprocessed data are then sent to data mining modules for analysis. Two frequently used data mining modules in network intrusion detection: clustering and classification are included in Fig. 15.1 for illustration purpose. Clustering module can be used to identify outliers and group similar network data into clusters. The classification module, using labeled data which may be created by human analysts or other data mining functions, assigns class labels to incoming data and sends alarms to network administrators when intrusions are predicted. The outputs of data mining modules are analyzed and summarized by human analysts and stored in knowledge base. The knowledge base should be updated periodically to include new types of attacks.

Data mining modules can be implemented by various methods. For example, clustering module can be implemented by k-means, SOM, EM, BIRCH, or other clustering algorithms. Several factors need to be considered when select data mining methods. These factors include the available technical expertise, the available resources (e.g., software and hardware), and the performance of data mining methods. A satisfactory data mining method should generate high accuracy, low false alarm rates, and can be easily understood by network administrators.

In this application, we apply the kernel-based MCMP model to the network intrusion detection. The performance of this model is tested using two network datasets. The first dataset, NeWT, is collected by the STEAL lab at University of Nebraska

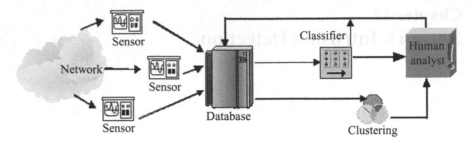

Fig. 15.1 Data mining for network intrusion detection system

at Omaha. A free version of Tenable NeWT Security Scanner is installed in a local area network node as the attacker and Ethereal version 0.10.1 is used as the data capturer in the victim machines. Tenable NeWT Security Scanner simulates the major network intrusions by generating attacks from one network node to the others and runs the same vulnerability checks using Nessus vulnerability scanner for the Microsoft Windows platform. The attack types are simulated by the sub-catalogs from Tenable NeWT Security Scanner and the normal data records consist of regular operations through networks, such as Internet browsing, ftp, and data files transferring. Each file collected from the network intrusion simulation contains the connection records traced from the raw binary data by Ethereal. Each connection record encapsulates the basic TCP/IP characteristics of all the IP traffic during the lifetime of a connection. Each record has 29 fields that are delimited by coma. Four types of network attacks are collected: denial-of-service (DOS); unauthorized access from a remote machine (R2L); unauthorized access to local root privileges (U2R); and probe. The definition and categorization of DOS, U2R, R2L, and Probe is the same as the KDDCUP-99 data. Because DOS, U2R, and R2L each have a small number of data records, we group them together into one class, named "other attack". Thus, NeWT data has three classes: probe, other attack, and normal records. The total number of data records is 34929, which include 4038 Probe, 1013 other attack and 29878 Normal. In order to apply the data mining technology such as MCMP in the original data set, non-numeric attributes are either dropped or transformed into numerical type. For example, we dropped the attributes contains IP address and time sequence information, such as "Source IP", "Destination IP", "First packet time", "Last packet time" and "Elapsed time" and transformed the "Connection status" from string to numeric. Each record ends up with 23 attributes. The attributes are separated by comma and the target attribute is the last column.

The second dataset is the KDDCUP-99 data set which was provided by DARPA in 1998 for the evaluation of intrusion detection approaches. A version of this dataset was used in 1999 KDD-CUP intrusion detection contest. After the contest, KDDCUP-99 has become a de facto standard dataset for intrusion detection experiments. KDDCUP-99 collects nine weeks of raw TCP dump data for a LAN simulating a typical U.S. Air Force LAN. Multiple attacks were added to the LAN operation. The raw training data was about four gigabytes of compressed binary TCP dump data from seven weeks of network traffic. A connection is a sequence of TCP

Table 15.1 NeWT data classification results (confusion matrix)

	Classified as		
	Probe	Other attacks	Normal
See5			
Probe	4033	5	0
Other attacks	14	994	5
Normal	663	492	28723
Running time	30 minutes		
MCMP			
Probe	4036	1	1
Other attacks	5	995	13
Normal	1181	413	28284
Running time	10–20 minutes		
MCMP with kernel			
Probe	4037	1	0
Other attacks	13	998	2
Normal	40	403	29435
Running time	10–25 minutes		

packets occurred during a specified time period and each connection is labeled as either normal or attack. There are four main categories of attacks: denial-of-service (DOS); unauthorized access from a remote machine (R2L); unauthorized access to local root privileges (U2R); surveillance and other probing. Because U2R has only 52 distinct records, we test KDDCUP-99 as a four-group classification problem. The four groups are DOS (247267 distinct records), R2L (999 distinct records), and Probe (13851 distinct records), and normal activity (812813 distinct records).

15.2 Classification of NeWT Lab Data by MCMP, MCMP with Kernel and See5

For the comparison purpose, See5 release 1.19 for Windows [169], a decision tree tool, was applied to both NeWT data and KDDCUP-99 data and the results of See5 were compared with the results of MCMP and Kernel-based MCMP. See5 is chosen because it is the winning tool of KDD 99 cup [164].

The 10-fold cross validation results of MCMP, See5, and MCMP with kernel on NeWT data are summarized in Table 15.1. Table 15.1 shows that all three methods achieve almost perfect results for Probe and excellent classifications for other attack. The difference of their performance is on Normal class. The classification accuracies of See5, MCMP, and MCMP with kernel for Normal are 96.13%, 94.66%, and 98.52%, respectively. A classifier with low classification accuracy for Normal

Table 15.2 Accuracy rates of nine classifiers for KDD99 data

	Overall (rank)	Probe (rank)	DOS (rank)	R2L (rank)	Normal (rank)
K-MCQP	.98997 (4)	.99235 (1)	.99374 (4)	.91291 (3)	.98888 (5)
SVM	.99218 (2)	.87192 (8)	.99666 (2)	.88088 (5)	.99300 (3)
BN	.98651 (6)	.91806 (6)	.96915 (8)	.91692 (2)	.99304 (2)
C4.5	.99560 (1)	.92080 (5)	.99659 (3)	.78078 (9)	.99683 (1)
Log.	.99202 (3)	.88181 (7)	.99684 (1)	.87487 (6)	.99258 (4)
CART	.98768 (5)	.94975 (4)	.99209 (5)	.83283 (8)	.98717 (6)
MC2QP	.97508 (7)	.97220 (3)	.99070 (6)	.89590 (4)	.97048 (7)
See5	.92280 (8)	.98339 (2)	.97766 (7)	.85686 (7)	.90516 (8)
NB	.88615 (9)	.81149 (9)	.96359 (9)	.93894 (1)	.86380 (9)

class will generate a lot of false alarms. The overall error rates of See5, MCMP, and MCMP with kernel are 3.38%, 4.62%, and 1.31%, respectively. MCMP with kernel achieves the lowest false alarm rate and highest overall classification accuracy.

15.3 Classification of KDDCUP-99 Data by Nine Different Methods

The KDD99 dataset was trained using nine classification algorithms: Bayesian Network, Naive Bayes, Support Vector Machine (SVM), Linear Logistic, C4.5, See5, Classification and Regression Tree (CART), MCMCQP (called MC2QP in short) (Model 10) and a version of MC2QP (Model 10) with kernel function, say K-MCQP. Some of these well-known algorithms have been recognized as the top algorithms in data mining [223]. Note that while the Naive Bayes classifier estimates the class-conditional probability based on Bayes theorem and can only represent simple distributions, Bayesian Network is a probabilistic graphic model and can represent conditional independences between variables. The nine classifiers were applied to KDD99 data using 10-fold cross-validation. The classification results shows that C4.5, Naive Bayes, Logistic and K-MCQP produce the best classification accuracies for Normal (0.996834), R2L (0.938939), DOS (0.996837) and Probe (0.992347), respectively. In terms of execution time, MC2QP (110 min) and K-MCQP (300 min) ranked third and fourth among the nine classifiers for KDD99 and outperformed the average of the other seven classifiers (498 min). Table 15.2 displays the accuracy rate of the nine classifiers for KDD99.

A successful network intrusion classification method needs to have both high classification accuracies and low false alarm rates. The false alarm rate is the percentage of classified events (Probe, DOS, R2L, other attacks and Normal) that are actually non-events. Although the accuracies and false alarm rates obtained by the nine classifiers may appear differently, the difference may not be statistically significant [102]. Therefore we further conducted paired statistical significance (0.05)

Table 15.3   False alarm rates and ranking scores of nine classifiers for KDD99 data

	Probe (rank)	DOS (rank)	R2L (rank)	Normal (rank)	Ranking score
K-MCQP	.045796 (7)	.00104 (1)	.011772 (6)	.050487 (1)	32
SVM	.014582 (1)	.008288 (5)	.002304 (1)	.190887 (8)	35
BN	.037866 (6)	.007216 (2)	.006851 (3)	.133756 (3)	38
C4.5	.032457 (5)	.014717 (6)	.003294 (2)	.202187 (9)	41
Log.	.021326 (3)	.02121 (8)	.010336 (4)	.170443 (6)	42
CART	.014859 (2)	.007672 (3)	.010414 (5)	.173647 (7)	45
MC2QP	.025514 (4)	.019469 (7)	.026839 (8)	.094168 (2)	48
See5	.054182 (8)	.007728 (4)	.069311 (9)	.141478 (4)	57
NB	.091323 (9)	.147005 (9)	.017724 (7)	.154311 (5)	67

tests to produce a performance score for each classifier. Note that a classifier with the lowest "ranking score" from statistical study achieves the best performance. Table 15.3 displays both the false alarm rate and ranking scores of the nine classifiers for KDD99. From Table 15.3, we observe that for the KDD99 data, K-BMCQLP, SVM and Bayesian Network are the top classifiers.

Chapter 16
Internet Service Analysis

16.1 VIP Mail Dataset

According to the statistic, as Chinese network advanced in the past few years, the total market size of Chinese VIP E-mail service has reached 6.4 hundred million RMB by 2005. This huge market dramatically enforced market competition among all E-mail service companies. The analysis for the pattern of lost customer accounts in hereby a significant research topic. This research can help decision-making in reducing the customer loss rate.

A dataset with 230 attributes, 5498 current records and 5498 lost records has been collected from a well-known Wall Street IPO Chinese Internet Co.

Original customer database is mainly composed of automated machine recorded customer activity journal and large amount of manually recorded tables, these data are distributed among servers located in different departments of our partnering companies, coving more than 30 kinds of transaction data charts and journal document, with over 600 attributes. If we were to directly analysis these data, it would lead to "course of dimensionality", that is to say, the drastic raise of computational complexity and classification error with data having big amount of dimensions [14]. Hence, the dimensionality of the feature space must be reduced before classification is undertaken.

With the accumulated experience functions, we eventually selected 230 attributes from the original 600 ones. Figure 16.1 displays the process which we used to perform feature selection of VIP E-mail dataset. We selected a part of the data charts and journal documents from the VIP E-mail System. Figure 16.1 displays the three logging journal documents and two email transaction journal documents, when the user log into pop3 server, the machine will record the user's log in into *pop3login*; similarly when the user log in to the smtp server, the machine will record it into *smtplogin*; when the user log in E-mail system through http protocol, the machine will record it into *weblogin*; when the user send E-mail successfully by smtp protocol, the system will record it into *smtpprcptlog* journal; when receiving a letter, it will be recorded into *mx_rcptlog* journal. We extracted 37 attribute from five journal document each, totaling 184 attribute to describe user logins and transactions.

Y. Shi et al., *Optimization Based Data Mining: Theory and Applications*,
Advanced Information and Knowledge Processing,
DOI 10.1007/978-0-85729-504-0_16, © Springer-Verlag London Limited 2011

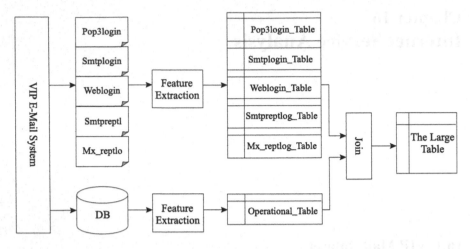

Fig. 16.1 An illustration of VIP E-mail dataset

From the database, shown in the left lower of Fig. 16.1, we extracted 6 features about customer complaint, 24 features about customer payment and 16 features about customer background information to form the *operational table*. Thus, 185 features from log files and 65 features from database eventually formed the Large Table. And the 230 attributes have depicted the feature of customer in detail. For the accumulated experience functions used in the feature extraction are confidential, further discussion of them exceeds the range of this chapter.

Considering the integrality of the records of customers, we eventually extracted two groups of them from a huge number of data, the current and the lost: 11996 SSN, 5998 respectively were chosen from the database. Combining the 11996 SSN with the 230 features, we eventually acquired the Large Table with 5998 current records and 5998 lost records, which will be the dataset for data mining.

16.2 Empirical Study of Cross-Validation

Cross-validation is frequently used for estimating generalization error, model selection, experimental design evaluation, training exemplars selection, or pruning outliers. By definition, cross-validation is the practice of partitioning a sample of data into subsamples such that analysis is initially performed on a single subsample, while further subsamples are retained "blind" in order for subsequent use in confirming and validating the initial analysis [175]. The basic idea is to set aside some of the data randomly for training a model, then the data remained will be used to test the performance of the model.

There are three kinds of cross-validation: holdout cross-validation, k-fold cross-validation, and leave-one-out cross-validation. The holdout method simply separated data into training set and testing set, taking no longer to compute but having a high variance in evaluation. k-Fold cross-validation is one way to improve over the holdout method. The data set is divided into k subsets and holdout method is repeated k times. Each time one of the k subsets is used for testing and the other $k - 1$ subsets are used for training. The advantage is that all the examples in the dataset are eventually used for both training and testing. The error estimate is reduced as k increasing. The disadvantage of this method is that it required high computation cost. Leave-one-out cross-validation takes k-fold cross-validation to its logical extreme, with k equal to N, the number of data points in the set.

In this section, a variance of k-fold cross-validation is used on VIP E-mail dataset, each time training with one of the subsets and testing with other $k - 1$ subsets. When we set k equals to 10, as it used to be, the process to select training and testing set is described as follows: first, the lost dataset (4998 records) is divided into 10 intervals (each interval has approximately 500 records). Within each interval, 50 records are randomly selected. Thus the total of 500 lost customer records is obtained after repeating 10 times. Then, as the same manner, we get 500 current customer records from the current dataset. Finally, the total of 500 lost records and 500 current records are combined to form a single training dataset, with the remaining 4498 lost records and 4498 current records merge into another testing dataset. The following algorithm is designed to carry out cross-validation:

Algorithm 16.1
REPEAT
Step 1. Generate the Training set (500 lost records + 500 current records) and Testing set (4498 lost records + 4498 current records).
Step 2. Apply the multiple-criteria programming classification models to compute $W^* = (w_1^*, w_2^*, \ldots, w_{230}^*)$ as the best weights of all 230 variables with given values of control parameters $(b, \sigma_\xi, \sigma_\beta)$.
Step 3. The classification Score[i] = $(x_i \cdot w^*)$ against of each observation has been calculated against the boundary b to check the performance measures of the classification.
END

Using Algorithm 16.1 to the VIP E-mail dataset, classification results were obtained and summarized. Due to the space limitation, only a part (10 out of the total 500 cross-validation results) of the results was summarized in Tables 16.1, 16.2, 16.3, and 16.4. The columns "lost" and "current" refer to the number of records that were correctly classified as "lost" and "current", respectively. The column "Accuracy" was calculated using correctly classified records divided by the total records in that class.

From Table 16.1 we can see that the average accuracy of 10 groups MCLP training sets is 80.94% on the lost user and 87.82% on the Current user, with the average accuracy of 10 groups testing sets 73.26% on the lost user and 83.81% on the current user. From Table 16.2 we can see that the average accuracy of 10 groups MCVQP training sets is 88.16% on the lost user and 92.26% on the Current user, with the average accuracy of 10 groups testing sets 79.80% on the lose user and 86.65% on

Table 16.1 Results of MCLP classification on VIP E-mail dataset

Cross-validation	Training set (500 lost dataset + 500 current dataset)				Testing set (4998 lost dataset + 4998 current dataset)			
	Lost	Accuracy	Current	Accuracy	Lost	Accuracy	Current	Accuracy
DataSet 1	408	81.6%	442	88.4%	3780	75.63%	4092	81.87%
DataSet 2	402	80.4%	439	87.8%	3634	72.71%	4183	83.69%
DataSet 3	403	80.6%	428	85.6%	3573	71.49%	4097	81.97%
DataSet 4	391	78.2%	454	90.8%	3525	70.53%	4394	87.92%
DataSet 5	409	81.8%	431	86.2%	3599	72.01%	4203	84.09%
DataSet 6	413	82.6%	443	88.6%	3783	75.69%	4181	83.65%
DataSet 7	399	79.8%	431	86.2%	3515	70.33%	4153	83.09%
DataSet 8	402	80.4%	433	86.6%	3747	74.97%	4099	82.01%
DataSet 9	404	80.8%	446	89.2%	3650	73.03%	4269	85.41%
DataSet 10	416	83.2%	444	88.8%	3808	76.19%	4218	84.39%

Table 16.2 Results of MCVQP classification on VIP E-mail dataset

Cross-validation	Training set (500 lost dataset + 500 current dataset)				Testing set (4998 lost dataset + 4998 current dataset)			
	Lost	Accuracy	Current	Accuracy	Lost	Accuracy	Current	Accuracy
DataSet 1	440	88.0%	460	92.0%	4009	80.21%	4294	85.91%
DataSet 2	436	87.2%	471	94.2%	3961	79.25%	4424	88.52%
DataSet 3	441	88.2%	457	91.0%	4007	80.17%	4208	84.19%
DataSet 4	454	90.8%	455	91.0%	4072	81.47%	4297	86.01%
DataSet 5	444	85.8%	562	92.4%	4073	81.49%	4367	87.37%
DataSet 6	449	89.8%	458	91.6%	4041	80.85%	4265	85.33%
DataSet 7	437	87.4%	465	93.0%	3899	78.01%	4397	87.98%
DataSet 8	440	88.0%	468	93.6%	3940	78.83%	4339	86.81%
DataSet 9	452	90.4%	453	90.6%	3999	80.01%	4316	86.35%
DataSet 10	430	86.0%	466	93.2%	3886	77.75%	4399	88.02%

the current user. From Table 16.3 we can see that the average accuracy of 10 groups MCCQP training sets is 86.04% on the lost user and 89.70% on the Current user with the average accuracy of 10 groups testing sets is 79.81% on the lose user and 86.22% on the current user. From Table 16.4 we can see that the average accuracy of 10 groups MCICP training sets is 84.94% on the lost user and 90.36% on the Current user, with average accuracy of 10 groups testing sets is 79.60% on the lose user and 86.14% on the current user. These results indicate that a good separation of the lost class and Current class is observed with these multiple-criteria programming classification models.

Table 16.3 Results of MCCQP classification on VIP E-mail dataset

Cross-validation	Training set (500 lost dataset + 500 current dataset)				Testing set (4998 lost dataset + 4998 current dataset)			
	Lost	Accuracy	Current	Accuracy	Lost	Accuracy	Current	Accuracy
DataSet 1	423	84.6%	446	89.2%	3978	79.59%	4270	85.43%
DataSet 2	438	87.6%	461	92.2%	3909	78.21%	4351	87.05%
DataSet 3	444	88.8%	454	91.8%	4146	82.95%	4310	86.23%
DataSet 4	428	85.6%	445	89.0%	4086	81.75%	4323	86.49%
DataSet 5	434	86.8%	453	90.6%	3984	79.71%	4232	84.67%
DataSet 6	424	84.8%	445	89.0%	3945	78.93%	4360	87.23%
DataSet 7	430	86.0%	447	89.4%	4022	80.47%	4252	85.07%
DataSet 8	433	86.6%	435	87.0%	3988	79.79%	4287	85.77%
DataSet 9	430	86.0%	452	90.4%	3950	79.03%	4323	86.49%
DataSet 10	418	83.6%	447	89.4%	3884	77.71%	4385	87.74%

Table 16.4 Results of MCIQP classification on VIP E-mail dataset

Cross-validation	Training set (500 lost dataset + 500 current dataset)				Testing set (4998 lost dataset + 4998 current dataset)			
	Lost	Accuracy	Current	Accuracy	Lost	Accuracy	Current	Accuracy
DataSet 1	420	84.0%	445	89.0%	3983	79.69%	4274	85.51%
DataSet 2	436	87.2%	461	92.2%	3920	78.43%	4348	86.99%
DataSet 3	445	89.0%	448	89.6%	4150	83.03%	4308	86.19%
DataSet 4	427	85.4%	441	88.2%	4078	81.59%	4318	86.39%
DataSet 5	426	85.2%	456	91.2%	3962	79.27%	4388	87.80%
DataSet 6	408	81.6%	448	89.6%	3780	75.63%	4290	85.83%
DataSet 7	425	85.0%	440	88.0%	3946	78.95%	4240	84.83%
DataSet 8	429	85.8%	451	90.2%	4102	82.07%	4190	83.83%
DataSet 9	416	83.2%	468	93.6%	3881	77.65%	4366	87.35%
DataSet 10	415	83.0%	460	92.0%	3983	79.69%	4332	86.67%

16.3 Comparison of Multiple-Criteria Programming Models and SVM

Because both the multiple-criteria programming classification model and SVM seek the solution by solve an optimization problem, so comparison between them seems to be much more convincible and convenient.

Following experiment compares the difference between MCLP, MCVQP, MC-CQP, MCIQP and SVM in free software LIBSVM 2.82, which can be downloaded from http://www.csie.ntu.edu.tw/~cjlin/libsvm/. The dataset described in Sect. 16.1 is used to these comparisons and five groups of results are summarized in Table 16.5.

Table 16.5 Comparison of the multiple-criteria programming models and SVM

Classification algorithms	Testing result (9996 records)	
	Records	Accuracy
MCLP	7850	78.53%
MCVQP	8319	83.23%
MCCQP	8298	83.02%
MCIQP	8284	82.87%
SVM	8626	86.29%

The first column lists all of the algorithms to be compared, including all of the multiple-criteria programming based algorithms and SVM. The second column is the number of the right classified records in the total testing dataset, and the third column is the percentage of the testing accuracy. From the results of Table 16.5 we can see that of the four multiple-criteria programming classification models, MCLP achieves the least testing accuracy, nearly 78.53%; MCVQP performs best, as high as 83.23%; MCCQP and MCIQP rank between MCLP and MCVQP, with the accuracy 83.02% and 82.87% respectively. And the accuracy of SVM on testing set is 86.29%. That is to say, SVM performs better that the multiple-criteria programming classification models on the VIP E-mail dataset. The reason maybe due to the fact that SVM considers the maximum classification between special points, which called support vectors; but the multiple-criteria programming classification models insist that all of the training samples should minimize overlapping layers of the classification and maximizing the distance between the classes simultaneously. This difference makes the multiple-criteria programming classification models much more sensitive on the outliers than SVM. Additionally, the kernel functions incorporated in SVM can classify the training sets in non-linear separation while the multiple-criteria programming classification models draw linear hyperplane between two classes. Fortunately, because of the VIP E-mail dataset is a linear separable one on the whole, this difference doesn't impact the accuracy of classification significantly.

Chapter 17
HIV-1 Informatics

17.1 HIV-1 Mediated Neuronal Dendritic and Synaptic Damage

The ability to identify neuronal damage in the dendritic arbor during HIV-1-associated dementia (HAD) is crucial for designing specific therapies for the treatment of HAD. To study this process, we utilized a computer based image analysis method to quantitatively assess HIV-1 viral protein gp120 and glutamate mediated individual neuronal damage in cultured cortical neurons. Changes in the number of neurites, arbors, branch nodes, cell body area, and average arbor lengths were determined and a database was formed (http://dm.ist.unomaha.edu/database.htm). We further proposed a two class model of multiple criteria linear programming (MCLP) to classify such HIV-1 mediated neuronal dendritic and synaptic damages. Given certain classes, including treatments with brain-derived neurotrophic factor (BDNF), glutamate, gp120 or non-treatment controls from our *in vitro* experimental systems, we used the two-class MCLP model to determine the data patterns between classes in order to gain insight about neuronal dendritic damages. This knowledge can be applied in principle to the design and study of specific therapies for the prevention or reversal of neuronal damage associated with HAD. Finally, the MCLP method was compared with a well-known artificial neural network algorithm to test for the relative potential of different data mining applications in HAD research.

HIV infection of the central nervous system (CNS) results in a spectrum of clinical and pathological abnormalities. Symptoms experienced can range from mild cognitive and motor defects to severe neurological impairment. This disease is known as HIV-1 associated dementia (HAD), and has affected approximately 20% of infected adults and 50% of infected children, before the era of highly active antiretroviral therapy [68, 84, 145, 154, 155]. The histopathological correlate of HAD is HIV-1 encephalitis (HIVE). HIVE is observed upon autopsy of many, but not all, patients suffering from HAD [87]. Damage to the dendritic arbor and a reduction in synaptic density are important neuropathological signatures of HIVE [83, 143, 144]. This type of damage to the neuronal network is thought to be an early event in the pathway leading to neuronal dropout via apoptosis. Notably, the best predictor of neurological impairment is not the level of viral expression; rather, it is the level

Y. Shi et al., *Optimization Based Data Mining: Theory and Applications*,
Advanced Information and Knowledge Processing,
DOI 10.1007/978-0-85729-504-0_17, © Springer-Verlag London Limited 2011

of immune activated mononuclear phagocytes (macrophages and microglia) in the brain [88]. HIV-1 infected and activated brain mononuclear phagocytes (MP) are believed to be the main mediators of neuronal injury in HAD. The mechanism by which this damage occurs remains unresolved, however, it is most likely induced by secretory factors from HIV-1 infected macrophages [82, 85, 86, 116, 138, 153].

Under steady-state conditions, MP act as scavengers and sentinel cells, nonspecifically eliminating foreign material, and secreting trophic factors critical for maintenance of homeostasis within the CNS microenvironment [67, 108, 135, 171, 233, 247]. Depending on the infection and/or activation state of these cells, MP also produces neurotoxic factors. It is believed that viral replication and/or immune activation ultimately results in MP production of neurotoxins such as glutamate and viral neurotoxic proteins, HIV-1 gp120 [27, 82, 85, 86, 116, 138, 151, 153, 173]. All of these factors could directly induce damage or may trigger secondary damageinducing mechanisms [82, 123]. However, a direct link between HIV-1 infection of brain macrophages and alterations in the dendritic arbor and synaptic density is unclear. Furthermore, while there is tremendous information regarding neuronal death/injury induced by HIV-1 or its related neurotoxins [27, 82, 85, 86, 116, 138, 151, 153, 173], there is no quantifiable data showing damage in the neuronal network, especially at the microscopic level, and its linkage to HIV-1 viral protein or each of the HIV-1 associated neurotoxins.

During the last two decades, assessments of neuronal dendritic arbor and neuronal network have been well-documented in the neuroscience and neuroinformatic research fields [11, 32, 43, 47, 48, 96, 131, 176, 190, 205]. These determinations characterize multiple features of neuronal morphology including the soma area, the area of influence, the length and diameter of the dendrites (internal or terminal), branching angles, axonal segments, the coverage factor, the complexity of tortuousity (represented by 3D bending energy), and the number of branch nodes, branches, bifurcations, and terminations. The measurements of neuronal morphology usually revolve around three major processes: laboratory experimentation (in vitro or in vivo), automatic or manual image analysis, and quantitative modeling and simulations through data sets [11, 43, 96, 131]. Thus, it is certainly feasible that the specific and quantifiable changes induced by HIV-1 viral protein or specific HIV-1 associated neurotoxins on the neuronal network and dendritic arbors be measured following this routine. Furthermore, new advances in applications of data mining technology motivate us to apply it to test the effect of pathogens and drug treatment on neuronal morphology.

To these ends, we have utilized a computerbased method to assess specific neuronal damage by uncovering the quantitative changes that occur in the neuronal network of cultured neurons. As a pilot study, we specifically considered four classes of treatments, in vitro, to describe the evolution from neurotrophins to neurotoxins. These classes include: (i) brain derived neurotrophic factor (BDNF) treatment (neurotrophin); (ii) non-treatment (control); (iii) glutamate treatment (neurotoxin); and (iv) HIV-1 envelope protein gp120 (neurotoxin). We measured several morphometric parameters that we stored in a database. We then applied a data mining technique, known as the two-class model of multiple criteria linear programming

(MCLP) [182, 183], to acquire knowledge related to neuronal damage found in dendrites and processes caused by the different treatments. Using the two-class models of MCLP, we classify individual neuron morphologies after neurons are treated with neurotrophic and/or neurotoxic factors. This could aid in successfully predicting which class or condition a neuron has been treated with according to its morphology. From this, we hope to understand early neuronal damage and how it relates to the transition between neurotrophic and neurotoxic activities that may occur during HAD. Furthermore, the analysis of these quantitative changes and utilization of MCLP may aid in the discovery of therapeutic options for HAD. We have also compared the MCLP method with a back-propagation algorithm of neural network to test the possibilities of extensive computer-based approaches investigating changes in neuronal morphologies related to HAD.

17.2 Materials and Methods

17.2.1 Neuronal Culture and Treatments

Major experimental steps include isolation of rat cortical neurons (RCN) and immunocytochemical detection of neural cells. RCN were cultured as previously described [248]. Briefly, the cortex was dissected from embryonic 17–18 day Sprague-Dawley rat pups. Individual cells were mechanically dissociated by titration in a Ca^{2+}/Mg^{2+}-free Hank's balanced salt solution (HBSS), 0.035% sodium bicarbonate, and 1 mM pyruvate (pH 7.4) after 30 minutes of 0.1% trypsin digestion. Trypsin was neutralized with 10% fetal bovine serum (FBS) and the cell suspension was washed three times in HBSS. Cells were re-suspended in neurobasal medium (Life Technologies, Grand Island, NY) containing 0.5 mM glutamine, 50 g/mL penicillin and streptomycin supplemented with B27, which is a neuronal cell culture maintenance supplement from Gibco, with glutamine and penicillin- streptomycin, and then plated in chamber-Tec slides at $2.5–5.0 \times 104$ cells/well. Cells were allowed to grow on the chamber-Tec slides for 3–9 days and 4–5 days old RCN were usually used for experiments. Earlier studies suggested that HIV macrophage conditioned media-treated neurons were characterized by a significant reduction in neuritic branching compared to control macrophage conditioned media-treated neurons [247].

Since HIV-1 macrophage-conditioned media (MCM) contain many factors, including HIV-1 proteins and other soluble neurotoxins, we furthered our study by focusing on the influence by BDNF (a representative of MP neurotrophic factors [185]), gp120 (neurotoxic HIV-1 viral protein [27], and glutamate (MP soluble neurotoxin [116, 242]. RCN were treated with BDNF at concentrations of 0.5 ng/mL, 5 ng/mL, and 50 ng/mL for 2–6 days; RCN were also treated for 2–6 days with HIV-1 viral protein $gp120_{IIIB}$ (0.1, 1, and 5 nM) (Advanced Biotechnologies Incorporated, Columbia, MD) or glutamate (10, 100, and 1000 μM). These different concentration levels were first tried to observe dose-dependent effects, which were assessed using the one-way analysis of variance (ANOVA). To build our data mining models, we then focused on data from high concentration treatments (for BDNF ≥ 5 ng/mL, for glutamate ≥ 100 uM, and for gp120 ≥ 1 nM).

17.2.2 Image Analysis

In order to analyze neuronal outgrowth, neurons labeled with antibodies for MAP-2 were analyzed by acquiring random TIFF images from the immunostained slides. These images were then imported into Image-Pro Plus, v. 4.0 (Media Cybernetics, Silver Spring, MD). Contrast and brightness were enhanced to provide the best possible images for data collection. Individual neurons were arbitrarily selected to measure the seven neuronal attributes and treatment conditions were blinded to the operator. While time-consuming, manual tracings were performed so that the process of measurement was under skilled human control, allowing for more reliable results. Analysis began with a manual trace of the outline of each neuron (cell bodies and tree processes). From this trace, a digital measurement of the length of the precise path of each neurite was gathered, as well as the area of the cell body. All data were taken from images of the same magnification ($10\times/25$ eyepieces, $20\times/0.50$n objectives). Length and surface were measured in pixels instead of metric units (with fixed pixel side of 13.3 μm). Each digital image covered the same amount of area on the slide.

We have measured seven parameters of a rat cortical neuron including: (i) the number of neurites, (ii) the number of arbors, (iii) the number of branch nodes, (iv) the average length of arbors, (v) the ratio between the number of neurites and the number of arbors, (vi) the area of cell bodies, and (vii) the maximum length of the arbors. Neurites, or branches, are the portions of the trees (excluding the single axon) between two consecutive bifurcations, between a bifurcations and a termination, or between the soma and a bifurcation or termination. Arbors are the groups of neurites extending from a single point on the cell body. Branch nodes are the points at which neurite segments branch into separate neurites. Note that, in binary trees, the number of neurites is always equal to the number of nodes plus the number of arbors. These three measures are thus obviously correlated. The area of a cell body represents the size of the soma. The maximum length of the arbors is the total length of the largest arbor in a neuron. Average arbor lengths and ratios between the number of neurites and the number of arbors were also calculated.

17.2.3 Preliminary Analysis of Neuronal Damage Induced by HIV MCM Treated Neurons

To first test the biological significance of neuronal damage in dendrites, RCN were treated with 20% crude MCM (secretory products) from control and HIV-1 infected macrophages. The results of these measurements (Fig. 17.1B) show that neurons treated with 20% MCM from HIV-1 infected monocyte-derived macrophages

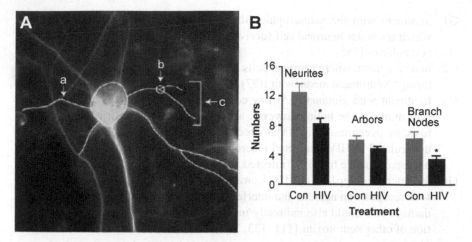

Fig. 17.1 Neuronal damage induced by secretory products from HIV-1 infected macrophages. (**A**) Image example (*a*: neurite; *b*: node; *c*: arbor). (**B**) Data (*HIV*) are compared to the group of neuronal cells treated with MCM from uninfected MDM (*Con*) and expressed as means ± SEM. * denotes $P < 0.01$. Experiments are representative of 3 replicate assays with MDM from 3 different donors

(MDM) ($n = 12$) significantly decreased both average neurite number ($P = 0.0087$, Student test, two-tailed p-value) and the amount of branching ($P = 0.0077$, Student t-test, two-tailed p-value) when compared to neurons treated with control MCM ($n = 12$). The change in the number of arbors was not quite significant ($P = 0.0831$, Student t-test, two-tailed p-value), however, when considering the ratios between the number of neurites and the number of arbors, there is a decrease. Furthermore, data shows that the mean area of the cell bodies did not change significantly between control (800.00 pixels) and infected cells (752.36 pixels) ($P = 0.2405$ Student t-test, two-tailed p-value), suggesting that damage occurred mainly in the processes. These observations suggest that HIV-1 infected MP secretory products cause neuronal damage by altering the complexity of the neuronal network. A neuronal antigen ELISA assay for the determination of changes in neuronal network by HIV-1 infected macrophage provided similar results [247].

As mentioned in "Neuronal Culture and Treatments", we furthered our study by focusing on the influence by four factors contained in HIV-1 MCM: BDNF, gp120, glutamate, and no treatment-steady state.

17.2.4 Database

The data produced by laboratory experimentation and image analysis was organized into a database comprised of four classes (G1–G4), each with nine attributes. The four classes are defined as the following:

G1: treatment with the neurotrophin BDNF (brain derived neurotrophic factor), which promotes neuronal cell survival and has been shown to enrich neuronal cell cultures [185].

G2: non-treatment, where neuronal cells are kept in the normal media used for culturing (Neurobasal media with B27).

G3: treatment with glutamate. At low concentrations, glutamate acts as a neurotransmitter in the brain. However, at high concentrations, it becomes neurotoxic by overstimulating NMDA receptors. This factor has been shown to be upregulated in HIV-1 infected macrophages [116, 242] and thereby linked to neuronal damage by HIV-1 infected macrophages.

G4: treatment with gp120, an HIV-1 envelope protein. This protein could interact with receptors on neurons and interfere with cell signaling, leading to neuronal damage or it could also indirectly induce neuronal injury through the production of other neurotoxins [111, 123, 246].

The nine attributes are defined as:

w_1 = the number of neurites,
w_2 = the number of arbors,
w_3 = the number of branch nodes,
w_4 = the average length of arbors,
w_5 = the ratio of the number of neurites to the number of arbors,
w_6 = the area of cell bodies,
w_7 = the maximum length of the arbors,
w_8 = the culture time (during this time, the neuron grows normally and BDNF, glutamate, or gp120 have not been added to affect growth), and
w_9 = the treatment time (during this time, the neuron was growing under the effects of BDNF, glutamate, or gp120).

The culture type and treatment time may affect neuron growth and its geometry; therefore, we recorded the two attributes relative to time, as well as the seven attributes on neuron geometry. The database contained data from 2112 neurons. Among them, 101 are from G1, 1001 from G2, 229 from G3, and 781 from G4. Figure 17.1 is a sample extracted from the HAD database (http://dm.ist.unomaha.edu/database.htm).

17.3 Designs of Classifications

By using the two-class model for the classifications on {G1, G2, G3, and G4}, there are six possible pairings: G1 vs. G2; G1 vs. G3; G1 vs. G4; G2 vs. G3; G2 vs. G4; and G3 vs. G4. The combinations G1 vs. G3 and G1 vs. G4 are treated as redundancies; therefore they are not considered in the pairing groups. G1 through G3 or G4 is a continuum. G1 represents an enrichment of neuronal cultures, G2 is basal or maintenance of neuronal culture and G3/G4 are both damage of neuronal cultures. There would never be a jump between G1 and G3/G4 without traveling through G2. So, we used the following four two-class pairs: G1 vs. G2; G2 vs. G3; G2 vs. G4; and G3 vs. G4. The hypotheses for these two-class comparisons are:

G1 vs. G2: BDNF should enrich the neuronal cell cultures and increase neuronal network complexity, i.e., more dendrites and arbors, more length to dendrites, etc.

G2 vs. G3: Glutamate should damage neurons and lead to a decrease in dendrite and arbor number including dendrite length.

G2 vs. G4: gp120 should cause neuronal damage leading to a decrease in dendrite and arbor number and dendrite length.

G3 vs. G4: This pair should provide information on the possible difference between glutamate toxicity and gp120-induced neurotoxicity.

As a pilot study, we concentrated on these four pairs of classes by a cross-validation process over 2112 observations. Since data sets from different batches of neuronal cultures contained varying numbers of observations, comparison between classes was based on the number of observations seen in the smaller class. This allowed for a balanced number of paired classes and avoided a training model over-fit from the larger class. Therefore, a varied cross-validation method was applied. To perform comparisons between class pairs, each smaller class G1 of {G1 vs. G2}, G3 of {G2 vs. G3}, G4 of {G2 vs. G4} and G3 of {G3 vs. G4} were partitioned into five mutually exclusive subsets S_1, S_2, S_3, S_4, and S_5, each containing 20% of the data. In the iteration j, where the subset S_j ($j = 1, \ldots, 5$), served as test data, the four remaining subsets, containing 80% of the data, were used as the training data. To form the final training set, equal amounts of data from the larger class were then randomly selected to match the training data from the smaller class. The remaining data from the larger class and the test data (S_j) from the smaller class were then teamed up as the final test set. In this way, five sets training and test data for each class pair were formed, each training set containing the same amount of data from each class in the pair. Given a threshold of training process that can be any performance measure, we have carried out the following steps:

Step 1: For each class pair, we used the Linux code of the two-class model to compute the compromise solution $W^* = (w_1^*, \ldots, w_9^*)$ as the best weights of all nine neuronal variables with given values of control parameters (b, ξ^*, β^*).

Step 2: The classification score $MCLP_i = (x_i \cdot W^*)$ against each observation was calculated against the boundary b to check the performance measures of the classification.

Step 3: If the classification result of Step 2 was not acceptable (i.e., the given performance measure was smaller than the given threshold), different values of control parameters (b, ξ^*, β^*) were chosen and the process (Steps 1–3) repeated.

Step 4: Otherwise, for each class pair, $W^* = (w_1^*, \ldots, w_9^*)$ was used to calculate the MCLP scores for all A in the test set and conduct the performance analysis.

According to the nature of this research, we define the following terms, which have been widely used in performance analysis:

TP (True Positive) = the number of records in the first class that has been classified correctly;

FP (False Positive) = the number of records in the second class that has been classified into the first class;

TN (True Negative) = the number of records in the second class that has been classified correctly; and

FN (False Negative) = the number of records in the first class that has been classified into the second class.

Then we have four different performance measures:

$$\text{Sensitivity} = \frac{TP}{TP + FN}; \qquad \text{Positive predictivity} = \frac{TP}{TP + FP};$$

$$\text{False Positive Rate} = \frac{FP}{TN + FP}; \quad \text{and} \quad \text{Negative Predictivity} = \frac{TN}{FN + TN}.$$

The "positive", in this article, represents the first class label while the "negative" represents the second class label in the same class pair. For example, in the class pair {G1 vs. G2}, the record of G1 is "positive" while that of G2 is "negative". Among the above four measures, more attention is paid to sensitivity or false positive rates because both measure the correctness of classification on class-pair data analysis. Note that in a given class pair, the sensitivity represents the corrected rate of the first class and one minus the false positive rate is the corrected rate of the second class by the above measure definitions.

We set the across-the-board threshold of 55% for sensitivity [or 55% of (1-false positive rate)] to select the experimental results from training and test processes. There are three reasons for setting 55% as the threshold:

1. No standard threshold values have been well established;
2. The threshold of 55% means the majority has been classified correctly;
3. This is a pilot study because of the limited data set.

17.4 Analytic Results

17.4.1 Empirical Classification

Since the value ranges of the nine variables are significantly different, a linear scaling transformation needs to be performed for each variable. The transformation expression is

$$x_n = \frac{x_i - \min(x_1 \ldots x_n)}{\max(x_1 \ldots x_n) - \min(x_1 \ldots x_n)},$$

where x_n is the normalized value and x_i is the instance value [167].

All twenty of the training and test sets, over the four class pairs, have been computed using the above procedure. The results against the threshold are summarized in Tables 17.1, 17.2, 17.3, and 17.4. The sensitivities for the comparison of all four pairs are higher than 55%, indicating that acceptable separation among individual pairs is observed with this criterion. The results are then analyzed in terms of both

Table 17.1 Classification results with G1 vs. G2

Training	N1	N2	Sensitivity	Positive predictivity	False positive rate	Negative predictivity
G1	55 (TP)	34 (FN)	61.80%	61.80%	38.20%	61.80%
G2	34 (FP)	55 (TN)				
Training	N1	N2	Sensitivity	Positive predictivity	False positive rate	Negative predictivity
G1	11 (TP)	9 (FN)	55.00%	3.78%	30.70%	98.60%
G2	280 (FP)	632 (TN)				

Table 17.2 Classification results with G2 vs. G3

Training	N2	N3	Sensitivity	Positive predictivity	False positive rate	Negative predictivity
G2	126 (TP)	57 (FN)	68.85%	68.48%	31.69%	68.68%
G3	58 (FP)	125 (TN)				
Training	N2	N3	Sensitivity	Positive predictivity	False positive rate	Negative predictivity
G2	594 (TP)	224 (FN)	72.62%	99.32%	8.70%	15.79%
G3	4 (FP)	42 (TN)				

Table 17.3 Classification results with G2 vs. G4

Training	N2	N4	Sensitivity	Positive predictivity	False positive rate	Negative predictivity
G2	419 (TP)	206 (FN)	67.04%	65.88%	34.72%	66.45%
G4	217 (FP)	408 (TN)				
Training	N2	N4	Sensitivity	Positive predictivity	False positive rate	Negative predictivity
G2	216 (TP)	160 (FN)	57.45%	80.90%	32.90%	39.39%
G4	51 (FP)	104 (TN)				

positive predictivity and negative predictivity for the prediction power of the MCLP method on neuron injuries. In Table 17.1, G1 is the number of observations predefined as BDNF treatment; G2 is the number of observations predefined as nontreatment; N1 means the number of observations classified as BDNF treatment and N2 is the number of observations classified as non-treatment. The meanings of other pairs in Tables 17.1, 17.2, 17.3, and 17.4 can be similarly explained.

Table 17.4 Classification results with G3 vs. G4

Training	N3	N4	Sensitivity	Positive predictivity	False positive rate	Negative predictivity
G3	120 (TP)	40 (FN)	57.45%	80.90%	24.38%	75.16%
G4	39 (FP)	121 (TN)				
Training	N3	N4	Sensitivity	Positive predictivity	False positive rate	Negative predictivity
G3	50 (TP)	19 (FN)	72.46%	16.78%	40.00%	95.14%
G4	248 (FP)	372 (TN)				

In Table 17.1 positive and negative predictivity are the same (61.80%) in the training set. However, the negative predictivity of the test set (98.60%) is much higher than that of the positive predictivity (3.78%). The prediction of G1 in the training set is better than that of the test set while the prediction of G2 in test outperforms that of training. This is due to the small size of G1. In Table 17.2 for {G2 vs. G3}, the positive predictivity (68.48%) is almost equal to the negative predictivity (68.68%) of the training set. The positive predictivity (99.32%) is much higher than the negative predictivity (15.79%) of the test set. As a result, the prediction of G2 in the test set is better than in the training set, but the prediction of G3 in the training set is better than in the test set.

The case of Table 17.3 for {G2 vs. G4} is similar to that of Table 17.2 for {G2 vs. G3}. The separation of G2 in test (80.90%) is better than in training (65.88%), while the separation of G4 in training (66.45%) is better than in test (39.39%). In the case of Table 17.4 for {G3 vs. G4}, the positive predictivity (80.90%) is higher than the negative predictivity (75.16%) of the training set. Then, the positive predictivity (16.78%) is much lower than the negative predictivity (95.14%) of the test set. The prediction of G3 in training (80.90%) is better than that of test (16.78%), and the prediction of G4 in test (95.14%) is better than that of training (75.16%).

In summary, we observed that the predictions of G2 in test for {G1 vs. G2}, {G2 vs. G3} and {G2 vs. G4} is always better than those in training. The prediction of G3 in training for {G2 vs. G3} and {G3 vs. G4} is better than those of test. Finally, the prediction of G4 for {G2 vs. G4} in training reverses that of {G3 vs. G4} in test. If we emphasize the test results, these results are favorable to G2. This may be due to the size of G2 (non-treatment), which is larger than all of other classes. The classification results can change if the sizes of G1, G3, and G4 increase significantly.

Chapter 18
Anti-gen and Anti-body Informatics

18.1 Problem Background

Antibodies bind their antigen using residues, which are part of the hypervariable loops. The properties of the antibody-antigen interaction site are primarily governed by the three-dimensional structure of the CDR-loops (complementarity determining region). The mode of antibody binding corresponding antigen is conservation. Antibody structure is rearranged to recognize and bind antigen with at least moderate affinity. Different types of antibody combining sites have been studied such as: cavity or pocket (hapten), groove (peptide, DNA, carbohydrate) and planar (protein) [214]. Much effort has focused on characters of antibody structure, antibody-antigen binding sit and mutation on the affinity and specificity of the antibody [13, 41, 45, 113, 189, 214, 224].

According to [62], in functional studies on antibody-antigen complexes, a few residues are tight bound among a number of contact residues in antibody-antigen interface. The distance between antibody's interface residue and antigen surface is the one of antigen-antibody binding characters. In this chapter, we set three type of interaction distance range between antibody residue and antigen surface. The residue belong to these distance range in antibody structure is predicted by Multiple Criteria Quadratic Programming (MCQP), LDA, Decision Tree and SVM to study correlation between characters of antibody surface residue and antigen-antibody interaction.

A kind of complex structure is designated as the basic complex structure and selected from the PDB file library, which is the start point of our research. Antibody's heavy chain, light chain and corresponding antigen from the complex are collectively defined as this basic complex structure. After testing the selected basic structure, some missing residues in the heavy chain and light chain of these structures are detected. We fill these missing residues using (HyperChem 5.1 for Windows (Hypercube, FL, USA)). If the heavy chains and the light chains of antibody in two complex structures are exactly the same, one of the structures is defined redundant. After examination of all the basic structures, 37 non-redundant complex structures are extracted.

Y. Shi et al., *Optimization Based Data Mining: Theory and Applications*, 259
Advanced Information and Knowledge Processing,
DOI 10.1007/978-0-85729-504-0_18, © Springer-Verlag London Limited 2011

In the 37 complex structures obtained, 3 of the antigens are nucleic acid, 4 of the antigens are heterocomplex, and the rest are proteins.

The residues in the antibody structure can be divided into residues embedded in the structure and residues in the surface of the structure. The Accessible Surface Area (ASA) method is utilized in the recognition of surface residues. The ASA is calculated using the DSSP program. If the ASA in the chain structure is 25% larger than the value calculated by the residue alone, this residue can be regarded to surface residue of the antibody structure. After computation, 5253 surface residues are recognized from the 37 non-redundant complex structures.

Antibody function is accomplished by combination antibody-antigen interaction surface. There are many methods to identify interaction surface residues from protein-protein complex. Fariselli [70] indicated that two surface residues are regarded as interface residues, if distance between $C\alpha$ atom of one surface residue in one protein and $C\alpha$ atom of one surface residue in another protein is less then 12 Å. In this chapter, the coordinate of α atom in residue is taken as the coordinate of this residue. Distances between one surface residue in the antibody structure and every atom in antigen in this complex structure are calculated. If the distance is less than 12 Å the surface residue is considered to interaction with the antigen. According to [62], in functional studies on antibody-antigen complexes, a few residues are tight bound among a number of contact residues in antibody-antigen interface. They play an important role in antigen-antibody interaction. For the research of the interaction distance between antibody residue and antigen, different ranges are used here. If the calculated distance is less than the threshold value, it is defined that the distance between interface residue and antigen belong to this range. The distance 8 Å, 10 Å and 12 Å are selected as the threshold values to cope with range. There are 329, 508, 668 residues belong to distance ranges 8 Å, 10 Å, 12 Å respectively.

There are many studies on composing sequence feature of the target residue [120, 121]. In this chapter, sequence patch is used for composing sequence feature. If the sequence patch size is set to be 5, it means the join sequence feature is composed by the target residue and 5 front neighbors and 5 back neighbors (the total of 11 residues) in sequence. Sequence patch is about the neighbor relationship in sequence. Sequence feature is coded as follows: each residue is represented by a 20-dimensional basic vector, i.e. each kind of residue corresponds to one of twenty dimensions in the basic vector. The element of the vector having value one means that it belongs to that kind of residue. Only one position has value one and others has zero. This sequence feature will be coded as a 220-dimensional vector if the sequence patch size is set to 5.

18.2 MCQP, LDA and DT Analyses

A ten-fold cross validation is applied on the Evaluation measure of classification accuracy. The details of a ten-fold cross validation is discusses as follows. The dataset is split into ten parts. One of ten acts as the testing set and the other nine as the training set to construct the mathematical model. The process rotates for ten times with each part as a testing set in a single round. In the generalization test, the mean value of the accuracy in the ten-fold cross validation test is used as the accuracy measure of the experiment. If the utilized method is correct, then the extracted features can well explain the correlation between antibody-antigen interaction residue and antibody-antigen interaction surface. In this situation, the mean accuracy of the ten-fold cross validation should also be comparatively high. The classification accuracy is composed with two parts: the accuracy of correct prediction residue in the distance range and the accuracy of correct prediction residue out of the distance range.

Other indictors, such as prediction Type I error, Type II error and correlation coefficient, can be obtained from the cross validation test for analyzing the effectiveness of the method.

Type I error is defined as the percentage of predicting the residues in the distance range that are actually out of the distance range and Type II error is defined as the percentage of residues out of the distance range that are actually in the distance range.

Correlation coefficient falls into the range of $[-1, 1]$. The introduction of correlation coefficient is to avoid the negative impacts of the imbalance between different classes of data. For example, if two types of data take up a $4 : 1$ position in a single dataset, then the prediction of the type with a large size of data will be 80% accurate. If the same dataset is used in the testing, then it is meaningless in the case of prediction. When using the model constructed on prediction of the data, the correlation coefficient will be -1 if the prediction is completely contrary to the exact value, 1 if the prediction is correct, and 0 if the prediction is randomly produced. The Correlation coefficient is calculated as follows:

$$\text{Correlation coefficient} = \frac{(TP \times TN) - (FP \times FN)}{\sqrt{(TP + FN)(TP + FP)(TN + FP)(TN + FN)}}, \quad \text{where:}$$

TP (True Positive): the number of residues in the distance range that has been classified correctly;

FP (False Positive): the number of residues out of the distance that has been classified into the class of in the distance range;

TN (True Negative): the number of residues out of the distance range that has been classified correctly;

FN (False Negative): the number of records in the distance range that has been classified into out of the distance range.

We are concerned about the accuracy of correct prediction residue in the distance range, Type II error and Correlation coefficient in five indictors, because the accuracy of correct prediction residue in the distance range is accuracy of identification residue belong to distance range, Type II error is percentage of residues out of the distance range that are actually in the distance range and Correlation coefficient is a measure of predictions correlate with actual data.

We conducted numerical experiments to evaluate the proposed MCQP model. The result of MCQP is compared with the results of 2 widely accepted classification tools: LDA and See5. The following tables (Tables 18.1–18.12) summarize the

Table 18.1 The results of the ten-fold cross validation tests for the distance ranges 8 Å with MCQP

MCQP (distance range 8 Å)	Classification accuracy		Error rate		Correlation coefficient
	Residue in the range	Residue out of the range	Type I	Type II	
Sequence patch size 1	62.27%	79.42%	24.84%	32.21%	42.32%
Sequence patch size 2	73.29%	81.09%	20.51%	24.78%	54.55%
Sequence patch size 3	79.73%	84.21%	16.53%	19.40%	64.00%
Sequence patch size 4	80.19%	85.92%	14.94%	18.74%	66.22%
Sequence patch size 5	80.18%	89.16%	11.91%	18.19%	69.62%

Table 18.2 The results of the ten-fold cross validation tests for the distance ranges 8 Å with See5

See5 (distance range 8 Å)	Classification accuracy		Error rate		Correlation coefficient
	Residue in the range	Residue out of the range	Type I	Type II	
Sequence patch size 1	1.82%	99.72%	13.50%	49.61%	7.55%
Sequence patch size 2	16.41%	99.15%	4.94%	45.74%	27.71%
Sequence patch size 3	27.36%	98.58%	4.94%	42.43%	36.95%
Sequence patch size 4	29.48%	98.62%	4.48%	41.69%	38.89%
Sequence patch size 5	29.79%	98.88%	3.62%	41.52%	39.66%

averages of 10-fold cross-validation test-sets results of LDA, See5, and MCQP for each dataset.

When the distance range is 8 Å, the prediction results of MCQP method are listed in Table 18.1. With the increase of sequence path size from 1 to 5, the accuracy of prediction residue in the distance range increases from 62.27% to 80.18%, the accuracy of prediction residue out of the distance range increases from 79.42% to 89.16% and correlation coefficient increases from 42.32% to 69.62%. This result indicates that the increase in the sequence path size will help reform the prediction result.

The results of the ten-fold cross validation tests for See5 and LDA can be similarly explained as shown in Tables 18.2 and 18.3. There exists an extraordinary situation in Table 18.3. Although the overall performance is improved, the Type I error rate of LDA increased due to the decreases in classification accuracy in residue out the range when the sequence patch size increases. In the comparison of the three methods, MCQP method has shown strong advantages in the accuracy of correct prediction residue in the distance range, Type II error and correlation coefficient with the same Sequence path size.

When the distance range is 10 Å, the prediction results of MCQP method are listed in Table 18.4. With the increase of sequence path size from 1 to 5, the accuracy of correct prediction residue in the distance range increases from 63.11% to 80.34%,

Table 18.3 The results of the ten-fold cross validation tests for the distance ranges 8 Å with LDA

LDA (distance range 8 Å)	Classification accuracy		Error rate		Correlation coefficient
	Residue in the range	Residue out of the range	Type I	Type II	
Sequence patch size 1	29.79%	98.88%	3.62%	41.52%	39.66%
Sequence patch size 2	29.79%	98.88%	3.62%	41.52%	39.66%
Sequence patch size 3	29.79%	98.88%	3.62%	41.52%	39.66%
Sequence patch size 4	74.77%	91.73%	9.96%	21.57%	67.48%
Sequence patch size 5	75.68%	92.85%	8.64%	20.75%	69.56%

Table 18.4 The results of the ten-fold cross validation tests for the distance ranges 10 Å with MCQP

MCQP (distance range 10 Å)	Classification accuracy		Error rate		Correlation coefficient
	Residue in the range	Residue out of the range	Type I	Type II	
Sequence patch size 1	63.11%	80.07%	24.00%	31.54%	43.81%
Sequence patch size 2	71.78%	82.96%	19.18%	25.38%	55.09%
Sequence patch size 3	76.58%	84.69%	16.66%	21.66%	61.47%
Sequence patch size 4	79.27%	86.85%	14.23%	19.27%	66.31%
Sequence patch size 5	80.34%	89.53%	11.53%	18.01%	70.17%

Table 18.5 The results of the ten-fold cross validation tests for the distance ranges 10 Å with See5

See5 (distance range 10 Å)	Classification accuracy		Error rate		Correlation coefficient
	Residue in the range	Residue out of the range	Type I	Type II	
Sequence patch size 2	36.61%	97.62%	6.11%	39.37%	43.20%
Sequence patch size 3	50.39%	97.09%	5.46%	33.82%	53.69%
Sequence patch size 4	51.57%	97.34%	4.90%	33.22%	55.01%
Sequence patch size 5	52.76%	97.24%	4.98%	32.70%	55.83%

the accuracy of correct prediction residue out of the distance range increases from 80.07% to 89.53%, Type I error decreases from 24.00% to 11.53%, Type II error decreases from 31.54% to 18.01% and correlation coefficient increases from 43.81% to 70.17%. Similar to the results of distance range 8 Å, the prediction results are also improved when the sequence path size is increased.

The results of See5 and LDA are summarized in Tables 18.5 and 18.6. MCQP shows strong advantages against See5 and LDA in the accuracy of correct prediction residue in the distance range and Type II error rate. The correlation coefficients of MCQP and LDA are close.

Table 18.6 The results of the ten-fold cross validation tests for the distance ranges 10 Å with LDA

LDA (distance range 10 Å)	Classification accuracy		Error rate		Correlation coefficient
	Residue in the range	Residue out of the range	Type I	Type II	
Sequence patch size 1	64.76%	76.76%	26.41%	31.46%	41.82%
Sequence patch size 2	69.49%	82.98%	19.68%	26.89%	52.95%
Sequence patch size 3	72.05%	89.73%	12.48%	23.75%	62.77%
Sequence patch size 4	74.21%	92.03%	9.70%	21.89%	67.32%
Sequence patch size 5	75.79%	93.74%	7.63%	20.53%	70.68%

Table 18.7 The results of the ten-fold cross validation tests for the distance ranges 12 Å with MCQP

MCQP (distance range 12 Å)	Classification accuracy		Error rate		Correlation coefficient
	Residue in the range	Residue out of the range	Type I	Type II	
Sequence patch size 1	56.26%	76.19%	29.74%	36.47%	33.11%
Sequence patch size 2	60.05%	79.23%	25.70%	33.52%	40.02%
Sequence patch size 3	63.96%	80.07%	23.76%	31.04%	44.61%
Sequence patch size 4	64.39%	82.48%	21.39%	30.15%	47.66%
Sequence patch size 5	66.03%	84.36%	19.15%	28.71%	51.26%

When the distance range is 12 Å, the prediction results of MCQP method are listed in Table 18.7. With the increase of sequence path size from 1 to 5, the accuracy of correct prediction residue in the distance range increases from 56.26% to 66.03%, the accuracy of correct prediction residue out of the distance range increases from 76.19% to 84.36%, Type I error decreases from 29.74% to 19.15%, Type II error decreases from 36.47% to 28.71% and correlation coefficient increases from 33.11% to 51.26%.

The results of distance range 12 Å for See5 and LDA are shown in Tables 18.8 and 18.9. The results of LDA are better than the results of MCQP and See5 while MCQP outperforms to See5.

Following the prediction of distance, another experiment is designed to predict the class of antigen because different types antibody-antigen interface have different surface characters [214]. These are cavity or pocket (hapten), groove (peptide, DNA, carbohydrate) and planar (protein). In our research, antigens are divided into two classes (protein, non protein) for study. Feature for inference of antigen class is constructed by interface residues belong to different range. It is a 21-dimensional vector, which denotes 20 kinds of residue composition in the interface residues plus the number of interface residue belong to distance ranges 8 Å, 10 Å, 12 Å respectively. The following tables (Tables 18.10, 18.11, and 18.12) summarize the predication results of antigen class by MCQP, See5 and LDA for each dataset.

Table 18.8 The results of the ten-fold cross validation tests for the distance ranges 12 Å with See5

See5 (distance range 12 Å)	Classification accuracy		Error rate		Correlation coefficient
	Residue in the range	Residue out of the range	Type I	Type II	
Sequence patch size 1	40.72%	97.40%	5.99%	37.83%	46.27%
Sequence patch size 2	47.31%	97.62%	4.78%	35.06%	51.99%
Sequence patch size 3	55.24%	97.75%	3.91%	31.41%	58.54%
Sequence patch size 4	58.38%	97.49%	4.12%	29.92%	60.71%
Sequence patch size 5	60.48%	97.84%	3.45%	28.77%	62.87%

Table 18.9 The results of the ten-fold cross validation tests for the distance ranges 12 Å with LDA

LDA (distance range 12 Å)	Classification accuracy		Error rate		Correlation coefficient
	Residue in the range	Residue out of the range	Type I	Type II	
Sequence patch size 1	63.32%	73.94%	29.16%	33.16%	37.47%
Sequence patch size 2	65.42%	86.65%	16.95%	28.52%	53.28%
Sequence patch size 3	69.31%	90.88%	11.62%	25.24%	61.64%
Sequence patch size 4	73.05%	93.63%	8.02%	22.35%	68.14%
Sequence patch size 5	76.40%	94.33%	6.88%	19.74%	72.25%

Table 18.10 The results of predication antigen class for the distance ranges 8 Å with MCQP, See5 and LDA

Distance 8 Å	Classification accuracy			Error rate		Correlation coefficient
	Overall	Interface	Surface	Type I	Type II	
MCQP	88.93%	89.81%	90.00%	10.02%	10.17%	79.81%
See5	83.78%	93.33%	42.86%	37.97%	13.46%	41.92%
LDA	86.49%	90.00%	71.43%	24.10%	12.28%	62.52%

We compared results of three methods in different distance range. The result of distance 8 Å is the best and the result of MCQP is better than the results of See5 and LDA.

The vast quantities of existing immunological data and advanced information technology have boosted the research work on computational immunology. The distance between antibody's interface residue and antigen surface is the one of antigen-antibody binding characters to observe the circumstantialities of antibody-antigen interaction surface. It will help us to understand position of each interface residue relative to antigen in three-dimensional, which connect affinity of antibody-antigen interaction. Experimental data analysis by using machine learning methods

Table 18.11 The results of predication antigen class for the distance ranges 10 Å with MCQP, See5 and LDA

Distance 10 Å	Classification accuracy			Error rate		Correlation coefficient
	Overall	Interface	Surface	Type I	Type II	
MCQP	75.36%	77.62%	73.33%	25.57%	23.38%	51.00%
See5	70.27%	76.67%	42.86%	42.70%	35.25%	20.75%
LDA	59.46%	63.33%	57.14%	40.36%	39.09%	20.52%

Table 18.12 The results of predication antigen class for the distance ranges 12 Å with MCQP, See5 and LDA

Distance 12 Å	Classification accuracy			Error rate		Correlation coefficient
	Overall	Interface	Surface	Type I	Type II	
MCQP	76.07%	82.38%	70.00%	26.70%	20.11%	52.79%
See5	72.97%	80.00%	42.86%	41.67%	31.82%	24.62%
LDA	75.68%	76.67%	71.43%	27.15%	24.62%	48.16%

may help explain and provide significant insight into the complex phenomenon of antibody-antigen interaction.

In this research, we set three type of interaction distance range between antibody residue and antigen surface. Three data sets contain 329, 508, 668 antibody residues corresponding to different distance ranges (8 Å, 10 Å, 12 Å). Based on distance range and sequence patch size, 15 samples have been created in the research of prediction these residues from 5253 antibody residues with MCQP. We noticed that all performance measures improved when the sequence path size is increased in the same distance range. This results show that the distance between antibody's interface residue and antigen can be inferred from the antibody structures data. The result of MCQP is compared with the results of Decision Tree based See5 and Linear Discriminant Analysis. MCQP achieved comparable or better results than See5 and LDA.

18.3 Kernel-Based MCQP and SVM Analyses

In this subsection, we use the same data sets in Sect. 18.2 to perform kernel-based MCQP (see Chap. 9) and kernel-based SVM (see Chap. 2).

The results obtained with the MCQP method are presented in Table 18.13. As the distance range increases from 8 to 12 Å, the accuracy of class I decreases from 80.18% to 66.03%, the accuracy of class II decreases from 89.16% to 84.36%, the Type I error increases from 11.91% to 19.15%, the Type II error increases from 18.19% to 28.71%, and the correlation coefficient decreases from 69.62% to 51.26%. The prediction results are worse for larger distance ranges.

Table 18.13   Results of the 10-fold cross validation tests with MCQP

Distance (Å)	Classification accuracy (%)		Error rate (%)		Correlation coefficient (%)
	Class I	Class II	Type I	Type II	
8	80.18	89.16	11.91	18.19	69.62
10	80.34	89.53	11.53	18.01	70.17
12	66.03	84.36	19.15	28.71	51.26

Table 18.14   Results of the 10-fold cross validation tests with SVM

Distance (Å)	Classification accuracy (%)		Error rate (%)		Correlation coefficient (%)
	Class I	Class II	Type I	Type II	
8	34.40	99.30	1.99	39.78	44.30
10	43.30	99.20	1.81	36.37	51.26
12	47.40	99.50	1.04	34.58	54.95

The results obtained with the SVM are presented in Table 18.14. As the distance range increases from 8 to 12 Å, the accuracy of class I increases from 34.40% to 47.40%, the accuracy of class II decreases from 99.30% to 99.50%, the Type I error decreases from 1.99% to 1.04%, the Type II error decreases from 39.78% to 34.58%, and the correlation coefficient increases from 44.30% to 54.95%. The prediction results are better for larger distance ranges.

The results of the statistical analyses (Tables 18.13 and 18.14) show that there is a significant difference between the kernel-based MCQP and the kernel-based SVM. The results obtained with the MCQP become worse when the distance range is increased, and the results obtained with the SVM improve when the distance range is increased. Comparison of the two methods shows that the MCQP is superior in the accuracy of class I, Type II error, and correlation coefficient for the same distance range.

Chapter 19
Geochemical Analyses

19.1 Problem Description

World natural diamond production for 2004 is estimated at 156 million carats and it translated into 61.5 billion US dollars in worldwide jewelery sales [58]. Even though, the current level of demand for diamonds with high color and quality is still not being met by the world's producing diamond mines. Numerous companies are carrying out various phases of diamond exploration in Botswana, which is the world's leading producer of gem quality diamonds. Due to the extensive Kalahari sand cover (and Karoo basalts underneath), sophisticated and innovative sampling and geophysical techniques are required to locate undiscovered kimberlites [221].

The goal of this chapter is to build an analytical model for kimberlites identification. Two classification methods are applied to a dataset containing information about rock samples drilled in Botswana.

The dataset contains rock samples data from one region of Botswana. Original dataset has 5921 row of observations and 89 variables, and each observation describes detailed information of one rock sample about its position, physical and chemical attributes. These variables include numeric and character types. After consulting the experts, we deleted some rows missing important variables and exclude some variables, which are irrelevant, redundant, or correlated such as sample-id and horizon. Then some types of variables are transformed from character to binary to satisfy the requirements of models, such as color and shape.

After data transformation, the dataset includes 4659 observations and 101 variables.

Data classification is a two-step process [102]. In the first step, a model is built describing a predetermined set of data classed or concepts. This model is constructed by analyzing database tuples described by attributes. Each tuple is assumed to belong to a predefined class, as determined by one of the attributes, called the class label attributes. The data used to build the model is called training set. And this step can be called supervised learning. In the second step, the model is used to classification. And the predictive accuracy of the model is estimated. The data set used to classification in this step is called testing set. When the constructed model is proved to be stable and robust, then this model can be used to predict the new data.

Y. Shi et al., *Optimization Based Data Mining: Theory and Applications*, 269
Advanced Information and Knowledge Processing,
DOI 10.1007/978-0-85729-504-0_19, © Springer-Verlag London Limited 2011

Table 19.1 Two-class LDA analysis with cross-validation

Count	0	3301	387	3688
	1	195	769	964
%	0	89.5	10.5	100
	1	20.2	79.8	100

Cross-validation is done only for those cases in the analysis. In cross-validation, each case is classified by the functions derived from all cases other than that case.

87.5% of cross-validated grouped cases correctly classified.

89.5% of kimberlites are grouped correctly, and 79.8% of other rocks are grouped correctly.

Table 19.2 Two-class DT analysis

Observed	Predicted		Percent correct
	0	1	
0	3493	202	95%
1	157	807	84%
Overall percentage	0.78	0.22	92%

The kimberlites identification for this dataset can be regarded as a four-group classification problem based on the fact that there are four important kinds of rock in this dataset. We will apply two standard classification methods, known as Linear Discriminant Analysis (LDA), Decision tree and C-SVM (see Chap. 2) for this work.

19.2 Multiple-Class Analyses

19.2.1 Two-Class Classification

We use 1 for kimberlites and 0 for other rocks. LDA has the following steps:

(1) change chemical attributes into numeric,
(2) stepwise used for attributes selection,
(3) prior probabilities: all classes equal,
(4) display: Leave-one-out classification.

The results are given in Table 19.1. The Decision Tree results are shown in Table 19.2.

Ten-fold cross-validation method is select here to estimate the accuracy of decision tree here. The Support Vector Machine (SVM) results are given in Table 19.3.

Table 19.3 Two-class SVM analysis

Subset	Accuracy of kimberlites (%)	Accuracy of others (%)	Error Type I	Error Type II
1	76.0	87.3	0.24	0.13
2	69.3	90.7	0.31	0.09
3	33.1	99.0	0.67	0.01
4	81.0	86.2	0.19	0.14

Hold-out method is used to test the accuracy of this model.

For each subset we get the accuracy of prediction, take subset 1 as an example, 76.0% of kimberlites are grouped correctly, and 87.3% other rocks are grouped correctly.

Table 19.4 Three-class LDA analysis with cross-validation

Count	0	812	108	264	1184
	1	161	2261	89	2511
	2	22	203	739	964
%	0	68.6	9.1	22.3	100
	1	6.4	90.0	3.5	100
	2	2.3	21.1	73.7	100

Cross-validation is done only for those cases in the analysis. In cross-validation, each case is classified by the functions derived from all cases other than that case.

81.8% of cross-validated grouped cases correctly classified.

76.7% of kimberlites are grouped correctly, 90.0% of Stormberg basalts are grouped correctly, and 68.6% other rocks are grouped correctly.

19.2.2 Three-Class Classification

We consider three classes, namely as kimberlites (2), Stormberg basalts (1) and other rocks (0). The LDA has two steps which are similar the above. The results are listed in Table 19.4.

These results are also displayed in Fig. 19.1. The Decision tree results are presented in Table 19.5. The SVM results are shown in Table 19.6.

19.2.3 Four-Class Classification

We use kimberlites (2), Stormberg basalts (1), Ecca and Karoo seds (3) and others (0). The LDA results are listed in Table 19.7. The four-class LDA is pictured as Fig. 19.2. The Decision tree results are shown in Table 19.8. The SVM results are given in Table 19.9.

Canonical Discriminant Functions

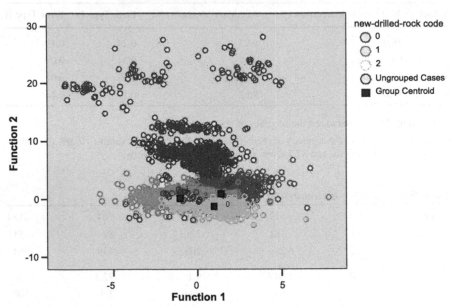

Fig. 19.1 Three-class LDA

Table 19.5 Three-class DT analysis

Observed	Predicted				Percent correct
	0	1	2	3	
0	522	74	41	23	79%
1	34	2422	55	0	97%
2	7	107	850	0	88%
3	0	0	125	399	76%
Overall percentage	0.12	0.56	0.23	0.09	90%

19.3 More Advanced Analyses

To gain more knowledge of the comparison for these methods, we conducted more advanced analyses. For this dataset are linearly inseparable, we used C-SVM with RBF kernel function [140] to classify the rock samples. (This model is called C-SVM[1].) There are two parameters (C for the objective function and GAMMA for the kernel function) selected for this model. The most common and reliable approach is to decide on parameter ranges, and to do an exhaustive grid search over the parameter space to find the best setting [33]. Figure 19.3 shows its process. It contains contour plots for training datasets. The different colors of the projected

Table 19.6 Three-class SVM analysis

| Subset | Accuracy | | |
	Kimberlites	Stormberg basalts	Others
1	87.4%	54.1%	76.3%
2	87.3%	60.6%	53.2%
3	74.5%	74.3%	39.6%
4	57.5%	62.8%	63.2%

Hold-out method is used to test the accuracy of this model.

For each subset we get the accuracy of prediction.

Table 19.7 Three-class SVM analysis with cross-validation

Count	0	423	75	126	29	653
	1	78	2226	90	117	2511
	2	59	192	709	4	964
	3	11	18	122	373	524
%	0	64.8	11.5	19.3	4.4	100
	1	3.1	88.6	3.6	4.7	100
	2	6.1	19.9	73.5	0.4	100
	3	2.1	3.4	23.3	71.2	100

Cross-validation is done only for those cases in the analysis. In cross-validation, each case is classified by the functions derived from all cases other than that case.

80.2% of cross-validated grouped cases correctly classified.

73.5% of kimberlites are grouped correctly, 88.6% of Stormberg basalts are grouped correctly, 71.2% of Ecca and Karoo seds and 64.8% other rocks are grouped correctly.

Table 19.8 Four-class DT analysis

| Observed | Predicted | | | | Percent correct |
	0	1	2	3	
0	589	43	5	23	89.2%
1	20	2368	123	0	94.3%
2	77	90	797	0	82.7%
3	2	0	123	399	76.1%
Overall percentage	14.8%	53.7%	22.5%	9.1%	89.1%

contour plots show the progression of the grid method's best parameter estimate. The final optimal parameter settings are $C = 128$ and GAMMA $= 0.0078125$.

Decision tree is assembled to SPSS 12 (SPSS Inc.), and it is easy to use the GUI to import the data and export the tree model. The depth of the tree is 3, and 36

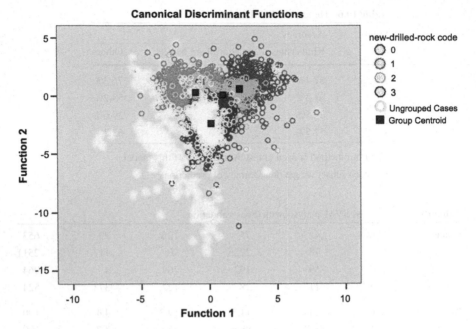

Fig. 19.2 Four-class LDA

Table 19.9 Four-class SVM analysis

Subset	Accuracy			
	Kimberlites	Stormberg basalts	Ecca and Karoo seds	Others
1	0.0%	99.1%	11.6%	77.3%
2	59.5%	0.2%	70.4%	71.4%
3	71.1%	77.4%	44.5%	43.6%
4	71.1%	59.2%	57.4%	28.9%

Table 19.10 The ten folds cross-validation accuracy for those methods

Methods	C-SVM[1]	Decision tree	Linear discriminant	C-SVM[2]
Cross-validation	95.66%	89.1%	80.1%	95.04%

nodes are created for this model. There are 12 rules to classify the rock samples. The accuracy is also estimated by 10-fold cross-validation method. Table 19.10 shows the accuracy result of both these methods compared with LDA.

The two main approaches take the comparable computation time with 2 minutes around, while the SVM has excellent accuracy compared with decision tree and linear discriminant. Still we find that the parameter selection for SVM takes a couple of hours. For reducing the computation time and computational complexity, a feature

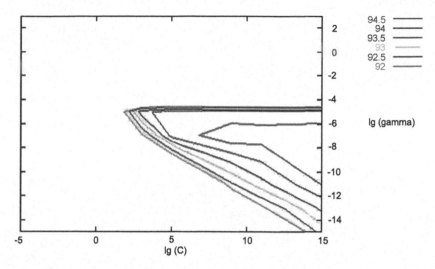

Fig. 19.3 Parameter selected by grid.py

Table 19.11 The accuracy after feature selection	Feature selected	101	88	44
	Cross-validation	95.66%	95.15%	95.04%

selection is needed. And this work can also help the geophysical experts to make right decision based on less rock sample attributes. In this article, we used F-score [38] as a feature selection criterion base on the simple rule that the large score is, the more likely this feature is more discriminative.

Based on the F-score rank, we selected 44 features and then apply C-SVM with RBF for training and prediction. Its accuracy is still above 95 percents (refer to Table 19.11). This model (C-SVM^2) takes less time on parameter selection and the best settings are $C = 64$, GAMMA $= 0.00195$. Through the ten-fold cross-validation, this compromise model is proved to be accurate and stable, so it can be applied to a new geochemical dataset.

A sophisticated and innovative method for Diamond-bearing kimberlites identification is needed for the diamond mining, especially in the area covered by extensive Kalahari sand. When a model is proved to be robust and effective for this work, it will be greatly helpful to the experts on kimberlites discovery. This article applies two methods to this new domain of application. Our results demonstrate that, both of these two methods have much higher prediction accuracy than 80 percents (the experts suggested). Which LDA is an effective method, the decision tree model is faster than SVM while SVM provides a higher accuracy.

Fig. 10.3 Response schematic of grade...

Table 10.11. The accuracy after feature selection

Feature selected	101	58	44
Cross-validation	95.63%	96.13%	95.03%

selection is needed. And this work can also help the geophysical experts to make right decision based on these rock sample attributes. In this article, we used F-score [8] as a feature selection criterion base on the simple rule that the large score is the more likely this feature is more discriminative.

Based on the F-score rank, we select 44 features and then apply C-SVM with RBF for training and prediction. Its accuracy is still above 95 percents (refer to Table 10.11). This model of C-SVM takes less time on parameter selection and the best settings are $C = 64$, GAMMA = 0.00195. Through the top-fold cross-validation, this comprehensive model is proved to be accurate and reliable, so it can be applied to a new geochemical district.

A sophisticated and innovative method for Diamond-bearing kimberlites identification is needed for the diamond mining, especially in the region covered by extensive Kalahari sand. When a model is proven to be robust and effective for this work, it will be greatly helpful to the exploration kimberlites discovery. This article applies two methods to this rock domain of exploration. Our results demonstrate that both of the two methods have much higher prediction accuracy than 80 percents (the 85% more suggestion). While LDA is an effective method, the decision tree model is fairer than SVM while SVM provides a higher accuracy.

Chapter 20
Intelligent Knowledge Management

20.1 Purposes of the Study

Since 1970s, researchers began systematically exploring various problems in knowledge management. However, people have been interested in how to collect, expand and disseminate knowledge for a long time. For example, thousands of years ago, Western philosophers studied the awareness and understanding of the motivation of knowledge [219]. The ancient Greek simply believed that personal experience forms all the knowledge. Researchers at present time pay more attention to management of tacit knowledge and emphasize on management of people as focusing on people's skills, behaviors and thinking patterns [211, 239].

Thanks to the rapid development of information technology, many western companies began to widely apply technology-based tools to organize the internal knowledge innovation activities. Thus it drove a group of researchers belonging to technical schools to explore how to derive knowledge from data or information. For instance, Beckman (1997) believes that knowledge is a kind of humans' logical reasoning on data and information, which can enhance their working, decision-making, problem-solving and learning performance. Knowledge and information are different since knowledge can be formed after processing, interpretation, selection and transformation of information [72].

In deriving knowledge by technical means, data mining becomes popular for the process of extracting knowledge, which is previously unknown to humans, but potentially useful from a large amount of incomplete, noisy, fuzzy and random data [102]. Knowledge discovered from algorithms of data mining from large-scale databases has great novelty, which is often beyond the experience of experts. Its unique irreplaceability and complementarity has brought new opportunities for decision-making. Access to knowledge through data mining has been of great concern for business applications, such as business intelligence [158].

However, from the perspective of knowledge management, knowledge discovery by data mining from large-scale databases face the following challenging problems.

First, the main purpose of data mining is to find hidden patterns as decision-making support. Most scholars in the field focus on how to obtain accurate models

Y. Shi et al., *Optimization Based Data Mining: Theory and Applications*, 277
Advanced Information and Knowledge Processing,
DOI 10.1007/978-0-85729-504-0_20, © Springer-Verlag London Limited 2011

and pay much emphasis on the accuracy of data mining algorithms. They halt immediately after obtaining rules through data mining from data and rarely go further to evaluate or formalize the result of mining to support business decisions [146]. Specially speaking, a large quantity of patterns or rules may be resulted from data mining. For a given user, these results may not be of interest and lack of novelty of knowledge. For example, a data mining project that classifies users as "current users, freezing users and lost users" through the use of decision tree classification algorithm produced 245 rules. Except for their big surprise, business personnel cannot get right knowledge from these rules [181]. The expression of knowledge should be not limited to numbers or symbols, but also in a more understandable manner, such as graphics, natural languages and visualization techniques. Knowledge expressions and qualities from different data mining algorithms differ greatly, and there are inconsistencies, even conflicts, between the knowledge so that the expression can be difficult. The current data mining research in expressing knowledge is not advanced. Hidden patterns obtained from data mining normally are just for a certain moment. The real databases, nevertheless, are constantly changing over time. It is hard to distinguish immediately to what extent the original knowledge needs to be updated. Furthermore due to the diversification of data storages in any organizations, a perfect data warehouse may not exist. It is difficult for data mining results based on databases or data warehouses to reflect the integration of all aspects of data sources. These issues lead to the situation that the data mining results may not be genuinely interesting to users and can not be used in the real world. Therefore, a "second-order" digging based on data mining results is needed to meet actual decision-making needs.

Second, many data mining techniques ignore domain knowledge, expertise, users' intentions and situational factors [163]. Note that there are several differences between knowledge and information. Knowledge is closely related to belief and commitment and it reflects a specific position, perspective or intention. Knowledge is a concept about operations and it always exists for "certain purposes". Although both knowledge and information are related to meaning, knowledge is in accordance with the specific situation and acquires associated attributes [156, 235]. From the culture backgrounds of knowledge, Westerners tend to emphasize on formal knowledge, while Easterners prefer obscure knowledge. It is also believed that these different kinds of knowledge are not totally separated but complementary to each other. In particular, they are closely linked in terms of how human and computer are interacted in obtaining knowledge. Because of the complexity of knowledge structure and the incrementality of cognitive process, a realistic knowledge discovery needs to explore interactively different abstraction levels through human-computer interaction and then repeat many times. Keeping the necessary intermediate results in data mining process, guiding role of human-computer interaction, dynamic adjusting mining target, and users' background knowledge, domain knowledge can speed up the process of knowledge excavation and ensure the effectiveness of acquired knowledge. Current data mining tools are unable to allow users to participate in excavation processes actually, especially for second-order excavation. In addition, both information and knowledge depend on specific scenarios, and they are relevant with the dynamic creation in humans' social interaction. Berger and Luckmann [17] argued that interacting people in certain historical and social scenario share information derived from social knowledge. Patterns or rules generated from data mining must be combined with specific business context in order to use in the enterprise. The context here includes relevant physics, business and other externally environmental and contextual factors, which also covers cognition, experience, psychology and other internal factors of the subject. It is the key element to a complete understanding of knowledge, affecting people's evaluation about knowledge. A rule may be useful to enterprises in a certain context, for a decision maker, at a certain time, but in another context it might be of no value. Therefore, context is critical for data mining and the process of the data mining results. In the literature, the importance of context to knowledge and knowledge management has been recognized by a number of researchers [28, 31, 56, 60, 90]. Though people rely on precise mathematical expressions for scientific findings, many scientific issues cannot be interpreted by mathematical forms. In fact in the real world, the results of data mining needs to effectively combine with the company reality and some non-quantitative factors, in particular, to consider the bound of specific context, expertise (tacit knowledge), users' specific intentions, domain knowledge and business scenarios, in order to truly become actionable knowledge and support business decisions [240].

Third, data mining process stops at the beginning of knowledge acquisition. The organizations' knowledge creation process derived from data should use different strategies to accelerate the transformation of knowledge in different stages of the knowledge creation, under the guidance of organizational objectives. Then a spiral of knowledge creation is formed, which creates conditions for the use of organizational knowledge and the accumulation of knowledge assets. At present, data mining process only covers knowledge creation part in this spiral, but does not involve how to conduct a second-order treatment to apply the knowledge to practical business, so as to create value and make it as a new starting point for a new knowledge creation spiral. Therefore, it cannot really explain the complete knowledge creation process derived from data. There is currently very little work in this area. In the ontology of data mining process, the discovered patterns are viewed as the end of the work. Little or no work involving the explanation of knowledge creation process at organizational level is studied in terms of implementation, authentication, internal process of knowledge, organizational knowledge assets and knowledge recreation. From the epistemological dimension, it lacks a deep study about the process of data–information–knowledge–wisdom and the cycle of knowledge accumulation and creation is not revealed. A combination of organizational guides and strategies needs to decide how to proceed with the knowledge guide at the organizational level so that a knowledge creation process derived from data (beyond data mining process) and organizational strategies and demands can be closely integrated.

Based on the above analysis, the knowledge or hidden patterns discovered from data mining can be called "rough knowledge". Such knowledge has to be examined at a "second-order" in order to derive the knowledge accepted by users or organizations. In this paper, the new knowledge will be called "intelligent knowledge" and the management process of intelligent knowledge is called intelligent knowledge management. Therefore, the focus of the study has the following dimensions:

- The object of concern is "rough knowledge".
- The stage of concern is the process from generation to decision support of rough knowledge as well as the "second-order" analysis of organizational knowledge assets or deep-level mining process so as to get better decision support.
- Not only technical factors but also non-technical factors such as expertise, user preferences and domain knowledge are considered. Both qualitative and quantitative integration have to be considered.
- Systematic discussion and application structure are derived for the perspective of knowledge creation.

The purposes of proposing intelligent knowledge management are:

- Re-define rough knowledge generated from data mining for the field of knowledge management explicitly as a special kind of knowledge. This will enrich the connotation of knowledge management research, promote integration of data mining and knowledge management disciplines, and further improve the system of knowledge management theory in the information age.
- The introduction of expertise, domain knowledge, user intentions and situational factors and the others into "second-order" treatment of rough knowledge may help deal with the drawbacks of data mining that usually pays too much emphasis on technical factors while ignoring non-technical factors. This will develop new methods and ideas of knowledge discovery derived from massive data.
- From the organizational aspect, systematic discussion and application framework derived from knowledge creation based on massive data in this paper will further strengthen and complement organizational knowledge creation theory.

20.2 Definitions and Theoretical Framework of Intelligent Knowledge

20.2.1 Key Concepts and Definitions

In order to better understand intelligent knowledge intelligent knowledge management, basic concepts and definitions are introduced in this subsection.

The research of intelligent knowledge management relates to many basic concepts such as original data, information, knowledge, intelligent knowledge and intelligent knowledge management. It also associated with several relevant concepts such as congenital knowledge, experience, common sense, situational knowledge etc. In order to make the proposed research fairly standard and rigorous from the beginning, it is necessary to give the definition of these basic concepts. Moreover, the interpretation of these concepts may provide a better understanding of intrinsic meanings of data, information, knowledge, and intelligent knowledge.

Definition 20.1 *Data* is a certain form of the representation of facts.

The above definition that is used in this paper has a general meaning of "data". There are numerous definitions of data from different disciplines. For example, in computing, data is referred to distinct pieces of information which can be translated into a different form to move or process; in computer component or network environment, data can be digital bits and bytes stored in electronic memory; and in telecommunications, data is digital-encoded information [217, 218]. In information theory, data is abstractly defined as an object (thing) that has the self-knowledge representation of its state and the state's changing mode over time [251]. When it is a discrete, data can be expressed mathematically a vector of n-dimensional possible attributes with random occurrences. Without any physical or analytic processing to be done, given data will be treated as "original" in this paper. Therefore, original data is the source of processing other forms (such as information, rough knowledge, intelligent knowledge and others). From the perspective of forms, the data here includes: text, multimedia, network, space, time-series data etc. From the perspective of structure, the data includes: structured, unstructured and semi-structured data, as well as more structured data which current data mining or knowledge discovery can deal with. From the perspective of quantity, the data includes: huge amounts of data, general data and small amounts of data etc. Data, judging from its nature, is only the direct or indirect statements of facts. It is raw materials for people to understand the world. Therefore, the characteristics of the original data here include: roughness (original, roughness, specific, localized, isolated, superficial, scattered, or even chaotic), extensive (covering a wide range), authenticity and manipulability (process through data technology). After access to original data, appropriate processing is needed to convert it into abstract and universal applicable information. Thus, the definition of information is given as:

Definition 20.2 *Information* is any data that has been pre-processed to all aspects of human interests.

Traditionally, information is the data that has been interpreted by human using certain means. Both scientific notation and common sense share the similar concepts of information. If the information has a numerical form, it may be measured through the uncertainty of an experimental outcome [195], while if it cannot be represented by numerical form, it is assigned for an interpretation through human [59]. Information can be studied in terms of information overload. Shi [178] classified information overload by exploring the relationships between relevant, important and useful information. However, Definition 20.2 used in this paper is directly for describing how to get knowledge from data where information is an intermediate step between these two. It is assumed that the pre-processed data by either quantitative or qualitative means can be regarded as information. Based on the concepts of data and information, the definition of rough knowledge is presented as follows:

Definition 20.3 *Rough knowledge* is the hidden pattern or "knowledge" discovered from information that has been analyzed by the known data mining algorithms or tools.

This definition is specifically made for the results of data mining. The data mining algorithms in the definition means any analytic process of using artificial intelligence, statistics, optimization and other mathematics algorithms to carry out more advanced data analysis than data pre-processing. The data mining tools are any commercial or non-commercial software packages performing data mining methods. Note that data pre-processing normally cannot bring a qualitative change of the nature of data and results in information by Definition 20.2, while data mining is advanced data analysis that discovers the qualitative changes of data and turns information into knowledge that has been hidden from human due to the massive data. The representation of rough knowledge changes with a data mining method. For example, rough knowledge from association method is rules, while it is a confusion matrix for the accuracy rates by using a classification method. The purpose of defining data, information and rough knowledge is to view a general expression of data mining process. This paper will call the process and other processes of knowledge evolution as "transformations". The transformation from data (or original data) to rough knowledge via information is called the first transformation, denoted as T_1. Let K_R stand for the rough knowledge and D denote as date. Then the first type of transformation can be expressed as:

$$T_1 : D \to K_R \quad \text{or} \quad K_R = T_1(D).$$

As it stands, T_1 contains any data mining process that consists of both data pre-processing (from data to information) and data mining analysis (from information to rough knowledge). Here the main tasks of T_1 can include: characterization, distinction, relevance, classification, clustering, outlier analysis (abnormal data), evolution analysis, deviation analysis, similarity, timing pattern and so on. Technologies of T_1 include extensively: statistical analysis, optimization, machine learning, visualization theory, data warehousing, etc. Types of rough knowledge are potential rules, potential classification tags, outlier labels, clustering tags and so on.

Characteristics of rough knowledge can be viewed as:

(i) Determined source: from results of data mining analysis.
(ii) Part usability: the possibility of direct support for business may exist, but most can not be used directly.
(iii) Rough: without further refinement, rough knowledge contains much redundant, one-sided or even wrong knowledge. For example, the knowledge generated from over-training has high prediction accuracy rate about the test set, but the effect is very poor.
(iv) Diversity: knowledge needs to be shown by a certain model for decision-making reference. There are many forms of rough knowledge, for instance, summary description, association rules, classification rules (including decision trees, network weights, discriminant equations, probability map, etc.), clustering, formulas and cases and so on. Some representations are easy to understand, such as decision trees, while some manifestations have poor interpretability, such as neural networks.
(v) Timeliness: compared with humans' experience, rough knowledge derives from data mining process in a certain time period, resulting in short cycle. It

may degrade in the short term with environmental changes. In addition, there are conflicts sometimes between the knowledge generated from different periods. As a result, as the environment changes the dynamic adaptability can be poor.

While rough knowledge is a specific knowledge derived from the analytic data mining process, the human knowledge has extensively been studied in the field of knowledge management. The item knowledge has been defined in many different ways. It is generally regarded as individual's expertise or skills acquired through learning or experience [220]. In the following, knowledge is divided as five categories in terms of the contents. Then, these terms can be incorporated into rough knowledge from data mining results for our further discussion on intelligent knowledge.

Definition 20.4 Knowledge is called *specific knowledge*, denoted by K_S if the knowledge representations by human through the certain state and rules of an object.

Specific knowledge is a cognitive understanding of certain objects and can be presented by its form, content and value [251]. Specific knowledge has a strict boundary in defining its meanings. Within the boundary, it is knowledge; otherwise, it is not [234].

Definition 20.5 Knowledge is called *empirical knowledge*, denoted by K_E if it directly comes from human experience gained empirical testing.

Note that the empirical testing in Definition 20.5 is referred to specifically nontechnical, but practical learning process from which human can gain experience. If it is derived from statistical learning or mathematical learning, knowledge is already defined as rough knowledge of Definition 20.2. Empirical testing here can be also referred as intermediate learning, such as reading from facts, reports or learning from other's experiences. When these experiences are confirmed through a scientific learning, they will become "knowledge". Otherwise, they are still "experiences" [251].

Definition 20.6 Knowledge is called *common sense knowledge*, denoted as K_C if it is well known and does not need to be proved.

Common sense is the facts and rules widely accepted by most of humans. Some knowledge, such as specific knowledge or empirical knowledge can become common sense as they are gradually popularized. Therefore, it is also called "postknowledge" [251].

Definition 20.7 Knowledge is called *instinct knowledge*, denoted by K_H if it is innate as given functions of humans.

Instinct knowledge is heritage of humans through the biological evolution and genetic process. It does not need to be studied and proved. If instinct knowledge is viewed as a "root" of the knowledge mentioned above, then a "knowledge ecosystem" can be formed. In the system, instinct knowledge first can be changed into empirical knowledge after training and studying. Then, if empirical knowledge is scientifically tested and confirmed, it becomes specific knowledge. As the popularity of specific knowledge develops, it is common sense knowledge. However, the system is ideal and premature since the creation of human knowledge is quite complex and could not be interpreted as one system [251].

Definition 20.8 Knowledge is called *situational knowledge*, denoted as K_U if it is context.

The term context in this paper, associated with knowledge and knowledge activities, is relevant to conditions, background and environment. It includes not only physical, social, business factors, but also the humans' cognitive knowledge, experience, psychological factors.

Situational knowledge or context has the following characteristics:

(i) It is an objective phenomenon which exists widely, but not certain whether it is fully aware of or not.
(ii) It is independent of knowledge and knowledge process, but keeps a close interacts with knowledge and knowledge process.
(iii) It describes situational characteristics of knowledge and knowledge activities. Its function is to recognize and distinguish different knowledge and knowledge activities. To humans, their contexts depict personal characteristics of one engaging in intellectual activities [159].

Based on the above definitions of different categories of knowledge, a key definition of this chapter is given as:

Definition 20.9 Knowledge is called *intelligent knowledge*, denoted as K_I if it is generated from rough knowledge and/or specific, empirical, common sense and situational knowledge, by using a "second-order" analytic process.

If the data mining is said as the "first-order" analytic process, then the "second-order" analytic process here means quantitative or qualitative studies are applied to the collection of knowledge for the pre-determined objectives. It can create knowledge, now intelligent knowledge, as decision support for problem-solving. The "second-order" analytic process is a deep study beyond the usual data mining process. While data mining process is mainly driven by a series of procedures and algorithms, the "second-order" analytic process emphasizes the combinations of technical methods, human and machine interaction and knowledge management.

Some researchers in the field of data mining have realized its importance of handling the massive rules or hidden patterns from data mining [170, 213, 222]. However, they did not connect the necessary concepts from the filed of knowledge management in order to solve such a problem for practical usage. Conversely, researchers in knowledge management often ignore rough knowledge created outside humans as a valuable knowledge base. Therefore, as to bridge the gap of data mining and knowledge management, the proposed study on intelligent knowledge in the paper is new.

As discussed above, the transformation form information to rough knowledge T_1 is essentially trying to find some existing phenomenological associations among specific data. T_1 is some distance away from the knowledge which can support decision-making in practice. The "second-order" analytic process to create intelligent knowledge from available knowledge, including rough knowledge, can be realized in general by transformation, defined as follows:

$$T_2 : K_R \cup K \to K_1 \quad \text{or} \quad K_1 = T_2(K_R \cup K),$$

where $K = \rho(K_S, K_E, K_C, K_H, K_U)$ is a power set.

The above transformation is an abstract form. If the results of the transformation are written in terms of the components of intelligent knowledge, then the following mathematical notations can be used:

(i) Replacement transformation: $K_I = K_R$;
(ii) Scalability transformation: $K_I = \alpha K_R$, where $-\infty < \alpha < +\infty$;
(iii) Addition transformation: $K_I = K_R + K_I$;
(iv) Deletion transformation: $K_I = K_R - K_I$;
(v) Decomposition transformation:

$$K_I = \alpha_1 K_{R1} + \alpha_2 K_{R2} + \alpha_3 K_{R3} + \cdots, \quad \text{where } -\infty < \alpha < +\infty.$$

In the above, replacement transformation is a special case of scalability transformation, and they, together with addition and deletion transformations are parts of decomposition transformation.

The coefficients of $\{\alpha_1, \alpha_2, \alpha_3, \ldots\}$ in the decomposition represent the components of $K = \rho(K_S, K_E, K_C, K_H, K_U)$ distributed in the knowledge creation process.

The intelligent knowledge has the following characteristics:

(i) The process of intelligent knowledge creation fully integrates specific context, expertise, domain knowledge, user preferences and other specification knowledge, and makes use of relevant quantitative algorithms, embodying human-machine integration principle.
(ii) Since intelligent knowledge is generated from the "second-order" analytic process, it is more valuable than rough knowledge.
(iii) It provides knowledge to people who need them at the right time, under appropriate conditions.
(iv) The objective of intelligent knowledge is to provide significant inputs for problem-solving and support strategic action more accurately.

To explore more advanced issues in the meaning of knowledge management, intelligent knowledge can be further employed to construct a strategy of problem-solving by considering goal setting, specific problem and problem environment.

Restricted by the given problem and its environmental constraints, aiming at the specific objectives, a strategy of solving the problem can be formed based on related intelligent knowledge. To distinguish the strategy that has be used in different fields, the strategy associated with intelligent knowledge is called intelligent strategy.

If P is defined as the specific problems, E is for problem solving environment and G is goal setting, then the information about issues and environment can be expressed as $I(P, E)$. Given intelligent knowledge K_I, an intelligent strategy S is another transformation, denoted as:

$$T_3 : K_I \times I(P, E) \times G \to S \quad \text{or} \quad S = T_3(K_I \times I(P, E) \times G).$$

Transformation T_3 differs from T_2 and T_1 since it relates to forming an intelligent strategy for intelligent action, rather than finding knowledge. Achieving the transformation from intelligent knowledge to a strategy is the mapping from a product space of $K_I \times I(P, E) \times G$ to strategy space S.

Action usually refers to the action and action series of humans. Intelligent action (a high level transformation) is to convert an intelligent strategy into actionable knowledge, denoted as T_4:

$$T_4 : S \to K_A, \quad \text{or} \quad K_A = T_4(S).$$

Term K_A is denoted as actionable knowledge. Some K_A can ultimately become intangible assets, which is regarded as "wisdom" [235]. For example, much actionable knowledge produced by great military strategists in history gradually formed as wisdom of war. A smart strategist should be good at using not only his/her actionable knowledge, but also the wisdom from history [156]. When processing qualitative analysis in traditional knowledge management, people often pay more attention to how intelligent strategy and actionable knowledge generated from tacit knowledge and ignore their source of quantitative analysis, where intelligent knowledge can be generated from combinations of data mining and human knowledge. Intelligent strategy is its inherent performance, while actionable knowledge is its external performance. Transformation T_4 is a key step to produce actionable knowledge that is directly useful for decision support. Figure 20.1 is the process of transformations from data to rough knowledge, to intelligent knowledge and to actionable knowledge.

The management problems of how to prepare and process all of four transformations leads to the concept of intelligent knowledge management:

Definition 20.10 *Intelligent knowledge management* is the management of how rough knowledge, human knowledge can be combined and upgraded into intelligent knowledge as well as management issues regarding extraction, storage, sharing, transformation and use of rough knowledge so as to generate effective decision support.

Intelligent knowledge management proposed in this paper is the interdisciplinary research field of data mining and knowledge management. One of frameworks can be shown as Fig. 20.2.

Fig. 20.1 Data→ Rough knowledge → Intelligent knowledge → Actionable knowledge

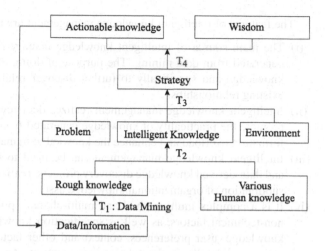

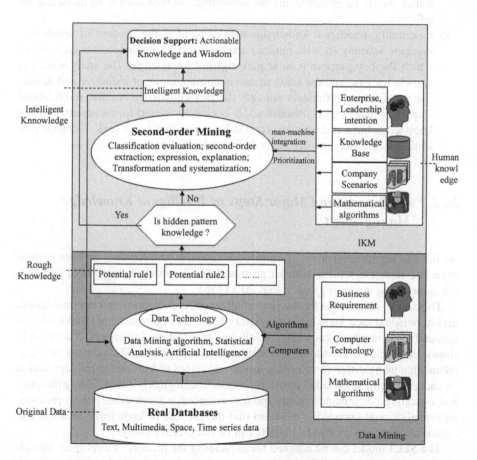

Fig. 20.2 A framework of Intelligent Knowledge Management (IKM)

The features of intelligent knowledge management are as follows:

(i) The main source of intelligent knowledge management is rough knowledge generated from data mining. The purpose of doing this is to find deep-seated knowledge and specifically to further discover relationships on the basis of existing relationships.

(ii) Intelligent knowledge management realizes decision support better, so as to promote the practicality of knowledge generated from data mining, reduce information overload and enhance the knowledge management level.

(iii) Intelligent knowledge management can be used to build organization-based and data-derived knowledge discovery projects, realizing the accumulation and sublimation of organizational knowledge assets.

(iv) It is a complex multi-method and multi-channel process. The technical and non-technical factors, as well as specification knowledge (expertise, domain knowledge, user preferences, context and other factors) are combined in the process of intelligent knowledge management. As a result, the knowledge found should be effective, useful, actionable, understandable to users and intelligent.

(v) Essentially, intelligent knowledge management is the process of combining machine learning (or data mining) and traditional knowledge management, of which the key purpose is to acquire rightful knowledge. The study source is knowledge base and the study means is combinations of inductive and deductive approaches. Ultimately not only the fact knowledge but also the relationship knowledge can be discovered. It is closely related to the organization of knowledge base and ultimate knowledge types that users seek. Adopted reasoning means may involve many different logical fields.

20.2.2 4T Process and Major Steps of Intelligent Knowledge Management

As the leading representative of knowledge creation process derived from experience, the Japanese scholar Nonaka proposed SECI model of knowledge creation, the value of the model is given in Fig. 20.3 [156].

This model reveals that through externalization, combination and internalization, highly personal tacit knowledge ultimately becomes organizational knowledge assets and turns into tacit knowledge of all the organizational members. It accurately shows the cycle of knowledge accumulation and creation. The concept of "Ba" means that using different strategies in various stages of knowledge transformation can accelerate the knowledge creation process. It can greatly enhance the efficiency and operating performance of enterprises' knowledge innovation. It also provides an organizational knowledge guide so that the process of knowledge creation and organizational strategies and demands can be integrated closely.

The SECI model can be adopted for explaining the process of intelligent knowledge management, especially the 4T process of transformation including data–rough

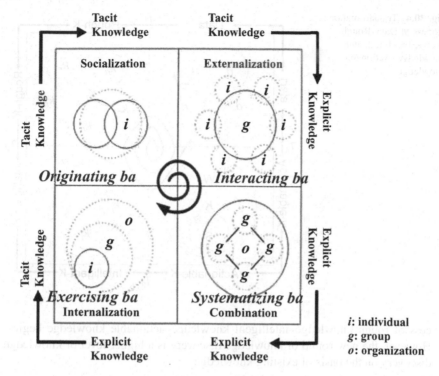

Fig. 20.3 Knowledge creation as the self-transcending process, source: [156]

knowledge–intelligent knowledge–actionable knowledge. From the organizational aspect, knowledge creation derived from data should be the process of knowledge accumulation like a spiral, which is shown in Fig. 20.4.

The transformation process includes:

- T_1 (from data to rough knowledge): after the necessary process of processing and analyzing original data, the preliminary result (hidden Pattern, rules, weights, etc.) is rough knowledge, as a result from a kind of primary transformation.
- T_2 (from rough knowledge to intelligent knowledge): on the basis of rough knowledge, given user preferences, scenarios, domain knowledge and others, the process carries out a "second-order" mining for knowledge used to support intelligent decision-making and intelligent action. The process carries out deep processing of the original knowledge, which is the core step in intelligent knowledge management.
- T_3 (from intelligent knowledge to intelligent strategy): in order to apply intelligent knowledge in practice, one must first convert intelligent knowledge into intelligent strategy through consideration of problem statement and solving environment. It is the process of knowledge application.
- T_4 (from intelligent strategy to actionable knowledge): once actionable knowledge is obtained, it can be recoded as "new data", which are either intangible assets or wisdom can be used as data source for decision support. The new pro-

Fig. 20.4 Transformation process of Data–Rough knowledge–Intelligent knowledge–Actionable knowledge

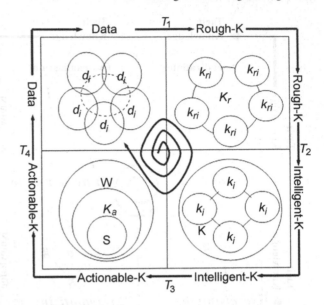

cess of rough knowledge–intelligent knowledge–actionable knowledge begins. However, the new round of knowledge discovery is a higher level of knowledge discovery on the basis of existing knowledge.

Therefore, it is a cycle, spiraling process for the organizational knowledge creation, and from the research review, the current data mining and KDD would often be halted when it is up to stage T_3 or T_4, leading to the fracture of spiral, which is not conducive to the accumulation of knowledge.

It also needs to be noted that in this process, different stages require different disciplines and technologies to support. Stage T_1 generally focuses on technical factors such as computer and algorithms, while stage T_2 needs expertise, domain knowledge, user preferences, scenarios, artificial intelligence for constrain and support. Stage T_3 needs a higher level of expertise to make it into actionable knowledge or even the intelligence. Stage T_4 generates new data primarily by computers, networks, sensors, records, etc. However, technical factors and non-technical factors are not totally separate, but the focus should be different at different stages.

20.3 Some Research Directions

Intelligent knowledge management can potentially be a promising research area that involves interdisciplinary fields of data technology, knowledge management, system science, behavioral science and computer science. The feature of intelligent knowledge management research is shown in Fig. 20.5. There are a number of research directions remaining to be explored. Some of them can be described as below.

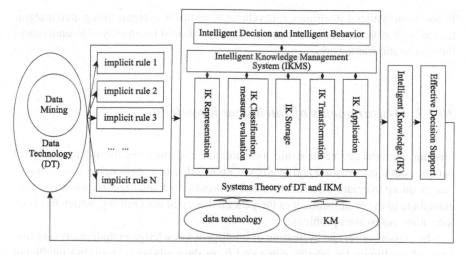

Fig. 20.5   Interdisciplinary feature of intelligent knowledge management

20.3.1   The Systematic Theoretical Framework of Data Technology and Intelligent Knowledge Management

Related to the above issues discussed in this paper, a general term, data technology, is used to capture a set of concepts, principles and theories of quantitative methodologies to analyze and transform data to information and knowledge. In accordance with the principles of system science, the following issues need to be raised and studied in the future:

(a) How to classify, describe and organize known data technologies, including data mining, artificial intelligence, statistics and others based on the treatment capacity and characteristics. How to effectively use the results of data mining as rough knowledge to discuss the logical relationship between intelligent knowledge and traditional knowledge structure? How to establish the mathematical model about intrinsic links between data technology and intelligent knowledge? These can be used to explain the characteristics of intelligent knowledge generated from the results of data analysis.

(b) From the perspective of knowledge creation derived from data, how to study the process of knowledge creation, establish knowledge creation theory derived from data and build a systematic framework of data mining and intelligent knowledge management.

As a class of "special" knowledge, the process management theory including extraction, transformation, application, innovation of intelligent knowledge in specific circumstances should be studied. In the study, not only knowledge itself, but also the tacit knowledge of decision-makers and users and other non-technical factors, such as domain knowledge, user preferences, scenarios, etc. should be considered.

In addition, artificial intelligence, psychology, complex systems, integrated integration as well as some empirical research methods should be employed to understand the systematic framework.

20.3.2 *Measurements of Intelligent Knowledge*

Finding appropriate "relationship measurement" to measure the interdependence between data, information and intelligent knowledge is a challenging task. This research on appropriate classification and expression of intelligent knowledge may contribute to the establishment of the general theory of data mining, which has been long-term unresolved problem.

The classification and evaluation of intelligent knowledge include analyzing features of intelligent knowledge generated from data mining, classifying intelligent knowledge, selecting appropriate indicators for different types of intelligent knowledge to do the effectiveness evaluation, and building a classification and evaluation system of intelligent knowledge.

In measure of intelligent knowledge, the theory and methods of subjective and objective measure of intelligent knowledge should be studied. From the point of view of distinction capacity, the mathematical method of the measurement for intelligent knowledge value should be more carefully studied. Measure is a "relationship". The establishment of intelligent knowledge measures can be very challenging.

Given applied goals and information sources, what data mining system must do is to evaluate the validity of intelligent knowledge structure. Results of the evaluation should not only quantify the usefulness of the existing intelligent knowledge, but also decide whether there is a need for other intelligent knowledge. The exploring of this area needs three aspects to conduct an in-depth study: (1) analysis of intelligent knowledge complexity; (2) analysis of the correlation between intelligent knowledge complexity and model effectiveness; (3) analysis of across heterogeneous intelligent knowledge effectiveness. In short, how to identify the results from data mining, and how to accurately measure the valuable intelligent knowledge and evaluate the quality of intelligent knowledge management are key issues of the application of knowledge management. The research of this area (Benchmark of Data Mining and Intelligent Knowledge) still remains unexplored, in need of deep exploration.

Furthermore, intelligent knowledge is viewed as a class of "special" knowledge, and the meaning and structure in mathematics and management of its preservation, transformation and application can be further studied.

The link between intelligent knowledge and intelligent action, intelligent decision-making and action, and how to apply intelligent knowledge to improve decision-making intelligence and decision-making efficiency can be interested research issues.

20.3.3 Intelligent Knowledge Management System Research

Based on the framework of the relevance and features of the intelligent knowledge management system, management information system and knowledge management system in this paper, both data mining and intelligent knowledge management are used to support enterprise management decision-making. For different industries, such as finance, energy policy, health care, communications, auditing with large-scale data infrastructures, an intelligent knowledge management system can be established, through the integration of data mining and intelligent knowledge management, to improve their knowledge management capability and overall competitiveness.

20.3.3 Intelligent Knowledge Management System Research

References

1. Alex, J.S., Scholkopf, B.: A Tutorial on SVR. Statistics and Computing, pp. 199–222. Kluwer Academic, Dordrecht (2004),
2. Alizadeh, F., Goldfarb, D.: Second order cone programming. Math. Program., Ser. B, **95**, 3–51 (2003)
3. Allwein, E.L., Schapire, R.E., Singer, Y.: Reducing multiclass to binary: A unifying approach for margin classifiers. J. Mach. Learn. Res. **1**, 113–141 (2001)
4. Altman, E.: Financial ratios, discriminant analysis and the prediction of corporate bankruptcy. J. Finance **23**(3), 589–609 (1968)
5. An, L.T.H., Tao, P.D.: Solving a class of linearly constrained indefinite quadratic problem by D. C. algorithms. J. Glob. Optim. **11**, 253–285 (1997)
6. Anderson, T.W.: An Introduction to Multivariate Statistical Analysis. Wiley, New York (1958)
7. Anderson, J.: Regression and ordered categorical variables (with discussion). J. R. Stat. Soc., Ser. B, Stat. Methodol. **46**, 1–30 (1984)
8. Angulo, C., Parr, X., Català, A.: K-SVCR, a support vector machine for multi-class classification. Neurocomputing **55**, 57–77 (2003)
9. Angulo, C., Ruiz, F.J., Gonzalez, L., Ortega, J.A.: Multi-classification by using tri-class SVM. Neural Process. Lett. **23**, 89–101 (2006)
10. Arie, B.D., Yoav, G.: Ordinal datasets. http://www.cs.waikato.ac.nz/ml/weka/ (2005)
11. Ascoli, G.A., Krichmar, J.L., Nasuto, S.J., Senft, S.L.: Generation, description and storage of dendritic morphology data. Philos. Trans. R. Soc. Lond. B, Biol. Sci. **356**, 1131–1145 (2001)
12. Bach, F.R., Lanckrient, G.R.G., Jordan, M.I.: Multiple kernel learning, conic duality and the SMO algorithm. In: Twenty First International Conference on Machine Learning, pp. 41–48 (2004)
13. Bath, T.N., Bentley, G.A., Fischmann, T.O., Boulot, G., Poljak, R.J.: Small rearrangements in structures of Fv and Fab fragments of antibody D1.3 on antigen binding. Nature **347**, 483–485 (1990)
14. Bellman, R.: Adaptive Control Processes: A Guided Tour. Princeton University Press, Princeton (1961)
15. Ben-Tal, A., Nemirovski, A.: Robust convex optimization. Math. Oper. Res. **23**(4), 769–805 (1998)
16. Ben-Tal, A., Nemirovski, A.: Robust solutions to uncertain programs. Oper. Res. Lett. **25**, 1–13 (1999)
17. Berger, P.L., Luckman, T.: The Social Construction of Reality. Doubleday, Garden City (1966)
18. Bhatt, R.B., Gopal, M.: On fuzzy-rough sets approach to feature selection. Pattern Recognit. Lett. **26**, 965–975 (2005)

19. Bi, J., Bennett, K.P.: Duality, geometry and support vector regression. In: Advances in Neural Information Processing Systems, pp. 593–600. MIT Press, Cambridge (2002)
20. Bie, T.D., Crisrianini, N.: Convex methods for transduction. In: Advances in Neural Information Processing Systems (NIPS-03), vol. 16 (2003)
21. Blake, C.L., Merz, C.J.: UCI Repository of Machine Learning Databases. University of California, Irvine. http://www.ics.uci.edu/~mlearn/MLRepository.html (1998)
22. Borwein, J.M.: Optimization with respect to partial orderings. Ph.D. Thesis, Oxford University, Jesus College (1974)
23. Boser, B., Guyon, I., Vapnik, V.N.: A training algorithm for optimal margin classifiers. In: Fifth Annual Workshop on Computational Learning Theory, pp. 144–152 (1992)
24. Bottou, L., Cortes, C., Denker, J.S.: Comparison of classifier methods: a case study in handwriting digit recognition. In: Proceedings of the International Conference on Pattern Recognition, pp. 77–82. IEEE Computer Society Press, New York (1994)
25. Boyd, S., Vandenberghe, L.: Convex Optimization. Cambridge University Press, Cambridge (2004)
26. Bradley, P., Mangasarian, O.: Feature selection via concave minimization and support vector machines. In: International Conference on Machine Learning. Morgan Kaufmann, San Mateo (1998)
27. Brenneman, D.E., Westbrook, G.L., Fitzgerald, S.P., et al.: Neuronal cell killing by the envelope protein of HIV and its prevention by vasoactive intestinal peptide. Nature **335**, 639–642 (1988)
28. Brezillion, P., Pomerol, J.: Contextual knowledge sharing and cooperation in intelligent assistant systems. Le Travail Humain **62**(3), 223–246 (1999)
29. Burges, C.J.C., Crisp, D.J.: Uniqueness of the SVM solution. In: Solla, S.A., Leen, T.K., Mller, K.R. (eds.) Proceedings of the Twelfth Conference on Neural Information Processing Systems. MIT Press, Cambridge (1999)
30. Candes, E., Wakin, M., Boyd, S.: Enhancing sparsity by reweighted l_1 minimization. J. Fourier Anal. Appl. **14**, 877–905 (2008)
31. CAP Ventures: The role of shared context in a knowledge-enabled environment. White paper, http://www.capv.com/bin/pdf/divine.pdf (2002)
32. Cesar, R.M., Costa, L.F.: Neural cell classification by wavelets and multiscale curvature. Biol. Cybern. **79**, 347–360 (1998)
33. Chang, C.C., Lin, C.J.: LIBSVM: a library for support vector machines. Software available at http://www.csie.ntu.edu.tw/cjlin/libsvm (2001)
34. Chang, M.W., Lin, C.J.: Leave-one-out bounds for support vector regression model selection. Neural Comput. **17**, 1188–1222 (2005)
35. Chen, X., Xu, F., Ye, Y.: Lower bound theory of nonzero entries in solutions of l_2–l_p minimization. Technical report, Department of Applied Mathematics, The Hong Kong Polytechnic University (2009)
36. Chen, X., Zhou, W.: Smoothing nonlinear conjugate gradient method for image restoration using nonsmooth nonconvex minimization. Preprint, Department of Applied Mathematics, The Hong Kong Polytechnic University (2008)
37. Chen, Y.: CMSVM v2.0: Chinese Modeling Support Vector Machines. Group of CMSVM, Training Center, China Meteorological Administration (2004)
38. Chen, Y.W., Lin, C.J.: Combining SVMs with various feature selection strategies. In: Feature Extraction. Foundations and Applications (2005)
39. Chopra, V.K., Ziemba, W.T.: The effect of errors in means, variances and covariances on optimal portfolio choice. J. Portf. Manag. **19**, 6–11 (1993)
40. Choquet, G.: Theory of capacities. Ann. Inst. Fourier **5**, 131–295 (1954)
41. Chothia, C., Lesk, A.M., Gherardi, E., Tomlinson, I.M., Walter, G., Marks, J.D., Lewelyn, M.B., Winter, G.: Structural repertoire of the human Vh segments. J. Mol. Biol. **227**, 799–817 (1992)
42. Chu, W., Keerthi, S.S.: New approaches to support vector ordinal regression. In: Proc. of International Conference on Machine Learning (ICML-05), pp. 145–152 (2005)

43. Coelho, R.C., Costa, L.F.: Realistic neuromorphic models and their application to neural reorganization simulations. Neurocomputing **48**, 555–571 (2002)
44. Cohen, W.W., Schapire, R.E., Singer, Y.: Learning to order things. J. Artif. Intell. Res. **10**, 243–270 (1999)
45. Collman, P.M., Laver, W.G., Varghese, J.N., Baker, A.T., Tulloch, P.A., Air, G.M., Webster, R.G.: Three-dimensional structure of a complex of antibody with influenza virus neuraminidase. Nature **326**, 358–363 (1987)
46. Conover, W.J.: Practical Nonparametric Statistics. Wiley, New York (1999)
47. Costa Lda, F., Manoel, E.T.M., Faucereau, F., Chelly, J., van Pelt, J., Ramakers, G.: A shape analysis framework for neuromorphometry. Netw. Comput. Neural Syst. **13**, 283–310 (2002)
48. Costa Lda, F., Velte, T.J.: Automatic characterization and classification of ganglion cells from the salamander retina. J. Comp. Neurol. **404**, 33–51 (1999)
49. Crammer, K., Singer, Y.: On the algorithmic implementation of multiclass kernel-based vector machines. J. Mach. Learn. Res. **2**, 265–292 (2002)
50. Crammer, K., Singer, Y.: Pranking with ranking. In: Proceedings of the Conference on Neural Information Processing Systems (NIPS) (2001)
51. Cristianini, N., Shawe-Taylor, J.: An Introduction to Support Vector Machines and Other Kernel-Based Learning Methods. Cambridge University Press, Cambridge (2000)
52. Data Mining Tools See5 and C5.0. http://www.rulequest.com/see5-info.html
53. Deng, N.Y., Tian, Y.J.: The New Approach in Data Mining—Support Vector Machines. Science Press, Beijing (2004)
54. Deng, N.Y., Tian, Y.J.: Support Vector Machine: Theory, Algorithms and Extensions. Science Publication, Beijing (2004)
55. Denneberg, D.: Fuzzy Measure and Integral. Kluwer Academic, Dordrecht (1994)
56. Despres, C., Chauvel, D.: A thematic analysis of the thinking in knowledge management. In: Despres, C., Chauvel, D. (eds.) The Present and the Promise of Knowledge Management, pp. 55–86. Butterworth–Heinemann, Boston (2000)
57. Devroye, L., Györfi, L., Lugosi, G.: A Probabilistic Theory of Pattern Recognition. Springer, Berlin (1996)
58. Diamond Facts: 2004/05 NWT Diamond Industry Report. http://www.iti.gov.nt.ca/diamond/diamondfacts2005.htm (2005)
59. Dictionary of Military and Associated Terms. US Department of Defence (2005)
60. Dieng, R., Corby, O., Giboin, A., Ribière, M.: Methods and tools for corporate knowledge management. Int. J. Hum.-Comput. Stud. **51**(3), 567–598 (1999)
61. Dietterich, T.G., Bakiri, G.: Solving multi-class learning problems via error-correcting output codes. J. Artif. Intell. Res. **2**, 263–286 (1995)
62. Dougan, D.A., Malby, R.L., Grunen, I.C.: Effects of substitutions in the binding surface of an antibody on antigen. Protein Eng. **11**, 65–74 (1998)
63. Dubois, D., Prade, H.: Fuzzy Sets and Systems: Theory and Application, pp. 242–248. Academic Press, New York (1980)
64. Eisenbeis, R.A.: Problems in applying discriminant analysis in credit scoring models. J. Bank. Finance **2**, 205–219 (1978)
65. El-Ghaoui, L., Lebret, H.: Robust solutions to least-square problems to uncertain data matrices. SIAM J. Matrix Anal. Appl. **18**, 1035–1064 (1997)
66. El-Ghaoui, L., Oustry, F.L., Lebret, H.: Robust solutions to uncertain semidefinite programs. SIAM J. Matrix Anal. Appl. **9**, 33–52 (1998)
67. Elkabes, S., DiCicco-Bloom, E.M., Black, I.B.: Brain microglia/macrophages express neurotrophins that selectively regulate microglial proliferation and function. J. Neurosci. **16**, 2508–2521 (1996)
68. Epstein, L.G., Gelbard, H.A.: HIV-1-induced neuronal injury in the developing brain. J. Leukoc. Biol. **65**, 453–457 (1999)
69. Ester, J.M., Kriegel, H., Xu, X.: Density-based algorithm for discovering clusters in large spatial databases with noise. In: Proceedings of 2nd International Conference on Knowledge Discovery and Data Mining, pp. 226–231 (1996)

70. Fariselli, P., Pazos, F., Valencia, A., Casadio, R.: Prediction of protein-protein interaction sites in heterocomplexes with neural networks. Eur. J. Biochem. **269**, 1356–1361 (2002)
71. Faybusovich, L., Tsuchiya, T.: Primal-dual algorithms and infinite-dimensional Jordan algebras of finite rank. Math. Program., Ser. B **97**, 471–493 (2003)
72. Feigenbaum, E.A.: The art of artificial of intelligence: themes and case studies of knowledge engineering. In: Fifth International Joint Conference on Artificial Intelligence, Cambridge, MA, pp. 1014–1029 (1977)
73. Fisher, R.A.: The use of multiple measurements in taxonomic problems. Ann. Eugen. **7**, 179–188 (1936)
74. Frank, E., Hall, M.: A simple approach to ordinal classification. In: Lecture Notes in Computer Science, vol. 2167, pp. 145–156. Springer, Berlin (2001)
75. Freed, N., Glover, F.: Simple but powerful goal programming models for discriminant problems. Eur. J. Oper. Res. **7**, 44–60 (1981)
76. Freed, N., Glover, F.: Evaluating alternative linear programming models to solve the two-group discriminant problem. Decis. Sci. **17**, 151–162 (1986)
77. Friedman, J., Hastie, T., Rosset, S., Tibshirani, R., Zhu, J.: Discussion of "Consistency in boosting" by W. Jiang, G. Lugosi, N. Vayatis and T. Zhang. Ann. Stat. **32**, 102–107 (2004)
78. Friess, C.T., Campbell, C.: The kernel adatron algorithm: a fast and simple learning procedure for support vector machines. In: Proceedings of 15th International Conference on Machine Learning (1998)
79. Fung, G., Mangasarian, O.: Proximal support vector machine classifiers. In: Proceedings of International Conference of Knowledge Discovery and Data Mining, pp. 77–86 (2001)
80. Fung, G., Mangasarian, O.L., Shavlik, J.: Knowledge-based support vector machine classifiers. In: NIPS 2002 Proceedings, Vancouver, pp. 9–14 (2002)
81. Fung, G.M., Mangasarian, O.L., Shavlik, J.: Knowledge-based nonlinear kernel classifiers. In: Schlkopf, B., Warmuth, M.K. (eds.) COLT/Kernel 2003. Lecture Notes in Computer Science, vol. 2777, pp. 102–113. Springer, Heidelberg (2003)
82. Gabuzda, D., Wang, J., Gorry, P.: HIV-1-associated dementia. In: Ransohoff, R.M., Suzuki, K., Proudfoot, A.E.I., Hickey, W.F., Harrison, J.K. (eds.) Chemokines and the Nervous System, pp. 345–360. Elsevier Science, Amsterdam (2002)
83. Garden, G.A., Budd, S.L., Tsai, E., et al.: Caspase cascades in human immunodeficiency virus-associated neurodegeneration. J. Neurosci. **22**, 4015–4024 (2002)
84. Gelbard, H.A., Epstein, L.G.: HIV-1 encephalopathy in children. Curr. Opin. Pediatr. **7**, 655–662 (1995)
85. Gelbard, H., Nottet, H., Dzenko, K., et al.: Platelet-activating factor: a candidate human immunodeficiency virus type-1 infection neurotoxin. J. Virol. **68**, 4628–4635 (1994)
86. Gendelman, H.E.: The neuropathogenesis of HIV-1-dementia. In: Gendelman, H.E., Lipton, S.A., Epstein, L.G., Swindells, S. (eds.) The Neurology of AIDS, pp. 1–10. Chapman and Hall, New York (1997)
87. Glass, J.D., Wesselingh, S.L., Selnes, O.A., McArthur, J.C.: Clinical neuropathologic correlation in HIV-associated dementia. Neurology **43**, 2230–2237 (1993)
88. Glass, J.D., Fedor, H., Wesselingh, S.L., McArthur, J.C.: Immunocytochemical quantitation of human immunodeficiency virus in the brain: correlations with dementia. Ann. Neurol. **38**, 755–762 (1995)
89. Goldfarb, D., Iyengar, G.: Robust portfolio selection problems. Math. Oper. Res. **28**, 1 (2003)
90. Goldkuhl, G., Braf, E.: Contextual knowledge analysis-understanding knowledge and its relations to action and communication. In: Proceedings of 2nd European Conference on Knowledge Management, IEDC-Bled School of Management, Slovenia (2001)
91. Grabisch, M., Sugeno, M.: Multi-attribute classification using fuzzy integral. In: Proceedings of the FUZZ-IEEE'92, pp. 47–54 (1992)
92. Grabisch, M., Nicolas, J.M.: Classification by fuzzy integral: performance and tests. Fuzzy Sets Syst. **65**, 255–271 (1994)
93. Grabisch, M.: A new algorithm for identifying fuzzy measures and its application to pattern recognition. In: Proceedings of 1995 IEEE International Conference on Fuzzy Systems (1995)

94. Grabisch, M.: The interaction and Möbius representation of fuzzy measures on finite spaces, k-additive measures: a survey. In: Fuzzy Measures and Integrals—Theory and Applications, pp. 70–93 (2000)
95. Graepel, T., et al.: Classification on proximity data with LP-machines. In: Ninth International Conference on Artificial Neural Networks, pp. 304–309. IEEE, London (1999)
96. Granato, A., Van Pelt, J.: Effects of early ethanol exposure on dendrite growth of cortical pyramidal neurons: inferences from a computational model. Dev. Brain Res. **142**, 223–227 (2003)
97. Gretton, A., Herbrich, R., Chapelle, O.: Estimating the leave-one-out error for classification learning with SVMs. http://www.kyb.tuebingen.mpg.de/publications/pss/ps1854.ps (2003)
98. Guo, H., Gelfand, S.B.: Classification trees with neural network feature extraction. IEEE Trans. Neural Netw. **3**, 923–933 (1992)
99. Guyon, I., Weston, J., Barnhill, S., Vapnik, V.: Gene selection for cancer classification using support vector machines. Mach. Learn. **46**, 389–422 (2002)
100. Gyetvan, F., Shi, Y.: Weak duality theorem and complementary slackness theorem for linear matrix programming problems. Oper. Res. Lett. **11**, 249–252 (1992)
101. Han, J., Kamber, M.: Data Mining Concepts and Techniques. Morgan Kaufmann, San Mateo (2002)
102. Han, J., Kamber, M.: Data Mining: Concepts and Techniques. Morgan Kaufmann, San Mateo (2006)
103. Hao, X.R., Shi, Y.: Large-scale MC2 Program: version 1.0, A C++ Program Run on PC or Unix. College of Information Science and Technology, University of Nebraska-Omaha, Omaha, NE 68182, USA, 1996
104. Har-Peled, S., Roth, D., Zimak, D.: Constraint classification: A new approach to multiclass classification and ranking. In: Advances in Neural Information Processing Systems, vol. 15 (2002)
105. Hartigan, J.A.: Clustering Algorithms. Wiley, New York (1975)
106. Hastie, T.J., Tibshirani, R.J.: Classification by pairwise coupling. In: Jordan, M.I., Kearns, M.J., Solla, S.A. (eds.) Advances in Neural Information Processing Systems, vol. 10, pp. 507–513. MIT Press, Cambridge (1998)
107. He, J., Liu, X., Shi, Y., Xu, W., Yan, N.: Classifications of credit card-holder behavior by using fuzzy linear programming. Int. J. Inf. Technol. Decis. Mak. **3**(4), 633–650 (2004)
108. Heese, K., Hock, C., Otten, U.: Inflammatory signals induce neurotrophin expression in human microglial cells. J. Neurochem. **70**, 699–707 (1998)
109. Herbrich, R., Graepel, R., Bollmann-Sdorra, P., Obermayer, K.: Learning a preference relation for information retrieval. In: Proceedings of the AAAI Workshop Text Categorization and Machine Learning, Madison, USA (1998)
110. Herbrich, R., Graepel, T., Obermayer, K.: Support vector learning for ordinal regression. In: Proceedings of the Ninth International Conference on Artificial Neural Networks, pp. 97–102. (1999)
111. Hesselgesser, J., Taub, D., Baskar, P., Greenberg, M., Hoxie, J., Kolson, D.L., Horuk, R.: Neuronal apoptosis induced by HIV-1 gp120 and the chemokine SDF-1alpha mediated by the chemokine receptor CXCR4. Curr. Biol. **8**, 595–598 (1998)
112. Horst, R., Pardalos, P.M.: Handbook of Global Optimization. Kluwer Academic, Dordrecht (1995)
113. Iba, Y., Hayshi, N., Sawada, I., Titani, K., Kurosawa, Y.: Changes in the specificity of antibodies against steroid antigens by introduction of mutations into complementarity-determining regions of Vh domain. Protein Eng. **11**, 361–370 (1998)
114. Jaakkola, T.S., Haussler, D.: Exploiting generative models in discriminative classifiers. In: Advances in Neural Information Processing Systems, vol. 11. MIT Press, Cambridge (1998)
115. Jgap: a genetic algorithms and genetic programming component by java (2008)
116. Jiang, Z., Piggee, C., Heyes, M.P., et al.: Glutamate is a mediator of neurotoxicity in secretions of activated HIV-1-infected macrophages. J. Neuroimmunol. **117**, 97–107 (2001)
117. Joachims, T.: Transductive inference for text classification using support vector machines. In: International Conference on Machine Learning (ICML) pp. 200–209 (1999)

118. Joachims, T.: SVMLight: Support Vector Machine. University of Dortmund. http://-ai. informatik.uni-dortmund.de/FORSCHUNG/VERFAHREN/SVM_LIGHT/svm_light.eng. html (November 1999)

119. Joachims, T.: Estimating the generalization performance of an SVM efficiently. In: Proceedings of the 17th International Conference on Machine Learning, San Francisco, CA, pp. 431–438. Morgan Kaufmann, San Mateo (2000)

120. Jones, S., Thornton, J.M.: Analysis of protein-protein interaction sites using surface patches. J. Mol. Biol. **272**, 121–132 (1997)

121. Jones, S., Thornton, J.M.: Prediction of protein-protein interaction sites using patch analysis. J. Mol. Biol. **272**, 133–143 (1997)

122. Kachigan, S.K.: Multivariate Statistical Analysis: A Conceptual Introduction, 2nd edn. Radius Press, New York (1991)

123. Kaul, M., Garden, G.A., Lipton, S.A.: Pathways to neuronal injury and apoptosis in HIV-associated dementia. Nature **410**, 988–994 (2001)

124. Keerthi, S., Shevade, S., Bhattacharyya, C., Murthy, K.: Improvements to Platt's SMO algorithm for SVM classifier design. Technical report, Dept. of CSA, Banglore, India (1999)

125. Kolesar, P., Showers, J.L.: A robust credit screening model using categorical data. Manag. Sci. **31**, 123–133 (1985)

126. Kou, G.: Multi-class multi-criteria mathematical programming and its applications in large scale data mining problems. PhD thesis, University of Nebraska Omaha (2006)

127. Kou, G., Liu, X., Peng, Y., Shi Y., Wise, M., Xu, W.: Multiple criteria linear programming to data mining: Models, algorithm designs and software developments. Optim. Methods Softw. **18**, 453–473 (2003)

128. Kou, G., Shi, Y.: Linux based multiple linear programming classification program: version 1.0. College of Information Science and Technology, University of Nebraska-Omaha, Omaha, NE 68182, USA, 2002

129. Kramer, S., Widmer, G., Pgahringer, B., DeGroeve, M.: Prediction of ordinal classes using regression trees. Fundam. Inform. **47**, 1–13 (2001)

130. Krebel, U.: Pairwise classification and support vector machines. In: Schölkopf, B., Burges, C.J.C., Smola, A.J. (eds.) Advances in Kernel Methods: Support Vector Learning, pp. 255–268. MIT Press, Cambridge (1999)

131. Krichmar, J.L., Nasuto, S.J., Scorcioni, R., Washington, S.D., Ascoli, G.A.: Effects of dendritic morphology on CA3 pyramidal cell electrophysiology: a simulation study. Brain Res. **941**, 11–28 (2002)

132. Kwak, W., Shi, Y., Cheh, J.J.: Firm bankruptcy prediction using multiple criteria linear programming data mining approach. Adv. Financ. Plan. Forecast., Suppl. **2**, 27–49 (2006)

133. Kwak, W., Shi, Y., Eldridge, S., Kou, G.: Bankruptcy prediction for Japanese firms: using multiple criteria linear programming data mining approach. Int. J. Bus. Intell. Data Min. **1**(4), 401–416 (2006)

134. Lanckriet, G., Cristianini, N., Bartlett, P., Ghaoui, L., Jordan, M.: Learning the kernel matrix with semidefinite programming. J. Mach. Learn. Res. **5**, 27–72 (2004)

135. Lazarov-Spiegler, O., Solomon, A.S., Schwartz, M.: Peripheral nerve-stimulated macrophages simulate a peripheral nerve-like regenerative response in rat transected optic nerve. GLIA **24**, 329–337 (1998)

136. Lee, Y., Lin, Y., Wahba, G.: Multicategory support vector machines. In: Computing Science and Statistics: Proceedings of the 33rd Symposium on the Interface, pp. 498–512 (2001)

137. LINDO Systems Inc: An Overview of LINGO 8.0. http://www.lindo.com/cgi/frameset.cgi? leftlingo.html;lingof.html

138. Lipton, S.A., Gendelman, H.E.: Dementia associated with the acquired immunodeficiency syndrome. N. Engl. J. Med. **16**, 934–940 (1995)

139. Liu, M., Wang, Z.: Classification using generalized Choquet integral projections. In: Proc. IFSA, pp. 421–426 (2005)

140. Mackay, D.: Introduction to Gaussian processes. In: Neural Networks and Machine Learning (1999)

141. Mangasarian, O.L., Musicant, D.R.: Successive overrelaxation for support vector machines. IEEE Trans. Neural Netw., **10**(5), 1032–1037 (1999)
142. Mangasarian, O.L., Musicant, D.R.: Lagrangian support vector machines. J. Mach. Learn. Res. **1**, 161–177 (2001)
143. Masliah, E., DeTeresa, R.M., Mallory, M.E., Hansen, L.A.: Changes in pathological findings at autopsy in AIDS cases for the last 15 years. AIDS **14**, 69–74 (2000)
144. Masliah, E., Heaton, R.K., Marcotte, T.D.: Dendritic injury is a pathological substrate for human immunodeficiency virus-related cognitive disorders. Ann. Neurol. **42**, 963–972 (1997)
145. McArthur, J.C., Sacktor, N., Selnes, O.: Human immunodeficiency virus-associated dementia. Semin. Neurol. **19**, 129–150 (1999)
146. Mcgarry, K.: A survey of interestingness measures for knowledge discovery. Knowl. Eng. Rev. **20**(1), 39–61 (2005)
147. Meng, D., Xu, C., Jing, W.: A new approach for regression: visual regression approach. In: CIS 2005, Part I. LNAI, vol. 3801, pp. 139–144. Springer, Berlin (2005)
148. Mikenina, L., Zimmermann, H.J.: Improved feature selection and classification by the 2-additive fuzzy measure. Fuzzy Sets Syst. **107**(2), 197–218 (1999)
149. Miranda, P., Grabisch, M.: Optimization issues for fuzzy measures. Int. J. Uncertain. Fuzziness Knowl.-Based Syst. **7**, 545–560 (1999)
150. Morrison, D.F.: Multivariate Statistical Methods, 2nd edn. McGraw-Hill, New York (1976)
151. Moses, A.V., Bloom, F.E., Pauza, C.D., Nelson, J.A.: HIV infection of human brain capillary endothelial cells occurs via a CD4 galactosylceramide-independent mechanism. Proc. Natl. Acad. Sci. USA **90**, 10474–10478 (1993)
152. Murofushi, T., Sugeno, M., Fujimoto, K.: Separated hierarchical decomposition of the Choquet integral. Int. J. Uncertain. Fuzziness Knowl.-Based Syst. **5**(5), 563–585 (1997)
153. Nath, A., Hartloper, V., Furer, M., Fowke, K.R.: Infection of human fetal astrocytes with HIV-1: viral tropism and the role of cell to cell contact in viral transmission. J. Neuropathol. Exp. Neurol. **54**, 320–330 (1995)
154. Navia, B.A.: Clinical and biologic features of the AIDS dementia complex. Neuroimaging Clin. N. Am. **7**, 581–592 (1997)
155. Navia, B.A., Jordan, B.D., Price, R.W.: The AIDS dementia complex: I. Clinical features. Ann. Neurol. **19**, 517–524 (1986)
156. Nonaka, I., Toyama, R., Konno, N.: SECI, Ba and Leadership: a unifying model of dynamic knowledge creation. In: Teece, D.J., Nonaka, I. (eds.) New Perspectives on Knowledge-Based Firm and Organization. Oxford University Press, New York (2000)
157. Ohlson, J.: Financial ratios and the probabilistic prediction of bankruptcy. J. Account. Res. **18**(1), 109–131 (1980)
158. Olson, D.L., Shi, Y.: Introduction to Business Data Mining. McGraw-Hill/Irwin, Englewood Cliffs (2007)
159. Pan, X.W.: Research on some key knowledge of knowledge management integrating context. Graduate University of Zhejiang University **9**, 23–28 (2005)
160. Pardalos, P.M., Rosen, J.B.: Constrained Global Optimization: Algorithms and Applications. Lecture Notes in Computer Science. Springer, Berlin (1987)
161. Pawlak, Z.: Rough sets. J. Comput. Inf. Sci. Eng. **11**, 341–356 (1982)
162. Peng, Y., Shi, Y., Xu, W.: Classification for three-group of credit cardholders' behavior via a multiple criteria approach. Adv. Model. Optim. **4**, 39–56 (2002)
163. Peng, Y., Kou, G., Shi, Y., Chen, Z.: A descriptive framework for the field of data mining and knowledge discovery. Int. J. Inf. Technol. Decis. Mak. **7**(4), 639–682 (2008)
164. Pfahringer, B.: Winning the KDD99 classification cup: bagged boosting. ACM SIGKDD Explor. Newsl. **1**(2), 65–66 (2000)
165. Platt, J.: Fast training of support vector machines using sequential minimal optimization. Technical report, Microsoft Research (1998)
166. Powell, M.J.D.: ZQPCVX a FORTRAN subroutine for convex quadratic programming. DAMTP Report NA17, Cambridge, England (1983)
167. Pyle, D.: Normalizing and Redistributing of Variables in Data Preparation for Data Mining. Morgan Kaufmann, San Francisco (1999)

168. Qi, Z.Q., Tian, Y.J., Deng, Y.N.: A new support vector machine for multi-class classification. In: Proceedings of 2005 International Conference on Computational Intelligence and Security, Xi'an, China, pp. 580–585 (2005)
169. Quinlan, J.: See5.0. http://www.rulequest.com/see5-info.html (2004)
170. Ramamohanarao, K.: Contrast pattern mining and application. In: IEEE Data Mining Forum (2008)
171. Rapalino, O., Lazarov-Spiegler, O., Agranov, E., et al.: Implantation of stimulated homologous macrophages results in partial recovery of paraplegic rats. Nat. Med. **4**, 814–821 (1998)
172. Rosenberg, E., Gleit, A.: Quantitative methods in credit management: A survey. Oper. Res. **42**, 589–613 (1994)
173. Ryan, L.A., Peng, H., Erichsen, D.A., et al.: TNF-related apoptosis-inducing ligand mediates human neuronal apoptosis: links to HIV-1 associated dementia. J. Neuroimmunol. **148**, 127–139 (2004)
174. Schölkopf, B., Smola, A.J.: Learning with Kernels—Support Vector Machines, Regularization, Optimization, and Beyond. MIT Press, Cambridge (2002)
175. Schneider, J.: Cross validation. http://www.cs.cmu.edu/~schneide/tut5/node42.html (1997)
176. Scorcioni, R., Ascoli, G.: Algorithmic extraction of morphological statistics from electronic archives of neuroanatomy. In: Lecture Notes Comp. Sci., vol. 2084, pp. 30–37 (2001)
177. Shashua, A., Levin, A.: Ranking with large margin principle: two approaches. Adv. Neural Inf. Process. Syst. **15**, 937–944 (2003)
178. Shi, Y.: Human-casting: a fundamental method to overcome user information overload. Information **3**(1), 127–143 (2000)
179. Shi, Y.: Multiple Criteria and Multiple Constraint Levels Linear Programming, pp. 184–187. World Scientific, Singapore (2001)
180. Shi, Y., He, J.: Computer-based algorithms for multiple criteria and multiple constraint level integer linear programming. Comput. Math. Appl. **49**(5), 903–921 (2005)
181. Shi, Y., Li, X.S.: Knowledge management platforms and intelligent knowledge beyond data mining. In: Shi, Y., et al. (eds.) Advances in Multiple Criteria Decision Making and Human Systems Management: Knowledge and Wisdom, pp. 272–288. IOS Press, Amsterdam (2007)
182. Shi, Y., Peng, Y., Xu, W., Tang, X.: Data mining via multiple criteria linear programming: applications in credit card portfolio management. Int. J. Inf. Technol. Decis. Mak. **1**, 131–151 (2002)
183. Shi, Y., Wise, M., Luo, M., Lin, Y.: Data mining in credit card portfolio management: a multiple criteria decision making approach. In: Koksalan, M., Zionts, S. (eds.) Multiple Criteria Decision Making in the New Millennium, pp. 427–436. Springer, Berlin (2001)
184. Shi, Y., Yu, P.L.: Goal setting and compromise solutions. In: Karpak, B., Zionts, S. (eds.) Multiple Criteria Decision Making and Risk Analysis and Applications, pp. 165–203. Springer, Berlin (1989)
185. Shibata, A., Zelivyanskaya, M., Limoges, J., et al.: Peripheral nerve induces macrophage neurotrophic activities: regulation of neuronal process outgrowth, intracellular signaling and synaptic function. J. Neuroimmunol. **142**, 112–129 (2003)
186. Sim, M.: Robust optimization. Ph.D. Thesis (2004)
187. Smola, A., Scholkopf, B., Rutsch, G.: Linear programs for automatic accuracy control in regression. In: Ninth International Conference on Artificial Neural Networks, Conference Publications London, vol. 470, pp. 575–580 (1999)
188. Sonnenburg, S., Ratsch, G., Schafer, C.: Learning interpretable SVMs for biological sequence classification. In: 9th Annual International Conference on Research in Computational Molecular Biology, pp. 389–407 (2005)
189. Stanfield, R.L., Fieser, T.M., Lerner, R.A., Wilson, I.A.: Crystal structures of an antibody to a peptide and its complex with peptide antigen at 2.8? Science **248**, 712–719 (1990)
190. Sterratt, D.C., van Ooyen, A.: Does morphology influence temporal plasticity. In: Dorronsoro, J.R. (ed.) Artificial Neural Networks—ICANN 2002, International Conference, Madrid, Spain. Lecture Notes in Computer Science, vol. 2415, pp. 186–191. Springer, Berlin (2002)

191. Sturm, J.F.: Using sedumi1.02, a Matlab toolbox for optimization over symmetric cones. Optim. Methods Softw. **11-12**, 625–653 (1999)
192. Sugeno, M., Fujimoto, K., Murofushi, T.: Hierarchical decomposition theorems for Choquet integral models. In: Proceedings of 1995 IEEE International Conference on Fuzzy Systems (1995)
193. Suykens, J.A.K., Wandewalle, J.: Least squares support vector machine classifiers. Neural Process. Lett. **9**, 293–300 (1999)
194. Tangian, A., Gruber, J.: Constructing quadratic and polynomial objective functions. In: Proceedings of the 3rd International Conference on Econometric Decision Models, Schwerte, Germany, pp. 166–194. Springer, Berlin (1995)
195. The American Heritage Dictionary of the English Language. Houghton Mifflin Company (2003)
196. The National Health Care Anti-Fraud Association. http://www.nhcaa.org/ (2005)
197. Thomas, L.C., Edelman, D.B., Crook, J.N.: Credit Scoring and Its Applications. SIAM, Philadelphia (2002)
198. Tian, Y.J.: Support vector regression machine and its application. Ph.D. Thesis, China Agricultural University (2005)
199. Tian, Y.J., Deng, N.Y.: Support vector classification with nominal attributes. In: Proceedings of 2005 International Conference on Computational Intelligence and Security, Xi'an, pp. 586–591 (2005)
200. Tian, Y.J., Deng, N.Y.: Leave-one-out bounds for support vector regression. In: International Conference on Intelligent Agents, Web Technologies and Internet Commerce, Austria, vol. 2, pp. 1061–1066 (2005)
201. Tian, Y.J., Deng, N.Y.: A leave-one-out bound for support vector regression. In: Hawaii International Conference on Statistics, Mathematics and Related Fields, pp. 1551–1567 (2006)
202. Tian, Y.J., Qi, Z.Q., Deng, Y.N.: A new support vector machine for multi-class classification. In: Proceedings of the 5th International Conference on Computer and Information Technology, Shanghai, China, pp. 18–22 (2005)
203. Tian, Y.J., Yan, M.F.: Unconstrained transductive support vector machines. In: Proceedings of the 4th International Conference on Fuzzy Systems and Knowledge Discovery, pp. 181–185 (2007)
204. Tsang, I.W., Kwok, J.T.: Distance metric learning with kernels. In: Proceedings of the International Conference on Artificial Neural Networks, Istanbul, Turkey (2003)
205. van Ooyen, A., Willshaw, D., Ramakers, G.: Influence of dendritic morphology on axonal competition. Neurocomputing **32**, 255–260 (2000)
206. Vapnik, V.N.: Statistical Learning Theory. Wiley, New York (1998)
207. Vapnik, V.N., Chapelle, O.: Bounds on error expectation for SVM. In: Advances in Large-Margin Classifiers (Neural Information Processing), pp. 261–280. MIT Press, Cambridge (2000)
208. Wang, Z., Guo, H.: A new genetic algorithm for nonlinear multiregressions based on generalized Choquet integrals. In: Proc. of FUZZ/IEEE, pp. 819–821 (2003)
209. Wang, Z., Klir, G.: Fuzzy Measure Theory. Plenum, New York (1992)
210. Wang, Z., Leung, K.-S., Klir, G.J.: Applying fuzzy measures and nonlinear integrals in data mining. Fuzzy Sets Syst. **156**, 371–380 (2005)
211. Wang, Z.T.: Knowledge System Engineering. Science Press, Beijing (2004)
212. Wapnik, W., Tscherwonenkis, A.: Theorie der Zeichenerkennung. Akademie Verlag, Berlin (1979)
213. Webb, G.: Finding the real pattern. In: IEEE Data Mining Forum (2008)
214. Webster, D.M., Henry, A.H., Rees, A.R.: Antibody-antigen interactions. Current Opinion in Structural Biology **4**, 123–129 (1994)
215. Weston, J., Watkins, C.: Multi-class Support Vector Machines. University of London Press, Egham (1998)
216. Weston, J., Mukherjee, S., Chapelle, O., Pontil, M., Poggio, T., Vapnik, V.: Feature selection for SVMs. In: Advances in Neural Information Processing Systems, vol. 13 (2001)

217. What is data?—A definition from the Webopedia. http://www.webopedia.com/TERM/D/data.html (2003)
218. What is data? A definition from Whatis.com. http://searchdatamanagement.techtarget.com/sDefinition/0,sid91_gci211894,00.html (2005)
219. Wiig, K.M.: Knowledge management: where did it come from and where it go? Expert Syst. Appl., **13**(1), 1–14 (1997)
220. Wikipedia: Knowledge (2008)
221. Williams, C., Coller, B., Nowicki, T., Gurney, J.: Mega Kalahari geology: challenges of kimberlite exploration in this medium (2003)
222. Wong, Andrew K.C.: Association pattern analysis for pattern pruning, pattern clustering and summarization. In: IEEE Data Mining Forum (2008)
223. Wu, X., Kumar, V., Quinlan, J.R., Ghosh, J., Yang, Y., Motoda, H., McLachlan, G.J., Ng, A., Liu, B., Yu, P.S., Zhou, Z.H., Steinbach, M., Hand, D.J., Steinberg, D.: Top 10 algorithms in data mining. Knowl. Inf. Syst. **14**, 1–37 (2008)
224. Xiang, J., Sha, Y., Prasad, L., Delbaere, L.T.J.: Complementarity determining region residues aspartic acid at H55 serine at tyrosines at H97 and L96 play important roles in the B72.3 antibody-TAG72 antigen interaction. Protein Eng. **9**, 539–543 (1996)
225. Xu, B.L., Neufeld, J., Schuurmans, D.: Maximum margin clustering. In: Advances in Neural Information Processing Systems (NIPS-04), vol. 17 (2004)
226. Xu, K., Wang, Z., Heng, P., Leung, K.: Classification by nonlinear integral projections. IEEE Trans. Fuzzy Syst. **11**, 187–201 (2003)
227. Xu, L.L., Schuurmans, D.: Unsupervised and semi-supervised multi-class support vector machines. In: Proceedings of the 20th National Conference on Artificial Intelligence, vol. 2 (2005)
228. Yan, N., Shi, Y.: Neural network classification program: version 1.0. A C++ program running on PC. College of Information Science and Technology, University of Nebraska-Omaha, Omaha, NE 68182, USA, 2002
229. Yan, N., Wang, Z., Shi, Y., Chen, Z.: Nonlinear classification by linear programming with signed fuzzy measures. In: Proceedings FUZZ/IEEE, vol. 11, pp. 187–201 (2006)
230. Yang, Z.X.: Support vector ordinal regression machines and multi-class classification. Ph.D. Thesis, China Agricultural University (2007)
231. Yang, Z.X., Tian, Y.J., Deng, N.Y.: Leave-one-out bounds for support vector ordinal regression machine. In: Neural Computing and Applications (2009)
232. Yang, Z.X., Deng, N.Y., Tian, Y.J.: A multi-class classification algorithm based on ordinal regression machine. In: International Conference on CIMCA2005 & IAWTIC 2005, Vienna, Austria, vol. 2, pp. 810–814 (2005)
233. Zeev-Brann, A.B., Lazarov-Spiegler, O., Brenner, T., Schwartz, M.: Differential effects of central and peripheral nerves on macrophages and microglia. GLIA **23**, 181–190 (1998)
234. Zeleny, M.: Knowledge of enterprise: knowledge management technology? Int. J. Inf. Technol. Decis. Mak. **1**(2), 181–207 (2002)
235. Zeleny, M.: Human Systems Management: Integrating Knowledge, Management and Systems. World Scientific, Singapore (2007)
236. Zhai, L.Y., Khoo, L.P., Fok, S.C.: In: Data Mining and Knowledge Discovery Approaches Based on Rule Induction Techniques, pp. 359–394. Springer, Heidelberg (2006)
237. Zhang, D., Tian, Y., Shi, Y.: A regression method by multiple criteria linear programming. In: 19th International Conference on Multiple Criteria Decision Making (MCDM), Auckland, New Zealand, Jan. 7–12 (2008)
238. Zhang, D., Tian, Y., Shi, Y.: Knowledge-incorporated MCLP classifier. In: Proceedings of Conference on Multi-criteria Decision Making (2008)
239. Zhang, L.L., Li, J., Shi, Y.: Study on improving efficiency of knowledge sharing in knowledge-intensive organization. In: WINE 2005, pp. 816–825 (2005)
240. Zhang, L.L., Li, J., Zheng, X., Li, X., Shi, Y.: Study on a process-oriented knowledge management model. Int. J. Knowl. Syst. Sci. **5**(1), 37–44 (2008)
241. Zhang, Z., Zhang, D., Tian, Y., Shi, Y.: Kernel-based multiple criteria linear program. In: Proceedings of Conference on Multi-criteria Decision Making (2008)

242. Zhao, J., Lopez, A.L., Erichsen, D., Herek, S., Cotter, R.L., Curthoys, N.P., Zheng, J.: Mitochondrial glutaminase enhances extracellular glutamate production in HIV-1-infected macrophages: linkage to HIV-1 associated dementia. J. Neurochem. **88**, 169–180 (2004)
243. Zhao, K., Tian, Y.J., Deng, N.Y.: Unsupervised and semi-supervised two-class support vector machines. In: Proceedings of the 7th IEEE ICDM 2006 Workshop, pp. 813–817 (2006)
244. Zhao, K., Tian, Y.J., Deng, N.Y.: Robust unsupervised and semi-supervised bounded C-support vector machines. In: Proceedings of the 7th IEEE ICDM 2007 Workshop, pp. 331–336 (2007)
245. Zhao, K., Tian, Y.J., Deng, N.Y.: Unsupervised and semi-supervised Lagrangian support vector machines. In: Proceedings of the 7th International Conference on Computational Science, pp. 882–889 (2007)
246. Zheng, J., Thylin, M., Ghorpade, A., et al.: Intracellular CXCR4 signaling, neuronal apoptosis and neuropathogenic mechanisms of HIV-1-associated dementia. J. Neuroimmunol. **98**, 185–200 (1999)
247. Zheng, J., Thylin, M.R., Cotter, R.L., et al.: HIV-1 infected and immune competent mononuclear phagocytes induce quantitative alterations in neuronal dendritic arbor: relevance for HIV-1-associated dementia. Neurotox. Res. **3**, 443–459 (2001)
248. Zheng, J., Thylin, M.R., Persidsky, Y., et al.: HIV-1 infected immune competent mononuclear phagocytes influence the pathways to neuronal demise. Neurotox. Res. **3**, 461–484 (2001)
249. Zhong, P., Fukushima, M.: A new multi-class support vector algorithm. In: Optimization Methods and Software (2004)
250. Zhong, P., Fukushima, M.: Second order cone programming formulations for robust multi-class classification. Neural Comput. **19**(1), 258–282 (2007)
251. Zhong, Y.X.: On the laws of information-knowledge-intelligence transforms. J. Beijing Univ. Posts Telecommun. **30**(1), 1–8 (2007)
252. Zhu, J., Rosset, S., Hastie, T., Tibshirani, R.: 1-norm support vector machines. In: Advances in Neural Information Processing Systems, vol. 16 (2004)
253. Zimmermann, H.J.: Fuzzy programming and linear programming with several objective functions. Fuzzy Sets Syst. **1**, 45–55 (1978)
254. Zou, H., Yuan, M.: The f_∞ norm support vector machine. Stat. Sin. **18**, 379–398 (2008)

Subject Index

Author Index

Y. Shi et al., *Optimization Based Data Mining: Theory and Applications*,
Advanced Information and Knowledge Processing,
DOI 10.1007/978-0-85729-504-0, © Springer-Verlag London Limited 2011